Pigments in Vegetables

PIGMENTS IN VEGETABLES

Chlorophylls and Carotenoids

Jeana Gross

An avi Book
Published by Van Nostrand Reinhold
New York

An AVI Book
(AVI is an imprint of Van Nostrand Reinhold)

Library of Congress Catalog Card Number 90-21039
ISBN 0-442-00657-8

Printed in the United States of America.

Van Nostrand Reinhold
115 Fifth Avenue
New York, New York 10003

Chapman and Hall
2-6 Boundary Row
London, SE1 8HN, England

Thomas Nelson Australia
102 Dodds Street
South Melbourne 3205
Victoria, Australia

Nelson Canada
1120 Birchmount Road
Scarborough, Ontario MIK 5G4, Canada

16 15 14 13 12 11 10 9 8 7 6 5 4 3 2 1

Library of Congress Cataloging-in-Publication Data

Gross, Jeana.
Pigments in vegetables: chlorophylls and carotenoids/Jeana Gross.
p. cm.
"An AVI book"—T.p. verso.
Includes bibliographical references and index.
ISBN 0-442-00657-8
1. Vegetables—Composition. 2. Chlorophyll. 3. Carotenoids.
4. Vegetables—Physiology. I. Title.
SB324.64.G76 1991
582.13'0419218—dc20 90-21039
CIP

Contents

Preface

Vegetables contain several classes of pigments: the green chlorophylls; the yellow, orange, and red carotenoids; the red, blue, or violet anthocyanins; and the red–violet betalains. This book covers the chlorophylls and the carotenoids, the two chief classes of vegetable pigments, and is the first major compilation of this kind.

Structure, physical and chemical properties, and analytical methods, including special methods adapted for certain vegetables, are discussed first, and are then followed by a discussion of physiological and biochemical aspects, including location, biosynthesis, and biochemical changes during plant development and senescence.

Both pigment classes have extremely important biological functions. The chlorophylls are of paramount importance in photosynthesis. The carotenoids, besides their photofunction, have a highly significant role in nutrition. Some of them have vitamin A value and, possibly, a role in the prevention of human cancer. The chlorophyll distribution table includes all of the currently available quantitative data on vegetables.

The discussion of carotenoid distribution describes the unique carotenoids found in vegetables and includes qualitative and quantitative distribution tables for the most common vegetables. These tables summarize all of the published data to date, but they are far from complete. However, application of recently available techniques should fill the enormous gaps, update old data, and permit an accurate assessment of the vitamin A value of the food. Such complete data would also make it possible to use carotenoid distribution in the taxonomy of vegetables.

The presentations of pigment distribution also consider the changes in pigment content during storage and processing.

The book also includes a presentation of factors affecting pigment synthesis in vegetables (e.g., phytohormones) and environmental factors (e.g., light, temperature), as well as fertilizers and pesticides.

Acknowledgments

I wish to thank Professor F. Lenz of the Institut für Obstbau und Gemüsebau der Universität Bonn, Germany, for the encouragement and support he gave me throughout the writing of this book, and for reading the section on chlorophyll.

I am also grateful for the warm support I received from Professor R. Ikan (Department of Organic Chemistry, Laboratory of Natural Products, The Hebrew University of Jerusalem, Israel).

Many thanks to all my colleagues around the world for providing relevant literature. I especially wish to thank Dr. Susanne Huyskens.

I am especially grateful to my husband, who displayed courageous patience while typing the reference list, and who by skillful manipulation of computers and enormous hard work literally saved this manuscript. Without his generous help publication of this book would not have been possible.

I also wish to acknowledge the partial financial support of the Deutsche Forschungsgemeinschaft.

Pigments in Vegetables

INTRODUCTION

> Par la science on se sent profondément en contact,
> en harmonie, avec la nature et ses secrets.
> Roger Martin du Gard—Les Thibault

In a broad sense, the term *vegetable* refers to any kind of plant life or plant product; in the narrow sense *vegetable* refers to the fresh edible portion of a herbaceous plant, whether consumed raw or cooked. The edible parts may be roots, stems, leaves and leafstalks, flowers, fruit, or seeds. The following listing includes only the most widespread species.

Root vegetables include beets, carrots, radishes, rutabagas, and turnips.

Stem vegetables include asparagus, of which the aerial stems are eaten, and kohlrabi, of which the edible portion is the enlarged overground stem.

Tuber vegetables have edible tubers that are underground stems. They include girasole (Jerusalem artichoke), potato, taro, and yam.

Leaf and leafstalk vegetables include Brussels sprouts, cabbage, endive, lettuce, parsley, spinach with edible leaves, and celery and rhubarb, which are cultivated for their edible leafstalks.

Bulb vegetables include garlic, leek, and onion.

Immature inflorescence vegetables include artichoke, broccoli, and cauliflower.

Fruits used as vegetables may be immature fruits such as green bean, cowpea, cucumber, eggplant, lima bean, okra, pea, summer squash, and sweet corn (maize). Mature fruit vegetables include pepper, pumpkin, and tomato. The seeds of beans, lima beans, peas, and pumpkins are edible.

This enumeration shows the heterogeneity of vegetables as derived primarily from the morphological diversity of their parts, in contrast to Gross (1987), which deals exclusively with the pigments of edible fruits.

On the other hand, as in fruits, the color of vegetables is due to green chlorophylls, the yellow to red carotenoids, and the red to violet anthocyanins, which are the main classes of plant pigments in nature. An additional minor pigment group is the betalains, of restricted distribution, which, for example, confer the deep purple color to red beets (*Beta vulgaris*). A great diversity in pigment changes during ontogeny of such various plant organs is thus to be expected.

Following pigment changes during ripening of fruits, different ways of pigment accumulation were observed. During ontogeny of different organs in vegetables, the changes are even more complex.

Because most vegetables are not consumed fresh, but are cooked, canned, frozen, dehydrated, or pickled, it is important to follow the pigment changes during these processes, considering that color is a quality characteristic. Pigments in vegetables are also nutritionally important. Orange-colored vegetables and those with high chlorophyll content (dark green leaves) are sources of provitamin A.

The chief function common to all these pigments in flowers and fruits, excluding chlorophylls, is to attract insects, birds, and other animals for pollination and seed dispersal. Some pigments (although this area is better investigated with flavonoids) may have a role in disease resistance. Pigments in other parts of the plant, especially underground parts such as stems and roots, seem to have no purpose other than to bestow color, or perhaps their function is not yet understood.

CHLOROPHYLLS

Of the three main natural classes of pigments, chlorophylls are the most widely distributed and the most important. Chlorophylls are the pigments responsible for photosynthesis, the fundamental life process that converts light energy into chemical energy. In the presence of light, carbohydrates and other organic compounds are produced from carbon dioxide and water, with molecular oxygen being liberated.

All green plants contain chlorophyll *a* and chlorophyll *b*. In higher plants chlorophyll *a* is the major pigment and chlorophyll *b* is an accessory pigment. Both are located in small subcellular organelles, the plastids, which are thus green and called *chloroplasts.*

In higher plants the chlorophyll ratio *a/b* is usually 3 to 1 and constitutes a parameter of the physiological status. The ratio also varies with growth conditions and environmental factors. Plant species exposed to sun tend to have a higher ratio (3.2 to 4) than shade plants (2.6 to 3.2) (Lichtenthaler, 1971; Lichtenthaler et al., 1981). The increased proportion of chlorophyll *b* in shade plants is due to its absorption properties. Being capable of absorbing strongly in the 450–480 nm range, it captures effective light at low intensity, partially filling the gap in the chlorophyll *a* spectrum.

It has been recently established that chlorophyll *a'*, a chlorophyll epimer, and pheophytin *a,* a chlorophyll derivative, are also present in green leaves (Watanabe et al., 1985b).

STRUCTURE

Chlorophylls are porphyrins containing the basic tetrapyrrole ring, of which one is reduced. The four rings are coordinated with the Mg^{2+} ion. A fifth isocyclic ring, ring E, is found near the third pyrrole ring. At the fourth ring, the propionic acid substituent is esterified with the diterpene alcohol phytol ($C_{20}H_{39}OH$), which is the hydrophobic side of the molecule, the rest of the molecule being hydrophilic.

Chlorophyll *b* differs from chlorophyll *a* only by having an aldehyde group (−CHO) in place of the methyl group at position 3 (see Figure 2).

Chlorophylls *a′* and *b′* are 10-epichlorophylls related to the parent pigments by inversion of the configuration at C-10, an asymmetric center at ring E (see Figure 2). They were discovered during sucrose column chromatography of plant extracts (Strain and Manning, 1942) as minor pigments accompanying the chlorophylls. They have the same absorption spectra as the parent pigments. Two decades later it was established that they are 10-epichlorophylls (Katz et al., 1968).

Chlorophylls *a* and *a′* are interconvertible. The process occurs slowly at room temperature, but rapidly at 95–100°C if the chlorophylls are heated in alcohol or pyridine. They may be prepared easily if green tissue is immersed for 1 min in boiling water; they are then extracted with acetone or methanol. The two epimers were considered artifacts produced in the course of handling plant extracts. Also, deep-frozen vegetable material (e.g., spinach) contains significant amounts of chlorophylls *a′* and *b′* (Scholz and Ballschmiter, 1981). Chlorophyll *a′* is a natural constituent of the photosynthetic membrane (Watanabe et al., 1985a).

PHYSICAL PROPERTIES

Chlorophylls are green in color because they absorb strongly in the red and blue regions of the visible spectrum. The small differences in the structures of the two chlorophylls produce differences in the absorption spectra, the maxima of chlorophyll *b* being shifted toward the green region of the spectrum. This produces the different green hues of the two pigments, blue–green for chlorophyll *a* and yellow–green for chlorophyll *b*. The absorption maxima of chlorophyll *a* and chlorophyll *b* are shown in Figure 1. With modern instruments these spectra have been reevaluated (see the subsection on chlorophyll determination). The position of the absorption maxima varies with the nature of the solvents. The chlorophylls are esters and soluble in most organic solvents.

In solution both chlorophylls are fluorescent. The absorption and fluorescence spectra in vivo of the chlorophylls are different, since chlorophylls are found as complexes with proteins within the photosynthetic biomembranes.

CHEMICAL PROPERTIES

The detailed chemistry of chlorophylls has been described in many reviews, including Willstaetter and Stoll (1913), Stoll and Wiedemann (1938), Aronoff (1966), Jackson (1976), Svec (1978), and Schwartz and Lorenzo (1990).

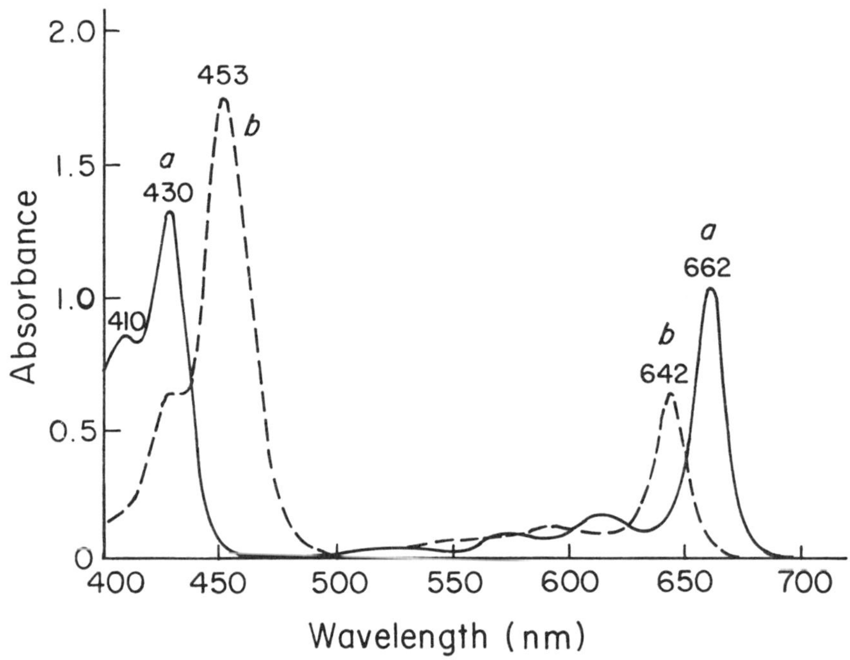

Figure 1. Absorption spectra of chlorophylls *a* and *b* in diethylether: chlorophyll *a* (——); chlorophyll *b* (- - -).

One of the most important characteristics of chlorophylls is their extreme lability. They are sensitive to light, heat, oxygen, and chemical degradation.

DERIVATIVES

Chlorophylls can be readily transformed, both in vivo and in vitro, into a series of derivatives (see Figure 2).

Chlorophyllides

The phytyl ester can be easily hydrolyzed to give chlorophyllide and phytol. The hydrolysis takes place under mild conditions either by acid or alkali. Generally, chlorophyllides are prepared enzymatically, the hydrolysis being catalyzed by chlorophyllase, an enzyme commonly found in green plant tissue. Leaves that are especially rich in chlorophyllase [e.g., sugar beet (Holden, 1961) or common cocklebur (*Xanthium pennsylvanicum*)] are used

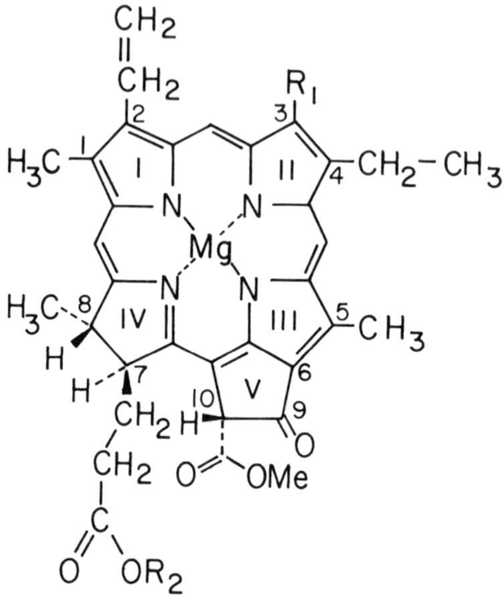

1

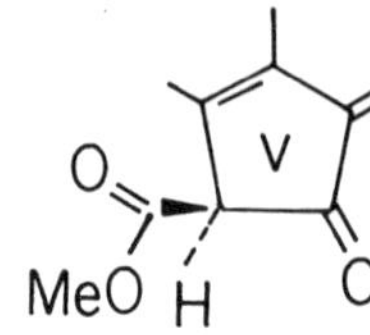

2

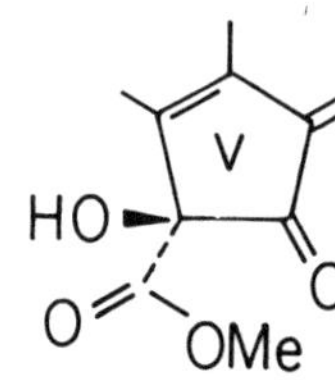

3

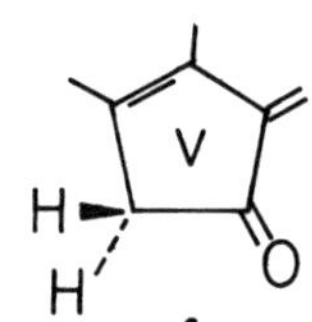

4

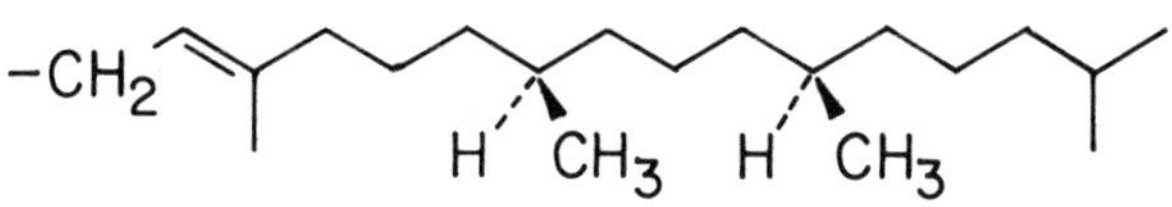

Phytyl group ($-C_{20}H_{39}$)

Compound	*Mg**	*R1*	*R2*	*Isocyclic Ring (V)*
Chlorophyll *a*	+	CH_3	Phytyl	1
Chlorophyll *b*	+	CHO	Phytyl	1
Chlorophyll *a′*	+	CH_3	Phytyl	2
Chlorophyll *b′*	+	CHO	Phytyl	2
Chlorophyllide *a*	+	CH_3	H	1
Chlorophyllide *b*	+	CHO	H	1
Pheophytin *a*	−	CH_3	Phytyl	1
Pheophytin *b*	−	CHO	Phytyl	1
Pheophorbide *a*	−	CH_3	H	1
Pheophorbide *b*	−	CHO	H	1
Chlorophyll *a*−1	+	CH_3	Phytyl	3
Pyropheophytin *a*	−	CH_3	Phytyl	4

*In pheophytins and pheophorhides, Mg is replaced by 2H.

Figure 2. Structural formulas and nomenclature of chlorophylls *a* and *b* and their various derivatives.

as enzyme sources (Pennington et al., 1964). The completion of the conversion to phytol-free derivatives is verified by checking their insolubility in petroleum ether.

Pheophytins *a* and *b*

Pheophytins are the magnesium-free derivatives of the chlorophylls. Pheophytins *a* and *b* are easily obtained from chlorophyll by the action of dilute acids, which remove the magnesium. The reaction lasts 1–2 min, and the concentration of HCl used is 13% (Pennington et al., 1964; Hynninen, 1973).

Pheophorbides *a* and *b*

Pheophorbides *a* and *b* are hydrolyzed chlorophylls without phytol (chlorophyllides) that have also lost the magnesium. Thus, they may be prepared from chlorophylls treated with concentrated acid (HCl 30%) or from acidified chlorophyllides (Pennington et al., 1964; Hynninen, 1973).

Allomerized Chlorophylls

Allomerized chlorophylls are oxidized chlorophylls with the oxygen atom being located at C-10, and hydroxychlorophylls being formed. The hydroxychlorophylls are called chlorophyll *a*-1 and chlorophyll *b*-1.

The process was named *allomerization* by Willstaetter as the oxidized products have absorption spectra identical to that of the parent compound. Chlorophyll is oxidized spontaneously by atmospheric oxygen when alcoholic solutions are left in the dark, as observed for the first time by Willstaetter and Stoll (1913).

Allomerized chlorophylls are prepared by passing oxygen for 72 hours through a methanolic solution of chlorophyll (Daley et al., 1973). The product was also observed during boiling of leaves (Holden, 1970b). Recently, it has been proved that chlorophyll allomerization proceeds via a free-radical mechanism, the autoxidation being inhibited by carotenoids (Hynninen, 1981).

Chlorophyll *a*-1 was recently identified in senescing excised leaves (Maunders et al., 1983).

Pyrochlorophylls

Pyroderivatives of chlorophylls or their derivatives are the compounds that have lost the carbomethoxy group $-COOCH_3$ at C-10 of the isocyclic ring, the group being replaced by hydrogen.

Chlorophyll *a,* methyl chlorophyllide *a,* pheophytin *a,* or methyl pheophorbide *a* when heated in pyridine at 100°C give rise to "pyro"-derivatives by decarbomethoxylation (Pennington et al., 1964).

ANALYTICAL METHODS

The analytical methods for chlorophyll analysis have been reviewed by Holden (1976) and more recently by Schwartz and Lorenzo (1990). In this book the analytical methods will be treated more extensively. We will refer also to intermediates that supervene in chlorophyll biosynthesis and to chlorophyll derivatives. These latter may be naturally found in some vegetables, or they may be formed in food during processing and storage.

Owing to the lability of the chlorophylls, a series of precautions must be taken to avoid their alteration. All manipulations must be carried out in dim light to avoid photobleaching; heating, oxidation, and enzymatic action must also be avoided.

Extraction

Plant material is extracted in a homogenizer where grinding and extraction occur simultaneously. Because chlorophylls are liposoluble, solvents such as acetone and methanol are used, which, being water-miscible, are adequate for fresh plant tissue. They break the chlorophyll–protein complex, noncovalent linkage and extract the pigments quantitatively. Undiluted solvents also block the chlorophyllase activity. To prevent pheophytin formation, the extract is neutralized. The pigments may be transferred to a water-free epiphase by adding water-immiscible solvents such as diethyl ether or petroleum ether.

Chlorophyll Determination

The chlorophylls and some of their derivatives, which absorb in the blue and red regions of the visible spectrum, are green. Also, both chlorophylls and some of their derivatives are fluorescent. Accordingly, chlorophylls and derivatives can be determined by spectrophotometric or spectrofluorometric methods.

Spectrophotometric Methods

Colorimetric methods are based on the Lambert–Beer law, which is a relationship between the concentration of a substance and its color. The measured absorption, extinction, or optical density is proportional to the specific absorption (extinction) coefficient, which is known, the length l (in centimeters) of the light path, and the concentration c. The specific absorption, $A^{1\%}_{1\,\text{cm}}$, or extinction, $E^{1\%}_{1\,\text{cm}}$ is the absorption of a 1% w/v solution in a 1-cm light path (values are given in reference tables). The absorption is usually measured at the wavelength of the absorption maximum of the substance.

The spectrophotometric determination for chlorophylls permits direct determination of both chlorophylls in the plant extract, without separation into individual pigments. The calculation is based on a two-equation system in which the absorption of each chlorophyll is corrected by subtracting the contributed absorption of the second pigment. Simultaneous equations for the various solvents used are given, and the respective absorption coefficients determined by Mackinney (1941) are used.

Because the absorption maxima vary with the solvent, different equations were derived for the various solvents used. For 80% acetone extracts, Arnon's (1949) equations are used:

$$\text{Chlorophyll } a \text{ (mg } l^{-1}) = 12.7A_{663} - 2.69A_{645}$$
$$\text{Chlorophyll } b \text{ (mg } l^{-1}) = 22.9A_{645} - 4.68A_{663}$$

For a diethyl ether extract the equations given are due to Smith and Benitez (1955):

$$\text{Chlorophyll } a \text{ (mg } l^{-1}) = 10.1A_{662} - 1.01A_{644}$$
$$\text{Chlorophyll } b \text{ (mg } l^{-1}) = 16.4A_{644} - 2.57A_{662}$$

Special equations have been derived for cases in which chlorophyll precursors or derivatives are present; their respective absorption coefficients, as previously determined, are used.

During food processing in canned and frozen vegetables, chlorophylls are easily converted to pheophytins, thereby affecting the color and, consequently, the food quality.

Numerous methods were developed to estimate the pigments of the vegetables before and after processing. Equations were derived for estimating the concentration of chlorophyll and percent conversion to pheophytin in acetone extracts of vegetables (Mackinney and Weast, 1940; Dietrich, 1958; Vernon, 1960). The method is based on the quantitative conversion of chlorophylls into pheophytins by adding an acid and reading the absorbances at two wavelengths. It was necessary to determine specific absorption coefficients, which may vary with the laboratory conditions and the vegetables analyzed. The method was adapted for determining chlorophyll and pheophytin in heated broccoli (Sweeney and Martin, 1958) and processed spinach (Tan and Francis, 1962), and chlorophyll stability as affected by pH in various vegetables such as peas, broccoli, Brussels sprouts, green beans, leaf spinach, and lima beans (Sweeney and Martin, 1961).

To follow the chlorophyll changes that occur during brining of green tissue, such as pickling cucumbers, researchers have developed a special method because the pigment mixture is complex, consisting of chlorophylls *a* and *b* and their derivatives: chlorophyllides, pheophytins, and pheophorbides (White et al., 1963). The method is based on the same principle—that is, on conversion of chlorophylls and chlorophyllides into pheophytins and pheophorbides, respectively, by adding acid and reading of absorbances at two wavelengths. The eight pigments are estimated by using a system of 16 equations.

Recently the analytical procedures for pigments of the photosynthetic biomembranes were reviewed and adapted (Lichtenthaler, 1987). Emphasis was placed on the spectral characteristics and the specific absorption coefficients of chlorophylls, pheophytins, and carotenoids, which are the basis of establishing equations for their determination. The data previously used in this field were based on old observations and old instrumentation, when the resolution of spectrophotometers was not as good as today. Therefore the specific absorption coefficients of the pigments were reevaluated and, in part, first established for organic solvents of different polarity and water content. This evaluation was carried out with a double-beam spectrophotometer, which allowed programmed automatic recording and printing of the proper wavelengths and absorbance values (Lichtenthaler and Wellburn, 1983). The former equations for diethyl ether (Smith and Benitez, 1955) are replaced by a new system of three equations. The third equation permits the simultaneous determination of the total carotenoids:

$$\text{Chl } a = 10.05A_{660.6} - 0.97A_{642.2}$$

$$\text{Chl } b = 16.36A_{642.2} - 2.43A_{660.6}$$

$$\text{Total carotenoids} = \frac{(1000A_{470} - 1.43 \text{ Chl } a - 35.87 \text{ Chl } b)}{205}$$

The pigment concentrations obtained by inserting the measured absorbance values are micrograms per milliliter plant extract solution.

The redetermined λ_{max} are at shorter wavelengths (660.6 and 642.2 nm) than so far reported for older instruments.

The absorption coefficients for *N,N*-dimethylformamide as solvent were also recently established (Moran, 1982; Inskeep and Bloom, 1985).

Spectrofluorometric Methods

Fluorometric methods are more sensitive than spectrophotometric methods. They permit accurate estimation at pigment concentrations of 10^{-2} to 10^{-3} of those required for spectrophotometric procedures. Such methods were used in early biosynthetic studies on protochlorophyll conversion to chlorophyll *a* (Virgin, 1955), and they continue to be used. The methods are suitable for determining the ratio of chlorophyll *a/b* when it deviates from the normal value of 3 (Boardman and Thorne, 1971). Fluorometric procedures were also used for estimating eight pigments (chlorophylls, chlorophyllides, pheophytins, and pheophorbides) in unsegregated mixtures (White et al., 1972; Bazzaz and Rebeiz, 1979). A combination of spectrophotometry and fluorometry was used for estimating zinc or copper pheophytins, chlorophylls, and pheophytins in mixtures (Jones et al., 1977a; White et al., 1977).

Increased resolving power is obtained by measuring the fluorescence emission spectra of photosynthetic tissue at 77 K (Papageorgiou, 1975). A rapid and simple method permits estimation of chlorophyll relative to its fluorescence emission intensity at 77 K when an adequate internal standard is used (Gershoni and Ohad, 1980). The differences among angiosperms in the biosynthesis and accumulation of monovinyl and divinyl protochlorophyllide during photoperiodic greening were determined by spectrofluorometric techniques (Carey and Rebeiz, 1985). High-performance liquid chromatography and fluorescence detection are widely used for resolution and quantification of chlorophylls and derivatives (Shioi and Sasa, 1982, 1983b; Shioi et al., 1983, 1984).

Low-temperature spectrofluorometry was widely used by Rebeiz and his group in biosynthetic studies of chlorophyll (Rebeiz, 1982).

Nondestructive Methods

In the previously described methods, the pigments were extracted after the destruction of the plant tissue. With nondestructive methods, the pigments may be estimated directly in the tissue, thus leaving the sample intact.

A color is characterized by three attributes: hue, lightness, and saturation. A three-dimensional representation of the color attributes is realized in what is known as a *color solid.* In the Munsell color solid, hue extends clockwise around a vertical axis, which represents lightness, and increases upward, and saturation extends horizontally. Each color is characterized by three coordinates.

A visual color rating is made by comparing with standardized color charts, atlases, fans, or special color comparators for specific fruits.

Physical methods imply the use of instruments, from simple to complex, for measuring light reflected from, or transmitted through, the sample. Measurements are, of course, more reliable because subjective factors do not interfere. On the other hand, evaluation by sensory panels, which reflects human visual perception, is the most adequate method for quality control.

The retina of the human eye has three types of color receptor molecules in the cone cells. Each pigment corresponds to a primary hue: red, blue, and green. The tristimulus colorimeters simulate the human eye by replacing the three receptor molecules with filters, one for each primary hue.

Reflectance or transmittance characteristics, whether obtained spectrophotometrically or with broadband filter photometers, normally are converted to the color system recommended by the Commission Internationale d'Eclairage, the CIE-*XYZ* coordinates, which improves the uniformity of the chromatic space and has been in use since 1976. But in food science, coordinates based on the Hunter solid are widely used: *L* (lightness), *a,* and *b,* where $+a$ represents red; $-a$, green; $+b$, yellow; and $-b$, blue. The ratio a/b is usually computed. It is negative for green plant material, approximately 0 for yellow material, and positive for orange or red material.

In green vegetables, the color changes occurring during thermal processing are usually followed by colorimetry both as routine quality control and to help in various investigations. The colorimetric parameter $-a/b$ was found to be highly correlated with the degree of conversion of chlorophyll to pheophytins in green peas (Gold and Weckel, 1959). The effect of pH adjustment on color retention in cut green beans and the effect of processing temperature on the color of spinach and pea puree were measured with Gardner or Hunterlab color difference meters (Sweeney and Martin, 1961; Tan and Francis, 1962; Buckle and Edwards, 1970b). The a/b ratio was used to follow the kinetics of the discoloration reaction in various vegetables (Hayakawa and Timbers, 1977). The effect of vacuum packaging on stored broccoli was studied by Wu and Salunkhe, following changes of the Hunter color values (Wu and Salunkhe, 1976).

Tristimulus colorimetric data have been tested and found suitable to determine the quality of dehydrated parsley samples in routine controls and to predict the level of chlorophyll degradation under different drying conditions (Berset and Caniaux, 1983). The blanching effect on the color of fresh and dried okra, before and after storage, and on green beans was followed with a Hunterlab colorimeter (Muftugil, 1986; Stone et al., 1986).

For rapid determination of chlorophyll in intact leaves, new improved portable colorimeters are used (Yadava, 1986; Marquard and Tipton, 1987). The readings correlated well with the extractable chlorophyll content. *N,N*-Dimethylformamide was used as an alternative solvent because it does not require grinding (Moran and Porath, 1980).

Alternatively, spectrophotometers are used for nondestructive color determination. For determination of reflectance and transmittance, the wavelength at which the highest correlation exists between measurements and color is chosen. This correlation was checked in citrus and soybean leaves by several authors (Benedict and Swidler, 1961; Turrel et al., 1961). For leaves of eight crops, including lettuce and spinach, this inverse proportional relation between reflectance and total chlorophyll and carotenoid content was optimal at 550 nm and more suitable for chlorophyll only (Thomas and Gausman, 1977).

For ripening tomatoes it was found that light transmittance techniques applied were more sensitive to small changes in color than the light reflectance technique. Measurements at the longest wavelength tested (595 nm) were highly correlated with visual color ratings during ripening (Jahn, 1975).

A reflectance spectrophotometer (Agtron) designed to measure green (546 nm) and red (640 nm) reflectance of a rotating object and to display the mathematical ratios of the two color modes in percentages is especially suitable for changes in developing fruits in which green and red are dominant colors, as in the tomato (Kader and Morris, 1978).

Separation of Plant Pigments

Separation by Partition Between Solvents

By 1864, as a result of studying chlorophyll spectroscopically, investigators concluded that chlorophyll is a mixture of two pigments. In 1873 the two pigments were isolated in relatively pure form by solvent partition between methanol and carbon disulfide (Humphrey, 1980). The method, which is based on a difference in solubility of the two pigments, is still used to separate chlorophylls and their derivatives, since the phytylated esters are epiphasic and the unphytylated pigments are hypophasic. Thus, their separation can be achieved by shaking an aqueous acetone chlorophyll solution with petroleum ether epiphase because the hydrolyzed pigments will remain in acetone. This separation is still used in the chlorophyllase assay, following the previous transformation of chlorophylls into pheophytins, which are photostable (Holden, 1976; Amir-Shapira et al., 1987).

A method for fractionation of chloroplast pigments by means of multiple liquid–liquid partition has been developed in order to avoid allomerization, which is considerable when conventional methods are used (Hynninen and Ellfolk, 1973).

Chromatography

Chromatography is the most important technique for separating chemical substances. Chromatography is based on a substance's relative adsorption rate on a stationary phase (the adsorbent) from a mobile phase, either liquid or gas. Partition chromatography, which involves liquid–liquid phases, is also used, the stationary phase being the water bound to the granules of the adsorbent. The solute molecules are partitioned between the stationary liquid phase and the mobile phase, which is a water-immiscible solvent.

The first separation of the leaf pigments on a $CaCO_3$ column developed with carbon disulfide into two green pigments and four yellow pigments was achieved by the Russian botanist Tswett (1906) and may be considered the beginning of chromatography. Modifications and applications of Tswett's method have been reviewed by Strain and Sherma (1972).

Column Chromatography. For separating chlorophylls and their derivatives by column chromatography, multiple adsorbents such as alumina, magnesia, polyethylene powder, powdered sucrose, starch, and cellulose are used. Organic adsorbents, particularly sucrose and cellulose, are preferred because they do not alter the chlorophylls. When leaf pigments were separated on a sucrose column developed with petroleum ether, β-carotene, being the least polar, was eluted first (Smith and Benitez, 1955). The other pigments were adsorbed in order of increasing polarity: chlorophyll *a,* lutein and zeaxanthin (together), chlorophyll *b,* violaxanthin, and neoxanthin being the most polar, on top.

The chlorophyll isomers, chlorophylls *a′* and *b′*, were discovered by chromatography on a sugar column on which they separated from chlorophylls *a* and *b* (Strain and Manning, 1942).

Pigment alteration occurring in frozen green vegetables, before and during cooking, was investigated by chromatography on a sugar column developed with petroleum ether containing increasing amounts of acetone (Sweeney and Martin, 1961). This procedure permitted the separation of the pheophytins (gray bands) into pheophytin *a,* following β-carotene, and pheophytin *b,* following chlorophyll *b.*

The pigments of thermally processed spinach were separated on a sugar–starch column (Tan and Francis, 1962). The protochlorophylls from pumpkin seed coats were separated by a combination of polyethylene and sugar-column chromatography (Houssier and Sauer, 1969).

Column chromatography is still widely used for preparative purposes when absorbents, such as diethylaminocellulose, Sephadex, or powdered sugar, are used. The chlorophyll source is primarily spinach leaves (Sato and Murata, 1978; Iriyama and Yoshiura, 1979; Iriyama et al., 1979). Highly purified chlorophylls from spinach leaves can be prepared in a short time by combined column chromatography on two kinds of cross-linked agar: DEAE-Sepharose and Sepharose CL. The former chromatography eliminates carotenoids, pheophytins, and chlorophyllides; the latter efficiently separates chlorophylls *a* and *b* (Omata and Murata, 1980, 1983).

Paper Chromatography. Paper chromatography of chloroplasts has been reviewed by Šesták (1980). It has been almost totally replaced by thin-layer chromatography, which is more sensitive. Two-dimensional paper chromatography was used to separate leaf pigments in the study of leaf color changes from spring to autumn (Sanger, 1971). Circular chromatography performed on Whatman No. 3 paper is still used for chlorophyll preparation (Burke and Aronoff, 1980).

Thin-Layer Chromatography. Thin-layer chromatography (TLC) techniques and precautions to be taken during pigment analysis are described by Stahl (1967). TLC methods for plastid pigments have been reviewed by Šesták (1975, 1982). Early methods based on partition chromatography on a mixture of silica gel and kieselguhr or on cellulose are still used to separate plastid pigments (Hager and Meyer-Bertenrath, 1966; Schneider, 1966).

Phytylated and nonphytylated chlorophylls of etiolated and green leaves were separated on silica gel H by adsorption chromatography (Seliskar, 1966).

Chlorophylls and their derivatives—pheophytins, pheophorbides, and chlorophyllides—were separated on cellulose (Bacon, 1965). Chlorophyll degradation products from processed pea puree (Buckle and Edwards, 1969) and of green and processed peas (Aczél, 1971a,b) were separated by this method.

Cellulose is suitable for separating chlorophyll derivatives and chlorophyll epimers (Sievers and Hynninen, 1977).

A recently described method using sucrose layers is suitable for analyzing C-10 epimeric chlorophylls and their oxidation products as well as chloroplast carotenoids (Sahlberg and Hynninen, 1984).

Chlorophylls and derivatives, as well as zinc and copper complexes of pheophytins, could be effectively separated by reversed-phase (r-p) TLC by using kieselguhr plates G impregnated with peanut oil (Jones et al., 1972a,b). More recently, chlorophylls were separated on oil-impregnated cellulose layers by reversed-phase chromatography (Scholz et al., 1981).

Separation of chlorophylls esterified with different C-20 diterpene alcohols was first achieved by TLC on silica gel that was previously buffered and impregnated with silver nitrate. The chlorophylls were first converted to pheophytins, which were then chromatographed. The separated pheophytins were saponified, and the alcohols were analyzed by gas chromatography and mass spectrometry (Rüdiger et al., 1976; Schoch et al., 1977).

A modern method, reversed-phase high-performance TLC, used C-18 bonded silica gel to separate chlorophylls and their derivatives (Susuki et al., 1987).

High-Performance (Pressure) Liquid Chromatography. Gradually, standard chromatographic methods are being replaced by high-performance liquid chromatography (HPLC), which is faster, more sensitive, and highly reproducible. In r-p liquid chromatography, which is more extensively used, the adsorbent is chemically bonded; hence, the elution is similar to that in the conventional partition system, with the more polar compounds eluted first. HPLC is widely used for analyzing plant pigments. The methods have been reviewed by several authors (Schwartz and von Elbe, 1982; Garrido-Fernández and Mínguez Mosquera, 1983; Lichtenthaler, 1984).

The first separation of chlorophylls by HPLC was achieved on a r-p column (Evans et al., 1975). The separation of spinach chloroplast pigments, chlorophylls, and carotenoids was successful, but for preparative use the separation time was 4.5 h (Eskins et al., 1977). The method was subsequently ameliorated and modified for etioplast analysis (Eskins and Harris, 1981).

A microscale method, using silica gel, separated chlorophylls, including the epimers *a′* and *b′*, pheophytins, and carotenoids (Iriyama et al., 1978). Chloroplast pigments, equivalent to 1 mg fresh weight of spinach leaves, were separated in 15 min by an isocratic HPLC method (Stransky, 1978).

Reversed-phase HPLC using octadecyl or octyl silylated silica gel as the stationary phase was first used by Braumann and Grimme (1981). It allowed the simultaneous separation of chlorophylls *a* and *b,* chlorophyll isomers, pheophytins *a* and *b,* and all the chloroplast carotenoids from a single sample within a short analysis time. This method was applied to pigment analysis of spinach chloroplasts; using the same method the pigments of chlorophyll-protein complexes from spinach thylakoids were analyzed (Braumann et al., 1982). A modified r-p HPLC was used to analyze the pigments in leaves, cotyledons, isolated chloroplasts, and subplastid fractions (thylakoids and envelopes) of spinach (Grumbach, 1983a).

A r-p HPLC method was developed enabling the separation of twelve pigments, chlorophylls and derivatives, to monitor changes in chlorophyll during thermal processing of spinach (Schwartz et al., 1981). For the first time pyropheophytins were detected in canned spinach. The failure to detect them earlier may be attributed to the lack of resolution of older methods and the identical spectra of pheophytins and pyropheophytins. The evidence for the identification of pyropheophytins is their lower polarity induced by the removal of the C-10 carbomethoxy group ($-CO_2CH_3$). This modification increases the retention time in r-p chromatography. The method was subsequently modified by substituting an isocratic solvent for the gradient elution system (Schwartz and von Elbe, 1983), and was then used to determine individual chlorophylls and their derivatives in thermal processed vegetables and to follow the kinetics of chlorophyll degradation to pheophytins and pyropheophytins. The method can be used to survey commercial products for the presence of these chlorophyll derivatives.

Chlorophylls and their derivatives have also been separated by other HPLC methods. See Burke and Aronoff (1981) and Falkowski and Sucher (1981).

Separating pheophytins and chlorophyllides esterified with different C-20 diterpene alcohols was successful by HPLC (Schoch, 1978; Schoch et al., 1978). A r-p HPLC method using an octadecyl silica column eluted with 100% methanol resolved five chlorophylls and four protochlorophyll species, esterified with different alcohols, within 22 min (Shioi and Sasa, 1982, 1983a; Shioi et al., 1983). However, the method did not allow separation of nonesterified chlorophylls, which was done with ion-suppression HPLC (Shioi et al., 1984).

Recently, polyethylene-based HPLC was used to separate mono- and divinyl chlorophyllides and other pigment mixtures (Shioi and Beale, 1987). Polyethylene proved to be a superior alternative phase to the known reversed-phase materials for chlorophyll separation.

An HPLC method for separating zinc and copper chelates of pheophytins *a* and *b* has also been developed (Schwartz, 1984). This method allows separation of allomerized compounds that are readily formed during the chelation reaction.

More and more HPLC methods are being applied to the analyses of chloroplast pigments in green vegetables (McMahon and Gladon, 1984; Khachik et al., 1986). The HPLC method developed by Wright and Shearer (1984) was used to investigate chlorophylls and carotenes of green vegetables grown in Japan (Izaki et al., 1986). This method separates carotenoids and chlorophylls and their degradation products in a single separation with a greater resolution than has previously been reported. A high-efficiency extraction, based on sonication and centrifugal filtration, removes cell debris from the initial pigment extract within 2 min. Such rapid extraction and cell debris removal minimize the formation of isomers and the contact time of the pigments with any remaining active enzyme.

Another new HPLC procedure allowed C-10 epichlorophyll (Chl *a*′) and pheophytin *a* to be detected in the leaves of 15 plant species (Watanabe et al., 1985b). Artifact formation has been rigorously excluded, and the column temperature of 7°C considerably enhanced the sensitivity.

The previously described methods for separation of chlorophylls and derivatives (Schwartz et al., 1981) were modified and applied to separate chlorophylls and pheophytins in fresh and frozen peas (Forni et al., 1988).

THE CHLOROPLAST

The chlorophylls are located in small, subcellular organelles called plastids, and are responsible for their green color and their name, *chloroplasts.* They are the site of photosynthesis. Plastids are dealt with in the comprehensive book by Kirk and Tilney-Basset (1978). They are found primarily in leaves and other green tissues, such as stems, immature fruits, and immature inflorescence vegetables.

The chloroplast is a lens-shaped body with diameter ~ 7μm and 3 μm thick. Its structure has been intensively investigated by electron microscopy. The chloroplast is surrounded by an envelope, which is a double membrane, that acts as a selective barrier to the transport of various metabolites into and out of the chloroplast. It also plays a role in the formation of new internal lamellae.

The chloroplast is filled with a matrix or *stroma.* Embedded within the stroma is a complex system of membranes or lamellae. The basic subunits of the lamellae are double membranes closed in on themselves to form flattened sacks called *thylakoids* (from the Greek *thylakos,* meaning sacklike). The thylakoids occur in regular packed stacks called *grana.* There are 5–20 thylakoids in a granum. In the chloroplasts of shade plants there are 100 thylakoids per granum (Meier and Lichtenthaler, 1981). Approximately every other thylakoid of the granum extends on one side or the other, thereby interlinking the grana by the stroma thylakoids. Due to these interthylakoid and intergranal connections, the thylakoid system constitutes a single, enormously complex, membrane-enclosed cavity. Although the concept that the thylakoid is a separate membrane-bound space is no longer thought to be true, it is still used for interpreting the chloroplast structure (Kirk and Tilney-Basset, 1978).

The stroma contains various particulate structures, such as ribosomes and strands of DNA. Chloroplasts have their own genome, and chloroplast ribosomes differ from cytosolic ribosomes. Starch grains are also usually present. Stroma components present in all plastids are the osmiophilic globules or plastoglobuli, which are apparent in sections fixed with osmium. They contain mostly the excess plastid quinones, but no chlorophylls.

Chemical Composition

The water content of the chloroplast is about 70–80%, and protein represents about 60% of the dry matter. There are two kinds of proteins: structural proteins, which are bound in the lipoprotein complex of the lamellae, and water-soluble proteins, the enzymes, which are found in the complex and in the stroma. The lipids bound in the lamellae amount to about 30–40%; the pigments, chlorophylls, and carotenoids, to 5–10% of the total dry matter.

In higher plants, chlorophylls *a* and *b* are present, and the ratio *a/b* is usually between 2.5 and 3.5.

The main carotenoids of chloroplasts in higher plants are β-carotene, lutein, violaxanthin, and neoxanthin. Chlorophylls and carotenoids exist in chloroplasts as chromoprotein complexes associated with the membrane structure.

Development

During leaf ontogenesis in plants growing under normal day–night conditions, chloroplasts develop from proplastids. *Proplastids* are small, colorless, undifferentiated plastids that occur in meristematic cells. They are the precursors of the more differentiated types of plastids, such as chloroplasts, etioplasts, and amyloplasts.

As the differentiation of the meristematic cell begins, the proplastids differentiate into chloroplasts. The ultrastructural changes involve the gradual formation of the thylakoids, mainly by invagination of the inner plastid membrane and their gradual stacking through invagination and protrusion processes.

Chloroplast development was followed in primary leaves of *Phaseolus vulgaris* by Whatley (1974). Leaf duration spanned 14 weeks from germination to senescence. For about 7 weeks the plastid remained a proplastid. After germination, development into a mature chloroplast lasted for 1 week. Changes in chloroplast number per cell during leaf development in spinach were followed by Possingham and Saurer (1969). A fivefold increase in the number of chloroplasts per cell took place over a growing period of 10 days, during which time the area of the leaves increased from 1 cm^2 to 50 cm^2. In young leaves, differentiated chloroplasts appeared to be undergoing division by constriction.

The characteristic plastids formed in etiolated leaves of plants grown in the dark are called *etioplasts*. Their most striking feature is the presence of usually one or more semicrystalline bodies formed of interconnected membraneous tubules in a regular pattern, and called a *prolamellar body,* because under illumination these membranes develop into thylakoids. All knowledge about chloroplast differentiation and chlorophyll biosynthesis is based on this conversion. Etioplasts are found in the inner seed coat cells of Cucurbitaceae fruits (see p. 47).

In addition to leaves, chloroplasts may be found in other plant organs, such as roots and tubers. Young carrot roots, grown from seeds, become green on continuous illumination (Wrischer, 1974). The proplastids develop gradually into chloroplasts over a period of 3 weeks. In illuminated potato tubers the ultrastructural changes are different. They involve the transformation of amyloplasts (mature plastids filled with starch grains) into chloroplasts (see p. 45).

Function

The function of the chloroplast is to capture light energy and use it for the conversion of carbon dioxide to carbohydrate. The photosynthetic process occurring in all oxygen-evolving plants can be represented by the equation

$$\underset{\text{carbon dioxide}}{6\,CO_2} + \underset{\text{water}}{6\,H_2O} \xrightarrow{\text{light}} \underset{\text{glucose}}{C_6H_{12}O_6} + \underset{\text{oxygen}}{6\,O_2}$$

In addition to carbohydrates, amino acids, proteins, and other organic compounds are also synthesized, and molecular oxygen is liberated. Photosynthesis is the source of all forms of food and of oxygen in the Earth's atmosphere.

The complex process of photosynthesis includes a photochemical or light stage and a dark stage that involves chemical reactions catalyzed by enzymes. In the light reactions, which take place in the thylakoid membrane system, the energy of the sun is captured in two relatively unstable molecules: reduced nicotinamide dinucleotide (NADPH) and adenosine triphosphate (ATP). In the dark reactions, which take place in the stroma of the chloroplast, ATP and NADPH supply the energy and reducing power needed to form glucose from carbon dioxide.

The photosynthetic pigments in the form of membrane–protein complexes are organized into two functional assemblies, called *photosystem I* (PS I) and *photosystem II* (PS II), that carry out two different light-dependent processes.

Most pigment complexes, about 99% of the total chlorophyll, function as antennae, light-harvesting complexes. They include chlorophylls *a* and *b*, other accessory pigments, and carotenoids, so nearly the whole region of the spectrum is exploited. The incident photons absorbed produce an electronic excitation that is transferred ultimately to the pigment complex of the reaction center, where the actual conversion of light to chemical energy occurs.

In PS I the reaction center contains chlorophyll *a* associated in a specific way with protein and has its absorption peak at 700 nm (P_{700}). The reaction center of PS II contains another form of chlorophyll *a* (P_{680}) and absorbs at a shorter wavelength. Recent evidence is that a chlorophyll *a′* dimer may constitute the PS I reaction center. This demonstration is the first of the association of chlorophyll *a′* with any functional site in vivo (Watanabe et al., 1985a).

Photosystem II is concerned with the removal of hydrogen from water, which is split into oxygen, protons, and electrons, and PS I is concerned with the reduction of NADP. Electrons flow from PS II to PS I through a series of intermediate electron carriers consisting of proteins and lipids, of which most are known. During this electron flow, protons move from outside the thylakoid membrane to within. ATP formation is linked to proton movement. The mechanism of photophosphorylation, in which ATP is formed from ADP, is not yet totally understood.

The Photosynthetic Unit

Photosynthetic membranes are formed from repeated assembled photosynthetic units, each consisting of antenna chlorophyll–proteins and reaction center complexes linked to one complete set of electron carriers. Even if the exact model is not well characterized, each reaction center of each photosystem is generally associated with an intermediate small antenna and a second, larger antenna that forms a separate entity.

At least four chlorophyll–protein complexes are found in a photosynthetic unit. Two have been well characterized: the P_{700} Chl *a*-protein complex, and the light-harvesting Chl *a/b*-protein complex (Thornber et al., 1979). The former represents the pigment molecules associated with the PS I reaction center; the latter, associated with PS II, represents the major light-harvesting Chl *a/b*-protein complex or antenna. Two additional complexes containing chlorophyll *a* are less well known, one of which is assumed to form the reaction center of PS II (Anderson et al., 1978). Several other, minor, light-harvesting and reaction-center chlorophyll proteins have been described in the literature (Lichtenthaler et al., 1981; Braumann et al., 1982; Bassi et al., 1985).

The P_{700} Chl *a*–protein complex is ubiquitous in the photosynthetic plant kingdom. The light-harvesting Chl *a/b*–protein complex transfers most of the absorbed light energy to the reaction center of PS II. It contains all the chlorophyll *b* of the chloroplast together with an equimolecular quantity of chlorophyll *a,* thus accounting for more than 50% of the total chlorophyll. Variations in its content, due to genetic or environmental factors, are fully explained by changes in both chlorophyll contents and chlorophyll *a/b* ratio.

Spectroscopic Properties of Chlorophyll in vivo

In vivo the chlorophylls are bound into lipoprotein membranes and occur there in several distinct spectroscopic forms. Chlorophyll *a* in living cells has a much broader absorption peak (~ 676 nm) than it does in extracts. At low temperature (77 K) there is a superior resolution that permits the detection of five spectral forms of Chl *a* with maxima at 662, 669, 677, 686, and 690 nm in the P_{700} Chl *a*–protein. In the light-harvesting Chl *a/b*–protein, four spectral forms, with maxima at 649, 650, 670, and 677 nm (the shorter peak belonging to Chl *b*), are found. The different forms are probably different aggregates formed in the thylakoid membrane.

The fluorescence emission spectra also give useful information on the state of organization of chlorophyll within the thylakoids. As with the absorption spectra, the resolution is superior at low temperatures. The chlorophyll–protein complexes can be identified by their specific fluorescence emission peaks at 77 K. These peaks are the Chl *a/b*–protein complex or light-harvesting complex of PS II (λ_F 686 nm), possibly the reaction center–antenna complex of PS II (λ_F 696 nm) and the reaction center–antenna complex of PS I (λ_F 730–740 nm) (Papageorgiou, 1975).

BIOSYNTHESIS

Chlorophyll biosynthesis has been extensively investigated. Recent progress in the field has been recorded by Bogorad (1976), Castelfranco and Beale (1981, 1983), and Leeper (1985).

Biochemical investigation of chlorophyll biosynthesis has been carried out with angiosperms and limited to only three families: Cucurbitaceae, Leguminosae, and Gramineae,thereby conferring an important role in science to the vegetables.

Research of the biochemical pathway was initiated around 1950 by Granick, who proposed a one-way biosynthetic pathway, which was generally accepted. The pathway from δ-aminolevulinic acid (ALA) to chlorophyll *a* is illustrated in Figure 3.

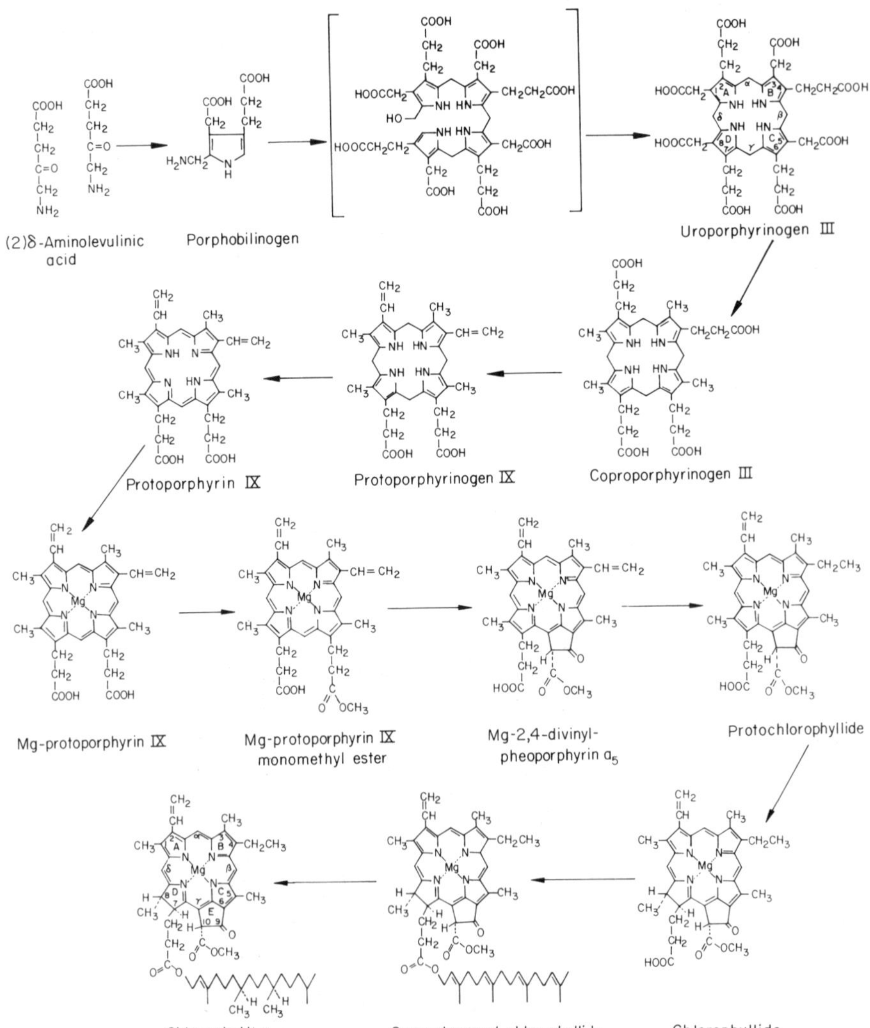

Figure 3. The pathway from ALA to chlorophyll *a* with the principal intermediates. (*From Castelfranco and Beale, 1983; reproduced, with permission, from the* Annual Review of Plant Physiology, *Volume 34, ©1983 by Annual Reviews Inc.)*

Dimerization of the first precursor ALA gives the monopyrrole porphobilinogen. Four molecules of porphobilinogen are condensed to give, first, the linear tetrapyrrole 1-hydroxymethylbilane, which cyclizes to uroporphyrinogen III (urogen), the first tetrapyrrole macrocycle. Subsequently, the side-chain substituents (acid acetic and propionic) are modified, and then the macrocycle is oxidated. Magnesium insertion, the first specific step in the chlorophyll pathway, is followed by the formation of a fifth ring, the cyclopentanone ring characteristic of chlorophylls. The fourth pyrrole ring is then reduced, and its propionic acid residue is esterified with a tetraisoprene alcohol, phytol. Thus, chlorophyll is formed.

Biosynthesis occurs entirely in the plastids. The various steps involved, each catalyzed by specific enzymes, are succinctly described as follows.

Step 1. Formation of δ-Aminolevulinic Acid

δ-Aminolevulinic acid is the key precursor of porphyrins, the source of the carbon and nitrogen in their molecules. Labeled ALA was enzymatically incorporated into porphobilinogen by spinach extracts, as demonstrated more than 30 years ago (Granick, 1954).

The classical path for ALA formation, the *Shemin pathway,* was accepted for many years. It involves the condensation of succinyl CoA with glycine, catalyzed by ALA synthetase. Only two uncommon plant tissues, a soybean callus culture and the greening skin of cold-stored potatoes, showed ALA synthetase activity (Wider de Xifra et al., 1971; Ramaswamy and Nair, 1973, 1974, 1976; Battle et al., 1975). Recent findings of ALA synthetase activity consider the Shemin pathway contributing to the synthesis of nonplastid tetrapyrroles (Kasemir, 1983).

In higher plants the generally accepted pathway for ALA formation is the *C-5 pathway* located in the plastids. ALA may be formed from the intact 5-carbon skeleton of either glutamate or α-ketoglutarate (Beale, 1978). The biochemical routes are shown in Figure 4.

Conversion of α-ketoglutarate to ALA involves a reduction to 4,5-dioxovalerate (DOVA), followed by a transamination. In the second route, the glutamate is reduced to glutamate-1-semialdehyde (GSA), which by an intramolecular transamination is transformed into ALA.

The formation of ALA from labeled C-5 precursors was first demonstrated in greening leaves of barley, red kidney beans, and greening cucumber cotyledons (Beale and Castelfranco, 1974). Similar results were obtained with maize leaves (Meller et al., 1975), greening cucumber cotyledons (Weinstein and Castelfranco, 1978), and other plants in which one or the other biosynthetic routes occurred.

Proofs for the first route from α-ketoglutarate via DOVA were the isolation of DOVA in maize leaves (Porra et al., 1980) and the demonstration of the involved enzymes. Two enzymes, partially purified from a maize leaf ex-

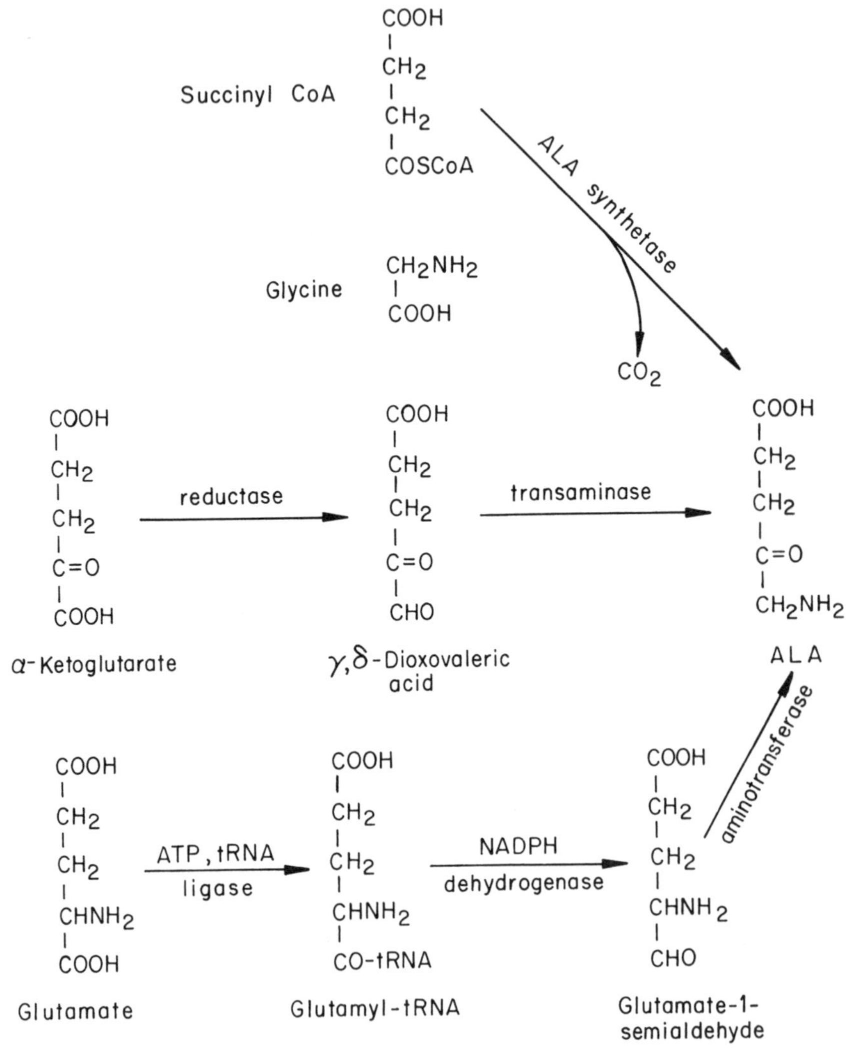

Figure 4. δ-Aminolevulinic acid (ALA) formation: Shemin pathway; C-5 pathway. (*From Castelfranco and Beale, 1983; Schneegurt and Beale, 1988; reproduced, with permission, from the* Annual Review of Plant Physiology, *Volume 34, © 1983 by Annual Reviews Inc.*)

tract, catalyzed the reduction of α-ketoglutaric acid to DOVA (Lohr and Friedmann, 1976). δ-Aminolevulinic acid transaminase was found in beans (Gassman et al., 1966). Recently, a specific aminotransferase for L-alanine and DOVA was found in cucumber seeds (Hayashi and Noguchi, 1983). The alanine: 4,5-dioxovalerate aminotransferase was found to transfer irreversibly the amino group from L-alanine to DOVA.

An enzyme activity lower than in cucumber seeds was detected in dry seeds of other plants, such as adzuki bean, broad bean, maize, peas, and *Cucurbita moschata.* The biosynthesis of α-amino[^{14}C]levulinic acid from [^{14}C]glutamate was catalyzed by chloroplast preparations from barley leaves (Gough and Kannangara, 1977). Subsequently the enzymes involved in this route, isolated from algae and higher plants, were fractionated, purified, and characterized (Wang et al., 1981; Weinstein et al., 1987). It appeared that in greening barley and other plants, as well as in algae, ALA is synthesized from glutamate by an unusual pathway. The glutamate (respectively the α-carbon of glutamate) is activated by ligation to tRNA before reduction (Huang et al., 1984; Kannangara et al., 1984; Weinstein et al., 1987). The RNA required for the biosynthesis of ALA was characterized by Schneegurt and Beale (1988).

Step 2. Pyrrole (Porphobilinogen) Formation

Porphobilinogen (PBG) is formed through an asymmetrical condensation of 2 molecules of ALA with elimination of 2 molecules of H_2O (see Figure 3). δ-Aminolevulinic acid dehydratase (porphobilinogen synthetase) has been widely studied. It is found in many bacteria, fungi, and higher plants. The higher plants studied are spinach leaves and pea seedlings (Gibson et al., 1955; Schneider, 1970), greening leaves of beans (Sluiters-Scholten et al., 1973), young corn (Schneider, 1973), and radish cotyledons (Shibata and Ochiai, 1976). Huault and Bruyant (1984) have purified the enzyme from radish seedlings.

The action mechanism of ALA dehydratase has also been studied. The molecule contributing to the propionic acid side chain is the one that is initially bound to the enzyme (Jordan and Seehra, 1980). Its subcellular location, recently investigated, showed that it is exclusively a plastid enzyme (Smith, 1988).

Step 3. Cyclic Tetrapyrrole Formation (Uroporphyrinogen III)

The intermediate step to cyclic tetrapyrrole formation is the formation of the linear tetrapyrrole 1-hydroxymethylbilane through head-to-tail condensation of four PBG molecules with elimination of NH_3, catalyzed by the enzyme porphobilinogen deaminase (hydroxymethylbilane synthetase). Cyclization occurs very rapidly in the presence of cosynthetase, the ring-closing and rearranging enzyme, with reversion of ring D to give uroporphyrinogen III. Of the four possible isomers, uroporphyrinogen III is the only significant isomer, the intermediate in chlorophyll synthesis. PBG deaminase has been found in spinach leaves, cosynthase in wheat germ (Bogorad, 1958a,b). Properties of PBG deaminase isolated from wheat germ and the leaves of Swiss chard and beans have been described by Frydman and Frydman (1970) and Bianchi and Stegwee (1979).

The mechanism of uroporphyrinogen III biosynthesis has been studied for years (Jordan and Sechra, 1979; Battersby et al., 1981). In recent works the purification and characterization of both enzymes from *Euglena gracilis* have been reported (Hart et al., 1984; Hart and Battersby, 1985).

Step 4. Porphyrin Side-Chain Modifications

This step involves the decarboxylation of the acetic and propionic acid residues attached to each of the four pyrrole rings. These transformations are catalyzed by two enzymes. Uroporphyrinogen decarboxylase catalyzes the decarboxylation of the acetic acid side chains to methyl groups, transforming uroporphyrinogen III, which possesses eight carboxy groups, to coproporphyrinogen III, a tetracarboxy compound. The intermediate hepta-, hexa-, and pentacarboxy analogues can be separated by r-p HPLC (Francis and Smith, 1984). Uroporphyrinogen decarboxylase has been found in numerous plants (Chen and Miller, 1974). From the six plant species tested (tobacco, cucumber, spinach, soybean, corn, and barley), tobacco leaf extract had the highest activity, about 9.5 times that of spinach and 1.6 times that of cucumber leaves.

The second enzyme, coproporphyrinogen oxidase, catalyzes the oxidative decarboxylation of two propionic acid side chains to vinyl groups (at rings A and B). Coproporphyrinogen III is thus transformed into protoporphyrinogen IX. Coproporphyrinogen oxidase activity has been detected in numerous plants, among them cucumber (Rebeiz et al., 1970).

Step 5. Oxidation of Protoporphyrinogen IX to Protoporphyrin IX

In this step six hydrogen atoms are removed from protoporphyrinogen, resulting in the formation of a stable chromophore—a planar system of conjugated double bonds that determines the spectroscopic properties of chlorophylls. Because porphyrinogen tends to dehydrogenate spontaneously in light in the presence of molecular oxygen, the reaction was not considered to be enzymatic. Only recently has protoporphyrinogen oxidase been observed in plants (Jacobs et al., 1982). The enzymatic oxidation occurring in plants has been studied. High oxidation rates were demonstrated in etioplasts, chloroplasts, and mitochondria from young shoots; much lower rates were obtained in chloroplasts from older barley or mature spinach and in mitochondria from potatoes (Jacobs and Jacobs, 1984). Protoporphyrinogen oxidase from etiolated barley was purified and partially characterized for the first time from a plant by Jacobs and Jacobs (1987).

Step 6. Magnesium Chelation of Protoporphyrin IX to Mg Protoporphyrin IX

Protoporphyrin may chelate with metals; in the animal kingdom it chelates with iron (Fe) to form hemoglobin, in plants with magnesium to form Mg protoporphyrin IX. The chelation of magnesium into protoporphyrin IX, catalyzed by Mg protoporphyrin chelatase, constitutes the first specific step in chlorophyll biosynthesis. The in vitro insertion of magnesium into exogenous protoporphyrin first succeeded with whole cells of *Rhodopseudomonas spheroides* (Gorchein, 1972), and later with a cell-free system from higher plants, namely greening cucumber cotyledons (Smith and Rebeiz, 1977).

Chelatase properties were studied by several authors (Castelfranco et al., 1979; Pardo et al., 1980; Fuesler et al., 1981). With HPLC the chelation products can be identified unequivocally (Richter and Rienits, 1980, 1982).

The intraplastidic localization of Mg chelatase was studied in intact developing cucumber chloroplasts. It is thought to be located in the chloroplast envelope, because its activity is inhibited by a membrane-impermeable organic mercurial (Füsler et al., 1984).

Step 7. Esterification of Mg Protoporphyrin IX

In this step the propionic acid side chain of position 6 is esterified to form Mg protoporphyrin IX monomethyl ester (MPE). The enzyme responsible for the formation of MPE is called *S*-adenosyl-L-methionine-magnesium protoporphyrin *O*-methyltransferase. It was found in bacteria, algae, and higher plants, such as developing *Zea mays* chloroplasts and beans (Radmer, 1966; Radmer and Bogorad, 1967). For a long time the sequence of Mg chelation and methylation was not exactly known because the reaction products are generally a mixture of Mg protoporphyrin and its ester. With HPLC, the specific chelation of magnesium was proved and verified that it precedes methylation. Mg protoporphyrin IX could serve as a substrate for the methylation step (Richter and Rienits 1980; Füsler et al., 1982).

Step 8. Isocyclic Ring Formation (Protochlorophyllide)

This step involves modification of the side chains at positions 4 and 6. The isocyclic ring, the fifth ring (E) in the molecule, is formed from the methylated propionic side chain at position 6 in ring C by β-oxidation and cyclization reactions. The product was called Mg-2,4-divinyl-pheoporphyrin a_5 (MgDVP), which is divinyl protochlorophyllide. It was found in the inner seed coats of *Cucurbita pepo,* along with monovinyl protochlorophyllide (Jones, 1966), and in etiolated barley and cucumber tissues (Belanger and Rebeiz, 1979).

The other modification—reduction of the vinyl group at position 4 to an ethyl group—was long a matter of controversy, the sequences of the two modifications not being established. The finding of MgDVP in the inner seed coats of pumpkins was proof that the formation of the isocyclic ring occurs first (Jones, 1966), whereas other experimental data showed that the reduction precedes ring formation. The problem was subsequently resolved by Rebeiz and co-workers, who showed that reduction might occur first at the Mg protoporphyrin stage, as will be further discussed.

The in vitro synthesis of MgDVP was demonstrated with developing chloroplasts of cucumber incubated with Mg protoporphyrin IX or MPE (Chereskin et al., 1982). The enzyme system responsible for isocyclic ring formation is called MPE oxidative cyclase. It was resolved into two enzymic components that were characterized by Wong and Castelfranco (1984). Its location does not seem to be in the chloroplast envelope, as with Mg chelatase, but deeper within the plastid (Füsler et al., 1984).

Step 9. Protochlorophyllide Reduction to Chlorophyllide

In this step—the greening process—the conversion of protochlorophyllide into chlorophyllide *a* occurs through photoreduction (reduction in the light) of ring D, namely, of the double bond between C_7 and C_8 as the dihydroporphyrin (chlorin) macrocycle is formed. The macrocycle contains the chromophore that imparts the green color to chlorophyll *a*. Upon illumination, the absorption maximum in vivo is shifted within 30 s from 635 nm to 678 nm. The process is complex, because various transient intermediate pigment-protein species with different spectroscopic properties are formed (Kasemir, 1983). The greening process was studied on etiolated seedlings that originate from seeds germinated in the dark, which are pale yellow rather than green, as are light-grown plants. When illuminated, they form chlorophyll.

The enzyme responsible for photoreduction is NADPH protochlorophyllide oxidoreductase (Griffiths, 1978). It is located in the prothylakoids of the etioplasts (Lütz et al., 1981).

Step 10. Esterification of Chlorophyllide *a*

In this last step of chlorophyll *a* biosynthesis, the 7-propionic acid residue is esterified with the C-20 isoprenoid alcohol, phytol ($C_{20}H_{39}OH$), to form chlorophyll *a*. Until recently, it was accepted that chlorophyllide *a* is esterified with phytol in a one-step reaction catalyzed by chlorophyllase, an enzyme with both biosynthetic and catabolic activity. In 1977 it was demonstrated that chlorophyllide *a* is esterified first with geranylgeraniol, which is transformed gradually to phytol by three hydrogenation steps in which the intermediates, chlorophyll dihydrogeranylgeranyl ester and tetrahydrogeranylgeranyl ester, are formed (Figure 5).

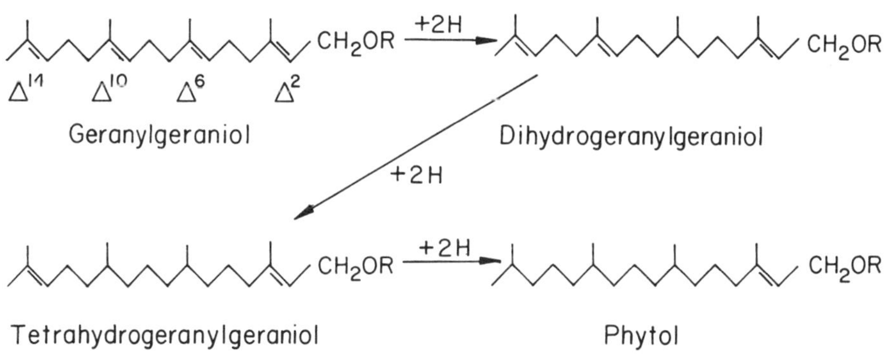

Figure 5. Intermediate esters during chlorophyll biosynthesis. (*From Rüdiger, 1986.*)

All of these chlorophyll esters have been separated and identified (Costes, 1966; Schoch et al., 1977; Schoch, 1978; Schoch and Schäfer, 1978). The pathway via ChlGG is the main pathway in greening etiolated seedlings, as demonstrated by the analysis of newly formed chlorophyll during deetiolation of oat and bean seedlings (Schoch et al., 1977; Schoch, 1978).

In vitro esterification of chlorophyllide by geranylgeranyl pyrophosphate was demonstrated in a cell-free system from maize shoots (Rüdiger et al., 1977). It was observed that the esterifying enzyme is different from chlorophyllase. The enzyme, named *chlorophyll synthetase,* responsible for the esterification of chlorophyllide was detected in etioplast membranes and characterized for the first time (Rüdiger et al., 1980). Subsequently, the enzyme was demonstrated in spinach chloroplasts to be located in the thylakoid membranes (Block et al., 1980; Soll and Schultz, 1981; Soll et al., 1983), and was also found in *Capsicum annuum* plastids (Dogbo et al., 1984). Chlorophyll synthetase may use geranylgeranyl pyrophosphate as well as phytyl pyrophosphate as substrates to form chlorophyll phytyl ester. It was demonstrated that two pathways exist in spinach chloroplasts for the reduction of geranylgeraniol to phytol. In the chloroplast envelope geranylgeranyl pyrophosphate (GGPP) is reduced to phytylPP (PPP), whereas in the thylakoid membranes the reduction of geranylgeraniol as a side chain of chlorophyll takes place (Soll et al., 1983). The pathway via ChlGG is the main pathway in greening etiolated seedlings, whereas the other pathway via PPP operates as the main pathway in green seedlings (Rüdiger, 1986).

Esterification may occur also at the protochlorophyllide level, the mechanism being the same as that acting in chlorophyll *a* formation. Four protochlorophylls esterified with different C_{20} alcohols are widely distributed in etiolated seedlings of higher plants, the relative amount of each protochlorophyll varying with the plant species and its age (Shioi and Sasa, 1983a,b).

With increasing age, successive inactivation of the hydrogenation step has been observed. The four protochlorophylls were not phototransformable; consequently, they are not direct precursors of chlorophyll, but are metabolized slowly during greening.

Step 11. Biosynthesis of Chlorophyll *b*

Chlorophyll *b* differs from chlorophyll *a* in that the 3-methyl group at ring B is oxidized to an aldehyde group. Chlorophyll *a* was considered the precursor of chlorophyll *b,* based on early pulse-labeling experiments. By incubating homogenates of soybean leaves with ^{14}C-labeled chlorophyll *a,* researchers obtained significant yields of ^{14}C-labeled chlorophyll *b* (Ellsworth et al., 1970). The accepted view favored a sequential scheme of chlorophyll *a,* assuming as precursor a special pool of more labile young chlorophyll *a* (Shlyk, 1971). A kinetic analysis of chlorophyllide *a* and chlorophyll *b* formation in etiolated mustard cotyledons suggested that chlorophyll *b* may be formed from chlorophyllide *a* instead of chlorophyll *a* (Ölze-Karow and Mohr, 1978).

Evidence continues to accumulate to indicate that chlorophyllide *b* is involved in the formation of chlorophyll *b*. Chlorophylls and chlorophyllides *a* and *b* were isolated from tea leaves chloroplasts following administration of [^{14}C]ALA. Relative pool sizes suggest that chlorophyllide *b* precedes chlorophyll *b* (Aronoff, 1981). An experiment on chlorophyll synthetase activity with etioplast membranes from etiolated oat seedlings showed that chlorophyllides *a* and *b* are good substrates for the enzyme (Benz and Rüdiger, 1981).

A chlorophyllide *b* pool has been detected in greening cucumber cotyledons and spinach leaves (Duggan and Rebeiz, 1982). The four esters from chlorophyllide *b*—geranylgeranyl ester to the phytyl ester—have all been detected in greening cucumber cotyledons. They can be separated and determined by HPLC (Shioi and Sasa, 1983c). The results suggest that chlorophyll *b* is formed by esterification of chlorophyllide *b* with geranylgeraniol followed by three hydrogenation steps of the alcohol moiety analogue to the pathway for chlorophyll *a* or protochlorophyll phytol formation.

An Alternative Chlorophyll Biosynthetic Pathway

The basic outline of the biosynthetic pathway just discussed is as follows:

$$8\,(\text{ALA}) \xrightarrow{-8H_2O} 4\,(\text{PBG}) \xrightarrow{-4NH_3} \underset{(\text{linear})}{\text{hydroxymethylbilane}}$$

$$\xrightarrow{-H_2O} \underset{(\text{cyclic})}{\text{uroporphyrinogen III}} \xrightarrow{-4CO_2} \text{coproporphyrinogen III}$$

$$\xrightarrow{-2CO_2} \text{protoporphyrinogen IX} \xrightarrow{-6H} \text{protoporphyrin IX}$$

$$\xrightarrow{Mg^{2+}} \text{Mg-protoporphyrin IX} \xrightarrow{\text{esterif.}} \text{Mg protoporphyrin IX monoester}$$

$$\xrightarrow{\text{ring formation}} \text{protochlorophyllide}$$

$$\xrightarrow{\text{reduction}} \text{chlorophyllide} \xrightarrow{\text{esterification}} \text{chlorophyll } a$$

This scheme, proposed by Granick (1950), represents the conversion of a simple amino acid into one chlorophyll *a* chemical species via a linear sequence of reactions.

In contrast to the apparent simplicity of the biosynthesis, chloroplast membranes show a very complex picture, the chlorophyll in situ showing a number of different spectroscopic forms in vivo determined by the environment (free, bound to various proteins, monomeric, dimeric, etc.). Recently, the assumption arose that, in addition to the physical heterogeneity, chemical heterogeneity may also be involved. Furthermore, the order of the sequence may not be as strict as described but may occur at any stage along the pathway.

The reinvestigation of the biosynthetic pathway was done by Rebeiz and co-workers, and the results have been reviewed (Rebeiz, 1982; Rebeiz et al., 1986).

Instead of the accepted single-branched pathway, a four-branched pathway was proposed, consisting of monovinyl (MV) and divinyl (DV) monocarboxylic routes and of monovinyl and divinyl esterified routes. The multibranched pathway is based on the fact that protochlorophyllide and protochlorophyllide ester are not synthesized from one another, but are formed independently from a common precursor (Mattheis and Rebeiz, 1977). Both protochlorophyllide and its ester can exist in two forms, depending on the nature of the side chain on ring B, which may be vinyl or ethyl. It was demonstrated that reduction of the vinyl group to an ethyl group might occur at any stage along the pathway, at first at the Mg protoporphyrin stage (see Figure 3).

The MV and DV analogues of all the intermediates were detected in higher plants (Belanger and Rebeiz, 1982; Wu and Rebeiz, 1984), due to differences in the fluorescence spectra at 77 K (Belanger and Rebeiz, 1980).

The pathway was later expanded to six biosynthetic branches, including two additional MV and DV isomeric monocarboxylic routes.

Subsequently it was found that higher plants differ markedly in their MV and DV protochlorophyllide biosynthetic capabilities, in darkness and in daylight. A survey of about 100 species revealed that plants can be divided into four greening groups based on whether they use acidic monovinyl or acidic divinyl routes (Rebeiz et al., 1986). Plants grown under field conditions may accumulate MV and/or DV protochlorophyllide either at night, in the dark (D), or in daylight (L); the possibilities are D DV/L DV, D MV/L DV, D DV/L MV, and D MV/L MV. The first group that accumulates predominantly divinyl protochlorophyllide at night and in daylight includes plants such as cucumber, mustard, and common purslane. By contrast, most of the higher plants that grow in temperate regions belong to the second group D MV/L DV, including cereals such as corn, wheat, barley, oat, and soybean. Cassava and apple belong to the D MV/L MV group.

The greening pattern, which varies widely among the plants, may be considered a taxonomic parameter for plant classification.

Based on these preliminary results, it appears that the DV greening pattern is more primitive than the MV pattern. Plants that appeared at the beginning of life—algae and gymnosperms—belong to the first group. Plants that appeared more recently belong mainly to the MV/MV or MV/DV, a probable indication that the monovinyl pathway may represent an evolution—adaptation to other light conditions (Rebeiz, 1988). This theory could be disputed.

DISTRIBUTION IN VEGETABLES

Chlorophylls *a* and *b* are ubiquitous in all the edible parts of vegetables, whether roots, stems, leaves, flowers, fruits, or seeds, at least at a certain developmental stage. Certainly, they are found in higher amounts in leaf vegetables and in lower amounts in immature fruits or immature inflorescence vegetables.

The analytical data found in the literature concerning the absolute chlorophyll content in vegetables are scarce.

Although the color of vegetables, their "greenness," is the result of the chlorophyll content, an extremely important qualitative attribute, commercially the results are expressed in terms of arbitrary color values. In investigations on chlorophyll changes during processing or storage, the percentage of chlorophyll loss, not absolute content, is generally determined. Applying HPLC the gaps found in the distribution tables should be completed; variations with species, cultivars, changes during ontogenesis, influence of environmental factors, can easily be followed.

Chlorophyll distribution in vegetables is shown in Table 1. For some vegetables, which were investigated more intensively, the results will be discussed in more detail.

Asparagus
(*Asparagus officinalis* L.) Liliaceae

The genus *Asparagus* belongs to the family of Liliaceae, with about 300 species. Best known is the garden asparagus *Asparagus officinalis,* cultivated as a green vegetable for its succulent spring stalks. The quality of the vegetable is assessed by its color (predominantly green), flavor, and tenderness—that is, its lack of fibrousness. These quality parameters were investigated in an early report in 200 samples of green asparagus grown in the United States (Kramer et al., 1949). The chlorophyll in the asparagus stalk is not uniformly distributed, but its concentration decreases from tip to base. This variation was demonstrated by separately analyzing different segments from tip to base. The results for two different lots are shown in Table 2 (on page 39).

After about 20 years, the chlorophyll *a* and *b* content in asparagus grown in the same region (New Jersey) was separately determined in the upper and lower stalk segments (Wang et al., 1971). The tip segment contained 139 μg (g fresh wt)$^{-1}$ chlorophyll *a* and 74 μg (g fresh wt)$^{-1}$chlorophyll *b*. The second section, 6.3–12.6 cm from the tip, contained 90 μg (g fresh wt)$^{-1}$ chlorophyll *a* and 47.5 μg (g fresh wt)$^{-1}$ chlorophyll *b,* results that prove the amelioration of the cultivar.

Purple asparagus contained 1.1 μg (g fresh wt)$^{-1}$ total chlorophyll, whereas green asparagus contained 30.2 μg (g fresh wt)$^{-1}$; both were grown in Romania (Dobreanu et al., 1983). In asparagus grown in Japan and analyzed with HPLC, the chlorophyll content is lower: 42.0 μg (g fresh wt)$^{-1}$ chlorophyll *a* and 13 μg (g fresh wt)$^{-1}$ chlorophyll *b* (Izaki et al., 1986). Green and white asparagus grown in Thailand had total chlorophyll content of 194–235 and 54 μg (g fresh wt)$^{-1}$,respectively (Ketsa and Uthaiburt, 1987).

(Text continues on page 39.)

Table 1. Chlorophyll Distribution in Various Vegetables

Common Name, Genus, Species	Cultivar	Part of Plant	*Chlorophyll Content* µg (g fresh wt)$^{-1}$ a	b	*Total*	µg (g dry wt)$^{-1}$ a	b	*Total*	*References*
Asparagus (*Asparagus*	Green	Stem			180–300				Kramer et al. (1949)
officinalis L.)					76.0				Nickerson et al. (1956)
		Upper stem	139.0	74.0					Wang et al. (1971)
		Lower stem	90.0	47.5					
			42.0	13.0					Izaki et al. (1986)
	Green				199–235				Ketsa and Uthaiburt (1987)
	White				54.0				
	Violet				1.1				Dobreanu et al. (1983)
	Green				30.2				
Beans, French snap	Green	Pod			20–80				Hoffman and Kanapaux (1955)
(*Phaseolus vulgaris* L.)					31.5				Nickerson et al. (1956)
					41.0				Walker (1964a)
					140.2				Walker (1964b)
								940.0	Groeschel et al. (1966)
					27–67			230–870	Mayland and Dean (1971)
			54.6	17.3					McMahon and Gladon (1984)
					167.9				Muftugil (1986)
Broccoli (*Brassica oleracea*		Head			70.5				Nickerson et al. (1956)
var *italica* Plenck.)		Head			91–150				Gilpin et al. (1959)
		Stems			35.0				
		Buds			170.0				Wu and Salunkhe (1976)
		Leaves			650				
	Corvet	Head				992	366		
	SG1					2016	775		Gross (1979)
	Gem					1177	536		
	Waltham 29				480				Wang (1979)
					226				Shewfelt et al. (1983)

			195.0	65.5		McMahon and Gladon (1984)
			106.0	33.0		Izaki et al. (1986)
			57.7	14.9	79.0	Khachik et al. (1986)
					740	Perrin and Gaye (1986)
	Corvet				166.9	Yang and Henze (1988a)
	SG1		424.9	137.8		Yang and Henze (1988b)
	Corvet		260.4	74.5		
Brussels sprouts (*Brassica oleracea* var *gemmifera* DC.)		Heads			87.5	Walker (1964b)
			47.0	9.0		Izaki et al. (1986)
			46.1	11.4	60.1	Khachik et al. (1986)
Cabbage (*Brassica oleracea* var *capitala* L.)		Head	75.3	25.3		McMahon and Gladon (1984)
			11.3	4.0	17.2	Khachik et al. (1986)
	Market					
	Prize		1200	430		
	Resistant					Rousos et al. (1986)
	Danish		1040	380		
			8.0	2.0		
	Green ball		22.0	6.0		
	Red		27.0	5.0		Izaki et al. (1986)
	Pointed		16.0	4.0		
Cauliflower (*Brassica oleracea* var *botrytis* L.)		Head	3.0	1.0		Izaki et al. (1986)
Carrot (*Daucus carota* L.)		Leaves	504	122		Izaki et al. (1986)
Celery (*Apium graveolens* L.)		Leaves	1143	255		
		Stalks	29.0	7.0		Izaki et al. (1986)
	Green		38.0	10.0		
	Yellow		8.0	2.0		
Chinese cabbage (*Brassica chinensis* L.)		Leaves			960	Wang (1983)
	Young		205.0	63.0		Izaki et al. (1986)
					1020	Ogura et al. (1987)
Chinese chive (*Allium tuberosum* Rottl.)		Leaves	589	178		Izaki et al. (1986)
		Flower stalks	250.0	59.0		

(continued)

Table 1. Continued

Common Name, Genus, Species	Cultivar	Part of Plant	Chlorophyll Content: μg (g fresh wt)$^{-1}$ a	b	Total	μg (g dry wt)$^{-1}$ a	b	Total	References
Chinese mustard (*Brassica campestris* var *chinensis* Czern.)		Leaves	210.0	28.0					Izaki et al. (1986)
Collards (*Brassica oleracea* var *viridis* L.)		Leaves	1009	216					Izaki et al. (1986)
Cress (*Lepidium sativum* L.)	Fine curled	Leaves	2610	730					Czuba and Ormrod (1974)
Cucumber (*Cucumis sativus* L.)		Fruit	56.4	25.9					Jones et al. (1962)
			51.8	25.9					White et al. (1963)
			58.1	24.3					McMahon and Gladon (1984)
			64.0	24.0					Izaki et al. (1986)
Garlic (*Allium sativum* L.)		Sprouts	90.0	28.0					Izaki et al. (1986)
Kale (*Brassica oleracea* var *acephala*)		Leaves	1370	464	1870				Khachik et al. (1986)
			1898	406					Izaki et al. (1986)
Kidney bean (*Phaseolus vulgaris* L.)			118.0	35.0					Izaki et al. (1986)
Kohlrabi (*Brassica caulorapa* L.)	Red	Leaves	90.0	20.0					Izaki et al. (1986)
	Green		80.0	10.0					
Lettuce (*Lactuca sativa* L.)									
butter head lettuce (*L. Sativa capitata*)	Mini	Head	334.0	62.0					
	salad-na		749.0	160.0					
crisp head lettuce			53.0	15.0					
stem lettuce (*L. sativa asparagina*)		Stem	83.0	24.0					
				71.0					Izaki et al. (1986)
leaf lettuce		Leaves	330.0						

	Red		125.0	26.0					
	Sunny		237.0	64.0					
	Green		517.0	154.0					
cos lettuce (*L. sativa longifolia*)			64.0	15.0					
	Grand Rapids		1100	250					Czuba and Ormrod (1974)
Lima beans (*Phaseolus lunatus* L.)		Seeds							
		Mature green				87–93	37–44		Abdul-Baki (1971)
		Dry				2.0–3.7	1.1–1.6		
Malabar spinach (*Basella alba* L.)		Leaves	672	195					Izaki et al. (1986)
Mung beans (*Phaseolus mungo* L.)		Sprouts	10.0						Izaki et al. (1986)
Mustard (*Brassica juncea* L.)		Leaves			1250				Kaur and Manjrekar (1975)
			620	146					Izaki et al. (1986)
Okra (*Abelmoschus esculentus* L.)		Pods			75–85				Fishbach (1943)
	Pusa Sawni		160.0	132.5					Singh and Dankhar (1980)
	Pusa Sawni				123.0				Gupta and Mukherjee (1981)
Onion (*Allium cepa* L.)		Leaves	424.0	86.0					Izaki et al. (1986)
Parsley (*Petroselinum crispum* Mill.)					1982				Yamauchi et al. (1980)
			1577	551	2185				Berset and Caniaux (1983)
			890	288					Izaki et al. (1986)
					2013				Goldschmidt (1988)
Pea (*Pisum sativum* L.)	Green	Seeds			99.0			383–407	Blair and Ayres (1943)
	Alderman								Wagenknecht et al. (1952)
					190.2				Walker (1964a)
	Edgell Freezer		98.0	33.8					Buckle and Edwards (1970a)
	Various				188–312				Aczél (1973)
			106.0	12.0					

(continued)

Table 1. Continued

Common Name, Genus, Species	Cultivar	Part of Plant	Chlorophyll Content μg (g fresh wt)$^{-1}$ a	b	Total	μg (g dry wt)$^{-1}$ a	b	Total	References
		Pods (immature)	37.0	8.0					Izaki et al. (1986)
Pepper (*Capsicum annuum* L.)		Fruit			135.0				Walker (1964a)
	Green		37.3	14.6					McMahon and Gladon (1984)
	Sweet		98.0	33.0					Izaki et al. (1986)
			55.0	21.0					
Perilla (*Perilla frutescens* L.)	Green	Leaves	2442	753					Izaki et al. (1986)
Savoy cabbage (*Brassica oleracea* var *sabauda* L.)		Head	10.0	2.0					Izaki et al. (1986)
Spinach (*Spinacia oleracea* L.)		Leaves			576.8				Dutton et al. (1943)
					1300				Kaur and Manjerkar (1975)
					1576				Yamauchi et al. (1985)
						6980	2490		Schwartz and von Elbe (1983)
			1380	440					Yamauchi et al. (1985)
			946	202	1270				Khachik et al. (1986)
			829	195					Izaki et al. (1986)
					1510				Ogura et al. (1987)
Turfed stone leek (*Allium ascalonicum* L.)		Leaves	309.0	95.0					Izaki et al. (1986)
Turnip (*Brassica campestris* var *rapa* L.)		Leaves	887.0	192.0					Izaki et al. (1986)
Watercress (*Nasturtium officinale* R. Br.)		Leaves	671	153					Izaki et al. (1986)
Welsh onion (*Allium fistulosum* L.)		Leaves	234.0	68.0					Izaki et al. (1986)
Zucchini squash (*Cucurbita pepo* var *melopepo* L.)		Fruit	73.4	30.8					McMahon and Gladon (1984)

Table 2. Effect of Plant Parts on Total Chlorophyll Content [μg (g fresh wt)$^{-1}$] of Raw Asparagus

Plant parts	*Lot 1*	*Lot 2*
Tip	124	129
First cut	59	66
Second cut	46	57
Third cut	—	46

Source: From Kramer et al. (1949).

Beans
(*Phaseolus* L.) Leguminosae

The genus *Phaseolus* is one of the most important of the Leguminosae family. The edible parts are the immature pods and the seeds. The snap bean or French bean (*P. vulgaris*) is one of the most popular vegetables in Europe and America. The importance of the green color of the pods as a quality parameter for fresh, frozen, and canned products was recognized over 40 years ago (Kramer and Smith, 1946).

The chlorophyll content was investigated in 10 "normal" maturing bean cultivars (Hoffman and Kanapaux, 1955). The results are given in Table 3.

Table 3. Chlorophyll Content [μg (g fresh wt)$^{-1}$] of Fresh Snap Bean Pods

Cultivar or breeding line	*Fall 1952*	*Spring 1953*	*Fall 1953*	*Spring 1954*	*Sieve size average*		*Season average*		*Cultivar average*
					3	*5*	*Spring*	*Fall*	
B2194	59	80	68	72	75	64	76	63	70
Wade	57	66	62	49	63	54	58	59	58
B3077	51	61	64	52	65	49	57	57	57
B2528	50	62	56	54	64	47	58	53	55
B2864	49	60	60	54	62	49	56	54	55
B2086	47	52	48	46	51	45	49	47	48
Seminole	43	49	49	50	53	42	49	46	48
EES No. 206	39	53	45	50	54	39	52	42	47
Logan	41	48	45	38	46	41	43	43	43
U.S. No. 5 Refugee	25	22	28	20	28	20	21	27	24
Average	46	55	53	49	56	45	52	49	51

Source: From Hoffman and Kanapaux (1955).

Discussion. It was found that five-sieve pods consistently contained less chlorophyll per unit of pod tissue than did three-sieve pods.

The difference between the means for all samples of the two sizes was 11 μg, the nearly mature pods having a lighter color than the younger smaller pods. Either the biosynthesis rate of chlorophyll slows down, or the degradation rate increases. Another possibility is that the increased sizes of seeds and pods have a dilution effect on the total chlorophyll content. The majority of varieties and breeding lines contained somewhat more chlorophyll in the spring, for which the lower temperature may be the determining factor.

Chlorophyll metabolism during ontogenesis of snap beans was investigated by Hall (1968). During maturation of the pod, a very rapid increase in total chlorophyll was followed by a rapid decrease with additional maturation. The total chlorophyll of 340 μg in 30-day-old pods increased to 1160 μg in the 40-day-old pods, decreased abruptly to 340 μg in the 46-day-old pods, and then continuously attained 20 μg in 70-day-old pods.

Whereas in normal-green beans during maturation, the pod becomes light brown or tan, in persistent-green beans, the pod remains green even in the dry pod and seed coat. Utilization of the persistent-green character in snap beans, for processing, has been suggested as a means of increasing color uniformity among pods of various physiological maturities (Dean, 1968).

The chlorophyll content of pods of various sieve size of normal- and persistent-green types was determined by Mayland and Dean (1971). The results are shown in Table 4. (The values for fresh weight were recalculated.) The normal-green cultivars were Tendercrop and White-seeded Tendercrop.

Table 4. Chlorophyll Concentration [$\mu g\ (g\ wt)^{-1}$] in Various-Sized Bean Pods and Cultivars

Line or cultivar	Pod sieve size									
	≤3		4		5		6		≥7	
	Dry	*Fresh*	*Dry*	*Fresh*	*Dry*	*Fresh*	*Dry*	*Fresh*	*Dry*	*Fresh*
XIda 69-3	930	47	790	61	670	52	590	47	390	44
Custer	870	67	650	54	520	49	390	42	320	40
Tendercrop	790	59	650	51	530	44	360	40	270	34
WS Tendercrop	850	67	580	48	490	49	380	41	290	37
XIda 266-5	820	60	640	49	480	44	350	41	270	38
XIda 68-5381	840	60	590	42	390	35	290	31	230	27
Mean	850	60	650	51	510	45	390	40	230	37

Source: From Mayland and Dean (1971).

Discussion. Chlorophyll concentration decreased as pod size increased in the permanent-green and normal-green cultivars.

Total chlorophyll per pod was influenced by pod length, its amount increasing from 158 μg pod^{-1} in sizes $\leq$ 3 to 296 μg pod^{-1} in sizes $\geq$ 7. Apparent chlorophyll synthesis occurred in field-grown green snap bean fruits up to 11-mm-diameter pods (sieve size 5 or 6), but a net decrease occurred with further maturation. Dry-mature persistent-green pods contained 300 μg chlorophyll on a dry-matter basis, whereas normal-green pods had none.

Lack of consistent agreement between visual and chemical methods to rank cultivars for pod color suggested differences in chlorophyll distribution in external versus internal tissue. Pods with chlorophyll concentrated in the exocarp appear darker than those with the same amount of chlorophyll distributed throughout the exocarp, mesocarp, endocarp, and developing embryos. Higher water content also gives the pods a darker appearance. Various other data have been reported on the chlorophyll content in green beans (see distribution in Table 1).

In research using HPLC, chlorophyll *a* and *b* were separately determined. The very different results obtained—54.6 μg (g fresh wt)$^{-1}$ chlorophyll *a* and 17.3 μg (g fresh wt)$^{-1}$ chlorophyll *b* in green bean pods grown in the United States, and 118 μg (g fresh wt)$^{-1}$chlorophyll *a* and 35 μg (g fresh wt)$^{-1}$ chlorophyll *b* in green beans grown in Japan—illustrate the influence of the environment (McMahon and Gladon, 1984; Izaki et al., 1986).

Changes in chlorophylls of green and bleached lima bean (*Phaseolus lunatus* L.) seeds during development and maturation have been investigated (Abdul-Baki, 1971). The chlorophylls were determined in mature-green seeds (5 weeks after pod formation) and in dry seeds and leaves. The results are given in Table 5.

Table 5. Chlorophyll Content [μg (g fresh wt)$^{-1}$] of Leaves and Seeds of the Lima Bean Cultivars G_2 and Green Fordhook 861

		G2		*Green Fordhook 861*	
Tissue	*Maturity stage*	*Chl* a	*Chl* b	*Chl* a	*Chl* b
Leaves	Fully expanded	5814	1909	6181	2360
Seeds	Mature green	93.3	44.0	86.6	37.0
Seeds	Dry; nonbleached	3.7	1.6	2.0	1.1
Seeds	Dry; bleached	0.08	0.1	tr	tr

Source: From Abdul- Baki (1971).
Note: tr = trace amounts.

Discussion. Chlorophyll content per seed increased during development, reached a maximum about 5 weeks after pod formation, then declined sharply during maturation.

Broccoli (*Brassica oleracea* var *italica* Plenck.) Cruciferae

Broccoli is a form of cabbage of the mustard family Brassicaceae, or Cruciferae, an annual plant that bears dense green clusters of flower buds at the end of the central axis and the branches. This immature flower head is eaten as a vegetable before the green buds have opened; fresh broccoli should be dark green with firm stalks and compact clusters.

One of the earliest analyses reports a chlorophyll content of 70.5 μg (g fresh wt)$^{-1}$ (Nickerson et al., 1956). Heads and stems of 12 samples, analyzed together, had a total chlorophyll content of between 90 μg and 150 μg (g fresh wt)$^{-1}$ (Gilpin et al., 1959). In three broccoli cultivars grown in Bonn, Germany, the chlorophyll content in the SG1 cv. was 2016 μg chlorophyll *a* and 775 μg (g dry wt)$^{-1}$ chlorophyll *b,* twice as high as in the Corvet cv. with 992 μg chlorophyll *a* and 336 μg chlorophyll *b,* whereas the third Gem cv. had a content in between, of 1177 μg chlorophyll *a* and 536 μg chlorophyll *b* (Gross, 1979).

Since broccoli has a short shelf life, rapidly turning yellow, there are numerous reports on the quality maintenance of fresh broccoli during storage. Implicitly the initial chlorophyll content was determined with very different results, which may illustrate the varying differences or the unevenness of the samples: buds, stems, and leaves contained 170, 35, and 650 μg (g fresh wt)$^{-1}$ total chlorophyll, respectively (Wu and Salunkhe, 1976). Other examples [in total chlorophyll μg (g fresh wt)$^{-1}$]: 480 (Wang, 1979), 266 (Shewfelt et al., 1983), 195 chlorophyll *a* and 65.5 chlorophyll *b* (McMahon and Gladon, 1984), and a high value of 740 (Perrin and Gaye, 1986).

In contrast, low values were obtained with HPLC methods, 79 μg (g fresh wt)$^{-1}$ in broccoli grown in America, which included, besides chlorophyll *a* and *b,* small amounts of pheophytins *a* and *b,* which may be artifacts (Khachik et al., 1986) and 139 μg (g fresh wt)$^{-1}$ in broccoli grown in Japan (Izaki et al., 1986).

The effect of cultivar and planting date on chlorophyll content was followed in two broccoli cultivars (Yang and Henze, 1988a). The results are shown in Table 6.

Discussion. Cultivar SG1, richer in chlorophyll, appeared to be affected by seasonal variations, whereas Corvet showed almost the same chlorophyll content, independent of planting date.

Data on chlorophyll in other *Brassica* species were provided by the comprehensive survey on chlorophyll, carotenes, and tocopherols in green vegetables grown in Japan (Izaki et al., 1986) (see distribution in Table 1).

Table 6. Effect of Cultivar and Planting Date on Chlorophyll Content [μg (g fresh wt)$^{-1}$] in Two Broccoli Cultivars

Cultivar	*Planting*	*Harvest*	*Chlorophyll*
SG1	7.06	6.08	483
	15.07	6.09	274
	5.08	1.10	217
Corvet	7.06	2.09	184
	15.07	15.09	155
	5.08	6.10	167

Source: From Yang and Henze (1988a).

For the Compositae family, it is of interest to mention the high chlorophyll content [1350 μg (g fresh wt)$^{-1}$] in lettuce grown in ideal conditions (Czuba and Ormrod, 1974) compared with other Japanese *Lactuca* species normally grown (Izaki et al., 1986). Again, here the developmental stage and nutrient status, as well as light and temperature, play a big role.

Peas
(*Pisum sativum* L.) Leguminosae

Pea is the common name for several species of herbaceous annuals of Leguminosae grown for their edible seeds borne in pods or legumes. The favorite market and garden pea cultivated everywhere in the world is a variety of *P. sativum*. Some varieties called sugar peas produce edible pods. The chlorophyll content in peas, especially the changes during processing, were thoroughly investigated. The variation of the chlorophyll content with cultivars grown in Hungary, wrinkled and smooth peas, are illustrated in Table 7 (Aczél, 1973).

Table 7. Total Chlorophyll Content [μg (g fresh wt)$^{-1}$] in Various Pea Cultivars

Cultivar	*Wrinkled*	*Cultivar*	*Smooth*
Express	233	M. Csudo	278
Rhein	312	Kobri	188
Debrecen	280	Perfection	280
Gloria	215		

Source: From Aczèl (1973).

The changes in chlorophyll content per organ during development and senescence in leaf, pod, and seeds were followed by Storey and Beevers (1978). In each organ, the level decreased after attaining full expansion, which occurred at 10 days postanthesis in leaf and at 12 and 22 days postanthesis in pods and cotyledons.

Potato
(*Solanum tuberosum* L.) Solanaceae

The common white starchy potato tuber presents an uncommon phenomenon; it turns green when stored in the cold and exposed to light. The greening is associated with bitterness and off-flavor. Chlorophyll synthesis is responsible for greening, and the bitterness is due to solanine, a toxic alkaloid (Larsen, 1949). Chlorophyll and solanine develop in the periderm and cortical parenchyma layers of the tubers. The same factors, such as light intensity and quality, storage duration, and age of tubers, influence the biosynthesis of both compounds. During a period of 600 h, the chlorophyll content of White Rose potatoes exposed to 140–205 lux of light at different temperatures was highest at 10°C (Larsen, 1950).

Chlorophyll and solanine accumulation in potatoes, as affected by cultivars, specific gravities, and light intensities, were investigated by Patil et al. (1971). The dependence on cultivars is illustrated in Table 8.

Table 8. Chlorophyll and Solanine Contents [μg (g fresh wt)$^{-1}$] of Different Potato Cultivars Exposed to White Fluorescent Light at 1,080 lux for 5 days at 21°C

Cultivar	*Chlorophyll*	*Solanine*
Bounty	24.3	703
Cascade	9.7	657
Kennebec	22.3	964
La Chipper	6.9	448
La Rouge	12.0	730
Norchip	14.0	792
Platte	7.2	59
Red Lasoda	15.7	700
Russet Burbank	20.8	690
Shurchip	18.1	445
Sioux	12.5	589

Source: From Patil et al. (1971).

The chlorophyll is about a hundredth of that found in leaves. At 540 lux and 2,160 lux the chlorophyll content was lower than that of tubers exposed to 1,080 lux and 1,620 lux light intensities.

Interestingly, the biosynthetic pathway of δ-ALA operative in greening potatoes was found to be the Shemin pathway, from succinate and glycine (see discussion of chlorophyll biosynthesis) (Ramaswamy and Nair, 1973, 1974, 1976).

Both chlorophyll *a* and *b* were formed in measurable amounts after 3 weeks of storage. After 9 weeks storage the total chlorophyll content was 26.7 μg (g fresh wt)$^{-1}$ peel at 0°C to 4°C, and 22.5 μg (g fresh wt)$^{-1}$ peel at 10°C to 15°C at a light intensity of 30 lux (Ramaswamy and Nair, 1974).

At the ultrastructural level, greening of potatoes involves the transformation of amyloplasts into chloroplasts. The cells of potato tubers contain numerous amyloplasts, whose function is to accumulate and store starch in the form of reserve starch granules. Upon exposure to light the amyloplasts are transformed into chloroplasts. The chloroplast thylakoids are formed from vesicles or membranous tubular extensions that arise by the invagination of the inner membrane of the amyloplast without an intermediate prolamellar body stage (Berger and Bergmann, 1967; Ohad et al., 1971; Anstis and Northcote, 1973; Badenhuizen and Salema, 1976; Zhu et al., 1984).

Vacuum packaging with polyethylene bags significantly inhibited light-induced greening and glycoalkaloid formation in three cultivars of potato tubers (Wu and Salunkhe, 1975).

Cucurbitaceae

The Cucurbitaceae family has about 100 genera and 700 species of food and ornamental plants. Of these, pumpkin, cucumber, watermelon, and squash are important food plants. Their fruits at the unripe stage contain chlorophylls *a* and *b* both in the pulp and peel.

The inner seed coats of certain Cucurbitaceae species are grayish green or olive–green due to the presence of large amounts of protochlorophylls. Protochlorophylls are generally found in the leaves of etiolated plants. They are the immediate precursors of chlorophylls, because under illumination the photoreduction of the D ring occurs and chlorophylls(ides) are formed, the leaves becoming green.

The presence of protochlorophyll in Cucurbitaceae was first discovered during early investigations on chlorophyll biosynthesis in plants (Noack and Kiessling, 1930; Rudolph, 1934). Later, two protochlorophylls *a* were reported in the same material (Seybold and Egle, 1939). According to Noack (1943), the thick fruit mesocarp prevents light from entering, thus inhibiting photochemical transformation of the protochlorophyll. When the fruit is opened, penetration of light causes transformation into chlorophyll. In dead seed coats of overmature fruits, protochlorophyll was not phototransformable. In fact, the pigment is nonphototransformable, and it is uncertain whether it is protein-bound (Virgin, 1981).

The pigments contained in the inner seed of *Cucurbita pepo* and *C. maxima* were identified spectroscopically as protochlorophyll and Mg vinyl pheoporphyrin a_5. The coats of *C. moschata* were found to contain Mg-free vinyl pheoporphyrin a_5 (Inada and Shibata, 1960).

When chromatographed on a sugar column, the pigments of the inner seed coats of *C. pepo* were resolved into two forms: true protochlorophyll and 2,4-divinyl pheoporphyrin a_5 methyl ester (Jones, 1966). The two protochlorophylls in the seed coats of *C. pepo* are present as protein complexes but are not photoconvertible.

The esterifying alcohols of the pumpkin (*C. pepo*) seed protochlorophylls were analyzed by gas chromatography–mass spectrometry (Ellsworth and Nowak, 1974). The results suggest that pumpkin seed protochlorophyll *a* is esterified with all possible C_{20} isoprenoid alcohols between, and including, geranylgeraniol and phytol, the latter forming 90% of the mixture of esterifying alcohols. The 2,4-divinyl protochlorophyll is esterified with farnesol (C_{15}) and all possible C_{20} isoprenoid alcohols, including geranylgeraniol and phytanol, phytol representing 50% of the mixture of esterifying alcohols. The composition of the esterifying alcohols differed in younger and older mature seeds.

Separation of protochlorophylls esterified with different alcohols from inner seed coats of three Cucurbitaceae species has been achieved by r-p HPLC (Shioi and Sasa, 1982). At least 17 protochlorophylls (mono and divinyl) esterified with different alcohols were separated from purified pigments of pumpkin (*C. moschata*) and balsam pear (*Momordica charantia*) seeds. The seeds of *Trichosanthes kirilowii* var. *japonica* contained 14 protochlorophyll species. The esterifying alcohols, analyzed by gas chromatography, were C_{13}, farnesol, C_{18}, geranylgeraniol, 6,7-dihydrogeranylgeraniol, 6,7,10,11-tetrahydrogeranylgeraniol, and phytol. The relative amounts of each pigment differed among the three species. In *M. charantia*, protopheophytin and nonesterified pigments, such as protochlorophyllide and protopheophorbide; were also identified. Its total protochlorophyll content was the highest—920 μg (g fresh wt)$^{-1}$, eight times that of *C. pepo* (Ellsworth and Nowak, 1974).

For larger-scale preparations of protochlorophyll pigments, column chromatography on polyethylene and sugar was used (Houssier and Sauer, 1969). From 2 kg of seeds, 20–50 mg protochlorophylls were obtained. In agreement with Inada and Shibata (1960), the content of various protochlorophylls depended on the pumpkin cultivar.

The ultrastructural changes in plastids from the inner seed coat of pumpkin seed (*C. pepo*) were followed at different stages during fruit development by Sundqvist and Ryberg (1979). The plastids changed greatly during development. In young seeds with low protochlorophyll content, regular prolamellar bodies were found, and starch grains filled most of the plastids. During development the starch grains decreased and the prolamellar bodies increased in size and lost their regularity. During maturation, the plastids accumulated plastoglobuli, probably containing protochlorophyll, and, the internal structure of the prolamellar tubular complex was eventually lost. The dry weight of seeds of the developmental stages varied from 0.04 g to 0.3 g. Such a series of seeds with decreasing water content, indicating increasing maturity, contained different amounts of protochlorophylls at the four stages: 0.20, 0.93, 82.7, and 500 μg (g fresh wt)$^{-1}$.

DEGRADATION

Chlorophyll metabolism is conspicuously visible by the vegetation's color transformation during seasonal changes. In spring and summer the greening process is due to biosynthesis of millions of tons of chlorophyll. In the autumn, an estimated 10^{11} kg of chlorophyll are degraded in a relatively short time during senescence of leaves and ripening of fruits and crops (Hendry and Stobart, 1986). Changes in chlorophyll content in leaves of *Acer campestre* during an entire season from April to November are shown in Figure 6.

As chlorophylls totally disappear, the green leaves become yellow, red,

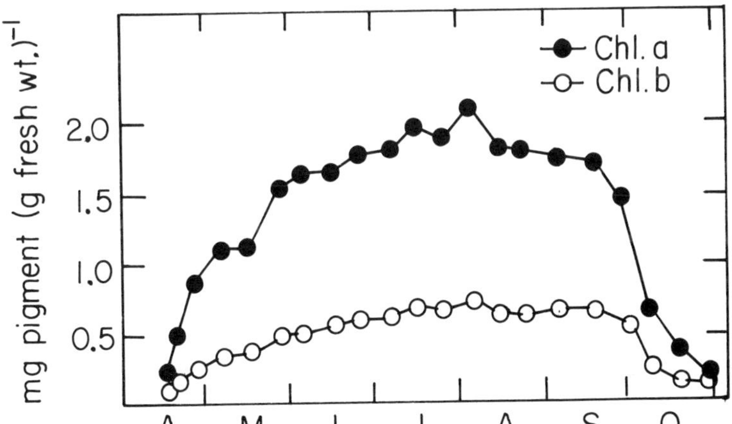

Figure 6. Changes in chlorophyll content of green leaves of *Acer campestre* during their development: A = April, M = May, ... (*From Kufner, 1980.*)

and, finally, brown due to the unmasking of the more stable carotenoids and, in some plants, to a de novo synthesis of anthocyanins.

Knowledge of chlorophyll degradation during leaf and fruit senescence is incomplete. These gaps in knowledge are emphasized in a few comprehensive reviews (Chichester and Nakayama, 1965; Simpson et al., 1976; Kufner, 1980). Moreover, a recent extensive and systematic review, which coordinates all the older and modern research on the subject, was entitled "Chlorophyll Degradation, a Biological Enigma" (Hendry et al., 1987).

The main difficulty in understanding the steps of chlorophyll degradation is that it occurs very rapidly and yields as end products colorless, low-molecular-weight compounds such as CO_2, NH_3, and H_2O. The overlapping of the products with degradation products of other substances adds to the difficulty.

Already in 1943 Seybold observed that, in autumn, leaf chlorophyll disappears without leaving traces (Seybold, 1943). Investigating chlorophyll degradation, Noack found that chlorophyll derivatives are degraded by hydrogen peroxide into the end products CO_2, NH_3, and H_2O (Noack, 1943). Egle (1960) applied these results to biological conditions in the leaf and proposed the following degradation pathway: the chlorophyll is first released from the chlorophyll–protein complex; then it is decomposed into chlorophyllides and phytol, the former being quickly oxidized by the aid of peroxides.

With further research, various derivatives of chlorophylls, which could be degradation products, were identified. Based on these results and on Egle's degradation pathway, Simpson et al. (1976) proposed similar hypothetical pathways. The degradation of chlorophyll *a* and its epimer proceeds via the pheophytins or via chlorophyllides, pheophorbides leading to the tetrapyrrolic chlorins, which are eventually oxidized in light to colorless compounds. A more recently proposed model for chlorophyll degradation involves chlorophyll *a*-1 as intermediate in addition to the cyclic tetrapyrrolic derivatives, namely chlorophyllides, pheophytins, and pheophorbides. These derivatives are transformed successively in linear tetrapyrroles and dipyrroles, then in corresponding monopyrroles and linear, nitrogen-containing components, and eventually in colorless simple molecules such as CO_2 and NH_3 (Hendry et al., 1987). The pathways are partially the reversal of some of the biosynthetic steps.

Leaf Senescence

The breakdown intermediates in senescing leaves have been the subject of many investigations. In earlier works there were indications that chlorophyll *a* disappears faster than chlorophyll *b* (Rudolph, 1934; Seybold, 1943; Egle, 1944). In later studies the changes in the content of all three groups of pigments present in leaves (chlorophylls, carotenoids, and anthocyanins) were followed in various species from their inception until decomposition (Goodwin, 1958a; Sanger, 1971; Park et al., 1973; Whitfield and Rowan, 1974; Kufner, 1980).

All curves representing changes in chlorophyll *a* and *b* content had, in most of the species, a similar form—a maximum attained in summer followed by a steep decrease to total disappearance in autumn. Nevertheless, the rate of chlorophyll breakdown varies considerably from species to species, the half-life being, for example, 91 days for *Prunus nigra,* 45 days for *Quercus ellipsoidalis,* and 70 days for *Acer campestre* (Goodwin, 1958a; Sanger, 1971; Kufner, 1980). Within a couple of weeks the chlorophyll content of ginkgo leaves decreased from 1.35 mg g^{-1} to 0.05 mg g^{-1} (Park et al., 1973).

As researchers had already found, chlorophyll *a* is more rapidly degraded than chlorophyll *b*. Consequently, the Chl *a/b* ratio is continuously shifted to lower values during leaf senescence.

The only breakdown products identified were low levels of pheophytin, with, occasionally, faint traces of pheophorbides and chlorophyllide (Sanger, 1971). In some yellow leaves, phytol-containing fragments derivable from the ring D, in amounts equivalent to those found in green leaves, have been recovered in about 12 species (Park et al., 1973). Some chlorophyll derivatives, mainly pheophytin, were identified in senescent tissues, but probably they were artifacts formed during extraction and chromatography (Goodwin, 1958a; Holden, 1970b).

The search for chlorophyll breakdown products continues, with the intensified use of in vitro model systems in which senescence is artificially induced. The progress of modern analytical methods, especially the very sensitive HPLC, permit detection of trace amounts, which may facilitate the task.

Pheophytin and pheophorbide have been found to be the first degradation products of chlorophyll in senescent tobacco culture (Schoch and Vielwerth, 1983). In senescing leaves of *Phaseolus vulgaris* and *Hordeum vulgare,* following excision from the plant, chlorophyll *a*-1 (the C-10 hydroxychlorophyll) and chlorophyll *b′* (the epimer chlorophyll *b*) were found (Maunders et al., 1983). Chlorophyll *a*-1 was not detected in healthy plants, in senescing attached leaves, or in chlorophyll–protein isolated from senescent tissues. No chlorophyll *b*-1 was detected, which illustrates the different susceptibilities of chlorophylls *a* and *b* to oxidation.

Chlorophyll catabolism in senescing *Citrus* fruit peels, autumnal China tree leaves (*Melia azedarach*), and detached dark held parsley (*Petroselinum sativum*) was investigated by Amir-Shapira et al. (1987). Chlorophyllide *a* and another dephytylated derivative accumulated in large amounts in senescent *Citrus* peels. Mature leaves of *Melia* contained chlorophyllide *a;* senescing leaves contained, in addition to traces of chlorophylls, mostly pheophorbide *a* and other polar dephytylated derivatives. Otherwise, in dark senescing parsley leaves, pheophytin *a* and other phytylated derivatives accumulated. The results suggest that chlorophyll can be degraded in other ways, depending on the plant system.

On the other hand, the suppositions and the results of earlier reports were confirmed: during senescence most of the chlorophylls disappeared without trace, and the mechanism is unknown.

Ultrastructural Changes During Leaf Senescence

The first changes observed in the chloroplasts of senescing leaves are the increase in size of the plastoglobules and the swelling of the thylakoids. The chloroplast volume diminishes considerably due to the disappearance of both the plastid stroma and the stroma thylakoids; grana thylakoids disappear more slowly. The number and size of plastoglobules increase continually during aging and become the predominant structure in the plastids of the yellow leaves (Barton, 1966; Ljubešić, 1968; Wrischer, 1978). Such plastids in the catabolic cells of senescent leaves, devoid of any function, are called "gerontoplasts," according to Sitte (1977).

Enzymes Involved in Chlorophyll Biodegradation

Research on the degradation steps at the enzymatic level has substantially progressed in recent years. Multiple kinds of enzymes are involved that catalyze either chlorophyll hydrolysis, removal of magnesium, modification of tetrapyrrole ring structure, and the final breakdown of the macrocyclic ring, the latter occurring also nonenzymatically. However, besides chlorophyllase and Mg dechelatase, none of the other enzymes has a specific function concerned with chlorophyll metabolism.

Chlorophyllase

The enzyme chlorophyllase catalyzes the hydrolysis of the ester linkage between the 7-propionic acid residue at ring D of the macrocyclic ring system and phytol, in chlorophylls and pheophytins.

Chlorophyllase was first described by Willstätter and Stoll (1913), who suggested that it catalyzes the first step in chlorophyll breakdown. For a long time, chlorophyllase was considered to have both biosynthetic and catabolic activity. But as shown previously, the enzyme responsible for the esterification during chlorophyll biosynthesis is chlorophyll synthetase, whereas chlorophyllase is, both in vitro and in vivo, a splitting enzyme (Holden, 1976).

Chlorophyllase was demonstrated in bacteria, algae, and higher plants. In higher plants its content varies in the various tissues, being higher in leaves and lower in roots and seeds. Its activity varies with the plant species. The enzyme was found to be located in the chloroplast of spinach leaves (Krossing, 1940) namely in the Chl-lipoprotein complex of spinach leaves chloroplasts (Ardao and Vennesland, 1960). Recent research established that chlorophyllase (chlorophyll chlorophyllide hydrolase) is an intrinsic membrane glycoprotein situated in thylakoid membranes (Terpstra, 1981). The enzyme has been purified and characterized (Lambers et al., 1985; Lambers and Terpstra, 1985).

Although a component of the thylakoid membrane in a normal functioning plant cell, chlorophyllase does not normally interact with its chlorophyll substrate, which is also situated within the same membrane. The intramembranous enzyme is kept in an inactive and stable conformation by the adjacent lipids (Lambers and Terpstra, 1985). A substantial degree of thylakoid disruption, either by mechanical means or by adding detergent, may be necessary to expose the chlorophyll molecules to degradation. Chloroplast fragments suspended in buffer form chlorophyllides, the hydrolysis being enhanced by detergents that act through disruption and partial solubilization of thylakoid membranes (Amir-Shapira et al., 1986). Chlorophyllase hydrolyzed only 40% of the chlorophyll in the chlorophyll–protein complexes of spinach. In the presence of detergent, all of the chlorophyll was hydrolyzed (Schoch and Brown, 1987).

Magnesium Dechelatase

Magnesium dechelatase is the enzyme responsible for the removal of the central magnesium ion. It was described in various systems, and shown to remove magnesium from chlorophyll and from chlorophyllide, so it is not clear which step is first (Ziegler et al., 1982; Schoch and Vielwerth, 1983). Cell-free extracts prepared from the leaves of etiolated red bean seedlings contain an enzyme that degrades Mg protoporphyrin monoester (Houghen et al., 1982). The product of the enzymatic breakdown has not been identified.

Chlorophyll Bleaching—Lipoxygenase, Peroxidases, and Oxidases

The bleaching of chlorophyll seems to be an oxidative process involving enzymes such as lipoxygenase (previously called lipoxidase), peroxidases, and oxidases. Strain first observed that chlorophylls *a* and *b* were oxidized to colorless substances in a system consisting of an aqueous extract from soybean, a fat, and oxygen (Strain, 1941). It was found that legume seeds and their extracts bleached chlorophyll in the presence of long-chain fatty acids (Holden, 1965). The most effective acids were the known substrates for lipoxygenase—the fatty acids linoleic and linolenic, which contain a *cis,cis*-1,4-pentadiene system. In the presence of molecular oxygen, lipoxygenase catalyzes their conversion into acids with a *cis,trans*-1,3-butadiene hydroperoxide system.

Chlorophyll appears to be bleached by co-oxidation during this chain reaction, which involves peroxidation of fatty acids and the breakdown of the hydroperoxide. Colored breakdown products were not obtained, because chlorophyll was bleached extremely rapidly. Chlorophyll *a* was bleached more rapidly than chlorophyll *b*.

Lipoxygenase was found to be present not only in legume seeds but also in various leaves (Holden, 1970a). The chlorophyll of various leaves was bleached in aqueous acetone solution at pH 6 in the dark. Leaves that bleached most were all rich in lipoxygenase, such as leaves of *Phaseolus mungo,* young and mature leaves of nettle (*Urtica dioica*), and leaves from barley and wheat seedlings. Lipoxygenase content was determined in various plant tissues. In addition to legume seeds, eggplant, potato, cauliflower, and avocado contained considerable activity (Pinsky et al., 1971).

The coupled oxidation of chlorophyll *a* with linoleic acid, catalyzed by lipoxygenase, was confirmed to be involved in bleaching (Orthoefer and Dugan, 1973; Imamura and Shimizu, 1974).

Studying the yellowing of postharvest spinach leaves and, in general, of leafy vegetables, it was concluded that first the chloroplast lipids are hydrolyzed by the lipolytic acyl hydrolase liberating unsaturated fatty acids; then the lipoxygenase catalyzes the formation of fatty acid hydroperoxides, which cause the degradation of chlorophylls (Yamauchi et al., 1986, 1987).

Besides lipoxygenase, chlorophyll bleaching was also ascribed to a peroxidase–hydrogen peroxide system, in the presence of phenolic compounds. Commercial horseradish peroxidase oxidizes chlorophyllides to yellow products, at higher rates in the presence of hydrogen peroxides and monophenols (Matile, 1980). Similarly, chlorophyll is effectively bleached by peroxidase extracted from orange flavedo (*Citrus sinensis*) in the presence of H_2O_2 in a reaction apparently coupled with peroxidation of certain phenols, resorcinol being the most effective (Huff, 1982). It was found that chlorophyll *a* is more rapidly destroyed than chlorophyll *b,* chlorophyllide *a* behaved similarly to chlorophyll *a,* and pheophytin was much less reactive than chlorophyll *a.*

It was suggested that the chlorophyll in parsley leaves may also be degraded by a peroxidase. The degradation was retarded by L-ascorbic acid (Yamauchi et al., 1980). Subsequently, the mechanism of chlorophyll degradation by peroxidase in parsley was elucidated as being a coupled oxidation involving first the oxidation of a phenolic compound, apigenin, a major flavonoid in parsley, which then degrades chlorophylls (Yamauchi and Minamide, 1985).

Besides lipoxygenase and peroxidases, bleaching activity was ascribed to an enzyme, called *chlorophyll oxidase.* Thylakoids of young barley chloroplasts possess a latent chlorophyll oxidase that catalyzes the bleaching of the endogenous chlorophylls in the presence of linolenic or other polyunsaturated fatty acids (Martinoia et al., 1982). However, chlorophyll oxidase is distinct from lipoxygenase in that pigment bleaching is not associated with the production of linolenate hydroperoxide. Moreover, the enzyme is active in the presence of oleic and even stearic acid—unlike lipoxygenase, which requires fatty acids containing a 1,4-pentadiene unit (Lüthy et al., 1984).

The first product of chlorophyll oxidase in barley thylakoids activated by linolenic acid was identified as 10-hydroxychlorophyll *a* (Chl *a*-1). It was formed within 2 min to 10 min, the quickest formation of this derivative ever reported (Schoch et al., 1984). On the other hand, under conditions of experimental senescence, the activation of the chlorophyll oxidase seems to be a slow process, the hydroxychlorophyll being detected only after several days (Maunders et al., 1983).

It has been suggested that the alteration of pigments in freshly isolated thylakoid membranes, aging in vitro, may be the result of a nonenzymatic process. The formation and subsequent accumulation of free radicals is the main factor involved in the destruction of pigments (Dupont and Siegenthaler, 1986).

Chlorophyll Turnover

Breakdown and synthesis of thylakoid membranes and their lipids in leaves occur during the natural daylight growth of plants. The turnover, however, is not visible, since the decomposition at night is recompensed by new synthesis during the day.

The biological half-time has been calculated to have values from 2.5 days to 7 days (Lichtenthaler and Grumbach, 1974). The daily turnover rate in developing radish seedlings amounts to about 15% per 24 h (Lichtenthaler et al., 1981). In young greening wheat leaves the turnover is a relatively rapid process, the half-time increasing with longer exposure to light, from 7 h to 58 h. However, in seedlings exposed to light for long periods, the turnover is progressively slower (days rather than hours) (Stobart and Hendry, 1984). Similar results were obtained with young barley leaves. The half-time of chlorophyll in 7-day-old leaves exposed to 24- and 48-h illumination, ranged from 15.8 h to 17 h and 43.6 h to 73 h, respectively (Hendry and Stobart, 1986).

Influence of Stress Conditions on Chlorophyll Degradation

Temperature

In nature, exposure to adverse temperature in either the high or the low range can initiate the onset of yellowing. Changes in chlorophylls following physical treatments of leaves were investigated systematically by Bacon and

Holden (1967). It was known 30 years ago that pheophytin is formed when green leaves are boiled (Campbell, 1937). In barley leaves heated for 5 min in distilled water 2% pheophytin *a* was formed at 61°C and 31% at 100°C. Pheophytin *b* was formed at higher temperatures, 3% at 75°C and 14% at 100°C. Chlorophyllide *a* was formed only at temperatures between 50°C and 85°C. After the leaves were boiled for 5 min, 19% of chlorophylls were converted to the epimer chlorophylls *a′* and *b′*.

Similar results were obtained with a suspension culture of tobacco at increased temperature in the dark. Chlorophyll *a* was transformed into pheophytin *a,* pheophorbide *a,* and 10-hydroxypheophorbide *a* (Schoch et al., 1984).

Chlorophyll degradation in vitro induced by heat stress was followed in citrus and parsley leaves (Amir-Shapira, 1986). The threshold for heat damage is different for the evergreen citrus and the herbaceous parsley leaves. Temperatures below 48°C did not cause damage in citrus, whereas for parsley the critical temperature was 42°C. In citrus, chlorophyllide- and pheophorbide-like pigments accumulated, whereas in parsley phytylated derivatives were detected.

The ratio of chlorophyll conversion to pheophytin in green vegetables and other plants, at different temperatures in buffered electrolyte systems, was studied by Haisman and Clarke (1975). Kinetic measurements showed that the conversion occurred above the threshold temperature of 50°C to 60°C, the rate varying with temperature, treatment duration, species, and pH. The beginning of chlorophyll degradation coincided with the disruption of the chloroplast structure at 60°C. Variation in chlorophyll degradation may be due to the subtle differences in molecular environment and is largely determined by the lipid–water interface.

Chilling effects (10°C for 7 days) were investigated in shoots of soybeans by Musser et al. (1984). Visible losses of chlorophyll were detectable within the first day of chilling, and regreening did not occur until the shoots were returned to 25°C. In citrus leaves, as a result of freezing treatment at −15°C, the cellular compartmentation was damaged and chlorophyll breakdown products, such as chlorophyllides, pheophorbides, and pheophytins, accumulated (Amir-Shapira, 1986).

Light Regime

Darkness is widely used in senescence studies to initiate and sustain the process in both attached and detached leaves. It has been demonstrated that senescence proceeds to virtual completion when the leaves are floated in water in the dark for 6 days. Proteolysis begins first; chlorophyll breakdown occurs only after 20 h (Thimann and Satler, 1979).

Opposite results were obtained with attached sycamore leaves (Maunders and Brown, 1983). In leaves held in darkness, breakdown was significantly less over a 14-day period than that occurring in leaves exposed to natural light. Chlorophyll *a* declined more rapidly than chlorophyll *b* in both situations, the stability of the latter being particularly increased in darkness. The data indicate that chlorophyll loss during senescence is, at least in part, the result of direct photochemical degradation of the pigment.

In vivo, however, even in full sunshine no appreciable chlorophyll loss occurs. In senescing attached leaves exposed to natural light, chlorophyll products, like those identified in model systems, are not formed. Possibly another mechanism, or a more rapid photodegradation, is involved (Maunders et al., 1983).

Photobleaching

Chlorophylls are sensitive to light, and photobleaching is one of the most important mechanisms of chlorophyll breakdown. Photobleaching does not occur in vivo as long as chlorophylls are found in intact thylakoid membranes and photoprotected by the endogenous carotenoids.

In vitro, chlorophyll solutions are irreversibly bleached by light in the presence of oxygen. The kinetics of the reaction and the intermediates were investigated by Jen and Mackinney (1970a,b). The reaction rates for chlorophylls *a* and *b* were of the same order of magnitude. Pheophytins and pheophorbides are more stable under light than are chlorophylls *a* and *b*. Their stability, as contrasted with chlorophylls and chlorophyllides, implies that Mg is essential for the excited triple state of the molecule. When the bleaching is interrupted, the red intermediates formed can be precipitated. They are more highly oxygenated, retain phytol, and have a N–Mg ratio of 4:1. Similar results were again found for these red intermediates, with a nitrogen–phytol–magnesium ratio of 4:1:1. After long exposure the N/Mg ratio remained unchanged, but the phytol was lost (Morris et al., 1973).

Pink pigments have been discovered in senescent leaves of *Festuca pratensis* and in yellowing autumnal leaves of a large number of species, such as barley, tobacco, and bean (Matile et al., 1987). They are considered to represent intermediary products of chlorophyll breakdown and may possibly be related to the red intermediates isolated during the in vitro bleaching. The color of these catabolites may be due to the opening of the porphyrin ring, with a modified chromophore. It is supposed that following chlorophyll breakdown by the action of chlorophyllase, the water-soluble porphyrins are transported from chloroplasts to vacuoles, where they undergo further cleavage and pink or blue fluorescent catabolites accumulate (Matile et al., 1989).

In some plants low-temperature damage occurs only in light. Chlorophyll degradation in leaves of cucumber (*Cucumis sativus*) is a photooxidative process with an optimum temperature of 1°C to 5°C (van Hasselt, 1972; van Hasselt and Strikwerda, 1976). Evidence for the role of singlet oxygen and superoxide in the breakdown of chlorophyll during chilling-enhanced photooxidation has been given by Wise and Naylor (1987). Chlorophylls *a* and *b* were degraded at essentially the same rate in detached leaves of cucumber, a chilling-sensitive plant. On the other hand, no evidence of photooxidative damage was observed in detached leaves of the chilling-resistant pea.

Degradation During Fruit Ripening

During fruit ripening the chlorophylls, which are present in all unripe fruit, break down following the chloroplast disintegration. There are few reports on chlorophyll changes during fruit ripening.

Chlorophyll degradation products have been followed in Hungarian wax peppers (*Capsicum frutescens*) at four ripening stages (Schanderl and Lynn, 1966). The first two stages (whitish green and dark green) represent those of increasing greenness. Then follows the color-break stage (the olive–orange one), when the onset of carotenoid biosynthesis occurs, and the last stage prior to complete ripeness (orange), when chlorophylls nearly disappear.

Pigment changes in five *Capsicum* cultivars at four stages of maturation and ripening were investigated by Rahman and Buckle (1980). The changes in chlorophyll *a* and *b* content during fruit ripening are given in Table 9.

Discussion of Table 9. In all five cultivars the chlorophylls decreased with ripening and were totally absent in the fully ripe fruits. Chlorophylls *a* and *b* were the only chlorophyll pigments.

In the pimento pepper (*Capsicum annuum* L.) 'Bighart', the total chlorophyll content was determined during growth and development (Basiouny and Biswas, 1981). The results are shown in Table 10.

Discussion of Table 10. At color break the rate of decrease in total chlorophyll became higher, but chlorophyll was still present in the red fruits, although masked by the other pigments.

In ripening cherry tomatoes (*Lycopersicum esculentum*) chlorophylls *a* and *b,* which represented 87% of the total pigment at the green stage, decreased to 76% at color break due mainly to the breakdown of chlorophyll *a,* which was present only in trace amounts in the ripe red fruit (Laval-Martin et al., 1975).

Total chlorophyll was determined in ripening Walter tomatoes at 10 ripening stages by Watada et al. (1976a). Tomatoes in stage 1, termed *immature-green,* lacked red internal color and gel-type placental tissue. Those in stages 2–4, termed *mature-green,* had increased amounts of red internal color and gel-type placental tissue. Those in stage 5, termed *breakers,* showed incipient ripening externally. Tomatoes in the remaining five stages had increasing amounts of red color. The results are given in Table 11.

Table 9. Chlorophyll Changes During Ripening in Five Capsicum Cultivars [μg (g fresh wt)$^{-1}$]

Cultivars		*Ripening stages*			
		Immature	*Mature*	*Half-ripe*	*Fully ripe*
	Color	Green	Green	Pink	Red
Long Red	Chl *a*	27	39	19	—
Cayenne	Chl *b*	14	18	10	—
	Color	Green	Green	Pink	Red
Pacific	Chl *a*	12	11	7	—
Bell	Chl *b*	6	6	2	—
	Color	Greenish yellow	Greenish yellow	Yellowish red	Red
Ram Horn	Chl *a*	5	4	2	—
	Chl *b*	3	2	1	—
	Color	Sulfury yellow	Sulfury yellow	Yellowish red	Red
College	Chl *a*	0.4	0.6	0.2	—
Gold	Chl *b*	0.2	0.4	tr	—
	Color	Green	Green	Greenish yellow	Yellow
Golden	Chl *a*	16	14	7	—
California	Chl *b*	8	8	4	—

Source: From Rahman and Buckle (1980).

Table 10. Changes in Chlorophyll Content During Growth and Development of Pimento Pepper Fruits

Harvest stage	*Total chlorophylls [μg (g fresh wt)$^{-1}$]*
Green	4460
Color break	4210
Red	3020
Overripe	1090

Source: From Basiouny and Biswas (1981).

Table 11. Chlorophyll Content of Walter Tomato Fruit During Ripening

Ripening stage	*Total chlorophylls [μg (g fresh wt)$^{-1}$]*
1	13.4
2	11.2
3	8.6
4	5.9
5	4.2
6	1.0
7	0.4
8	0.3
9	0.3
10	0.3

Source: From Watada et al. (1976a).

Discussion of Table 11. The average chlorophyll content decreased from 13.4 μg (g fresh wt)$^{-1}$ in immature-green fruit (stage 1) to 0.3 μg (g fresh wt)$^{-1}$ in partially ripe fruit (stage 8). Changes in the chlorophyll content after stage 6 were not significant, and a small amount (0.3 μg) of chlorophyll was still present in the ripe fruit (stage 10). The ratio Chl *a/b* decreased from 2 in the immature fruit to 1 as the fruit ripened.

Applying a new HPLC method, McMahon and Gladon separately determined chlorophylls *a* and *b* in ripening tomato fruits as a function of days after anthesis (McMahon and Gladon, 1984). The results were similar to those in Table 11. In 'Heinz 1350' tomatoes, chlorophylls *a* and *b* decreased continuously from 24.6 and 6.3 μg (g fresh wt)$^{-1}$, the initial contents after 15 days from anthesis. After 50 days, chlorophylls reached their lowest values of 4.6 and 1.3 μg (g fresh wt)$^{-1}$ and disappeared totally in fully ripe fruits after 55 days from anthesis.

Degradation During Processing and Storage of Green Vegetables

During processing and storage, the plant material containing chlorophyll undergoes changes or loss of color similar to that occurring in senescing plants. The purpose of preservation is to keep the material, as far as possible, in an unaltered condition. Consumers prefer cooked vegetables to have a bright green color because appearance is considered an index of freshness, palatability, and nutritional value. (An unappealing color is associated with spoiled food and will arouse the consumer's suspicion.)

The problem of maintaining the natural color of green vegetables, avoiding the indescribable chlorophyll changes, was the object of much investigation. As expected, the transformations are nearly the same as those occurring during senescence and are a consequence of the pigment's chemical properties, enzymes present, and physical conditions (e.g., temperature, moisture percentage).

Conversion of Chlorophylls to Pheophytins

Freezing Preservation and Blanching. The color change of the plant material from blue-green to dull olive-green is produced primarily by the conversion of chlorophylls *a* and *b* to their respective pheophytins. This is the most widespread alteration that takes place during cooking, thermal processing, freezing preservation, or storage of vegetables.

By 1937 it was recognized that the loss of green color in frozen Alderman peas stored at insufficiently low temperatures was produced by the slow conversion of chlorophylls into pheophytins, due to the action of the acid in the cell sap (Campbell, 1937). In frozen-pack peas and string beans, much of the chlorophyll changed to pheophytin (60–85%), and canned beans contained only pheophytin (MacKinney and Weast, 1940).

Nine fresh pea cultivars contained chlorophylls *a* and *b* ranging from 52 $\mu g\ g^{-1}$ to 75 $\mu g\ g^{-1}$ and from 19 $\mu g\ g^{-1}$ to 28 $\mu g\ g^{-1}$, respectively. Immediately after freezing, pheophytins *a* and *b* were detected (3–8 $\mu g\ g^{-1}$ and 3–5 $\mu g\ g^{-1}$, respectively). These levels were slightly higher after 6 months of storage at $-10°C$ (Forni et al., 1988).

Different acidities of peas and beans were suggested to be the cause for their different rates of chlorophyll conversion (Dietrich et al., 1957). The ratio of chlorophyll to pheophytin was proposed by the authors as a method for measuring color changes in green vegetables. A linear relationship between the color appearance of green beans to the percentage of chlorophyll conversion to pheophytin was found in frozen beans (Walker, 1964a).

In the early 1930s the importance of blanching was recognized, especially for green vegetables before freezing, as a means of preserving the natural green color and maintaining good quality. Blanching is one of the most important processes in the preparation of vegetables for freezing, canning, or dehydration (Lee, 1958). It involves treatment by means of some form of heat—steam or boiling water and, more recently, microwave and convection ovens.

The blanching process inactivates the enzymes present, shrinks the material, expels tissue gases, reduces any initial infection, and removes undesirable flavors. A great deal of research has been done to understand the chlorophyll degradation produced by blanching and to find optimal conditions that will induce minimal changes in different vegetables. Peas properly blanched by immersion in boiling water and then frozen at temperatures between −10°C and −21°C retained their characteristic bright green color for a long time (Diehl and Berry, 1933). In the assays on frozen and fresh string beans, it was observed that blanching reduced the degree of color change by "removing air and volatile and water-soluble constituents that would react with chlorophylls" (MacKinney and Weast, 1940).

The conversion of chlorophyll into pheophytin occurred in unblanched stored frozen peas, whereas the blanched lot remained unchanged (Wagenknecht et al., 1952). The effect of blanching on the chlorophylls conversion to pheophytins was studied in stored frozen beans (Walker, 1964b). As conversion increased with the blanching time, a period of 45 s to 1 min resulted in the most stable product.

In Brussels sprouts, blanching with water inactivated enzymes faster than blanching with steam and degraded less chlorophyll. Chlorophyll was more stable in water-blanched rather than steam-blanched samples stored 8 weeks at −7°C (Dietrich and Neumann, 1965).

For spinach the results were reversed; color was rated slightly higher on spinach that was blanched in steam (Sistrunk et al., 1977). The effect of modern blanching methods, such as microwave blanching and convection-oven blanching, were also investigated. Microwave-blanched asparagus had less color than either water- or steam-blanched, but no significant difference was found for green beans (Drake et al., 1981).

The effect of different types of blanching on the color and chlorophyll content in green beans was investigated by Muftugil (1986). Of the initial chlorophyll content, 88% was retained in water-blanched beans, 83% in steam-blanched, 97% in microwave-blanched, and 61% in beans blanched in convection ovens.

On the other hand, color measurements with a Hunterlab colorimeter indicated that blanching with water and steam resulted in a more intense greenness of the sample. Blanching is also an effective pretreatment of dehydrated okra for color retention (Stone et al., 1986).

Dehydration. In spinach processed by dehydration, 26% of the chlorophyll was converted to pheophytin during the process. In spinach blanched for 2 min and subsequently dehydrated, 44% of chlorophyll was converted to pheophytin, the percentage increasing with the length of blanching time (Dutton et al., 1943). It was demonstrated that both blanching and dehydration affect the chlorophyll content.

Chlorophyll degradation as a function of water activity in freeze-dried blanched spinach was studied by Lajolo et al. (1971, 1982). They found that conversion of chlorophyll to pheophytin occurs even at very limited water concentration.

Leafy vegetables (mustard greens, spinach, mint, and coriander leaves) available in northern India were dehydrated at a drying temperature between

63°C and 68°C to a low moisture content before storage at room temperature (Kaur and Manjrekar, 1975). Previous blanching in 2% sodium carbonate at 79°C helped to retain the green color of the dehydrated leaves. During storage for several months, the initial chlorophyll content decreased to 78% in spinach, 70% in mustard, and about 40% in mint and coriander. But the color of the latter is described as deep green, unaffected, after 3 months storage.

Conversion of chlorophylls to pheophytins occurred during drying of parsley leaves, which were dehydrated at temperatures between 80°C and 140°C (Berset and Caniaux, 1983). The pheophytin percentage was lower (3–7%) in samples dried at 80°C to 90°C and higher (12–15%) in samples dried at higher temperatures. On the other hand, samples dried at lower temperatures (60–110°C) and stored for 2 years contained a high percentage of pheophytins *a* and *b* (between 50% and 70%). The content of chlorophyll *a* varied between 13% and 60%, illustrating its higher instability compared with chlorophyll *b*, and the chlorophyll *a/b* ratio decreased accordingly, reaching values between 1 and 2.

Sterilization of Canned Vegetables. In canned green vegetables, the color change from blue-green to olive-green during the sterilization process by heat is very evident, as it is during cooking. This color change was rightly attributed to the conversion of chlorophyll to pheophytins in peas (Gold and Weckel, 1959) and in green beans (Westcott et al., 1955). In green peas there was a linear relationship between chlorophyll conversion and acidity, the latter increasing with heating and storage time.

The effect of processing temperature on pigments and color of spinach was similar (Tan and Francis, 1962). It was observed that a progressively smaller change of chlorophylls *a* and *b* to pheophytins *a* and *b* occurred as the processing temperature was raised. This was later confirmed by kinetic studies of thermal degradation of chlorophyll in spinach puree (Gupte et al., 1964). Data on chlorophyll changes in three green pea varieties during processing were reported by Aczél (1971a). During blanching, about 10% to 18% of the chlorophylls are converted to pheophytins. After sterilization (25 min at 120°C) chlorophyll was totally converted to pheophytin.

Chlorophyllides and Pheophorbides Formation

Pheophorbides were first detected during brining of cucumbers (Jones et al., 1961, 1962), suggesting that brining treatment favors chlorophyllase activity following the denaturation of the chlorophyll–lipoprotein complex by salt. The variations in chlorophylls and various chlorophyll derivatives formed during a brining period of 42 days were followed and reported by White et al. (1963). Conversion of chlorophylls to chlorophyllides and to pheophytins, and further conversion of chlorophyllides to pheophorbides, began immediately after the start of the brining process. Chlorophyllides *a* and *b* were found in measurable quantities in the first 7 days, and chlorophyllide *b* was found at higher concentrations, being probably more stable than chlorophyllide *a*. As the acidity increased rapidly, pheophorbides were formed at a rapid rate and represented the predominant pigments in the brined cucumber, with chlorophylls and chlorophyllides disappearing totally. The rapid rate of pheophorbide formation is evidence of high chlorophyllase activity and also of the ease with which magnesium in the chlorophylls and chlorophyllides may be replaced by hydrogen in a brine medium near a pH value of 6.0.

The effects induced in whole and sliced cucumbers by blanching and brining treatments were also studied (Jones et al., 1963). In unbrined cucumbers, blanching at 82°C induced the formation of relatively large quantities of chlorophyllides and appreciable quantities of pheophorbides. Blanching at 100°C mainly produced pheophytins and relatively small quantities of chlorophyllides and pheophorbides.

Brined, unblanched cucumbers contained mainly pheophorbides and some pheophytins. Blanching at 82°C did not essentially change the distribution. Blanching at 100°C partially inactivated the chlorophyllase so that, besides the main pigments, pheophytins (70% of the total) and pheophorbides (30%) were also found.

In the same report, the effect of blanching on other green vegetables, such as okra, turnip greens, and snap beans, was also investigated. Blanching at 82°C promoted formation of chlorophyllides and pheophorbides, in addition to pheophytin, in okra and turnip greens. The higher conversion of chlorophyll to chlorophyllides in okra than in turnip greens may have been due to the higher chlorophyllase concentration in okra. Blanching at 100°C inactivated chlorophyllase in okra and turnip greens so that chlorophyllides and pheophorbides were not formed. In snap beans only pheophytins were formed, a proof that snap beans are devoid of chlorophyllase. This report agrees with those of MacKinney and Weast (1940) and Walker (1964a,b).

Chlorophyllides and pheophorbides were also identified in frozen unblanched peas during storage (Buckle and Edwards, 1970a). Pheophorbides were identified among the chlorophyll degradation products in salted seisai (Japanese pickled vegetables) and showed a TLC pattern similar to that obtained from other salted vegetables (Shoko and Susumi, 1983).

A recent investigation confirmed the early results on chlorophyllase activities and color changes of fresh green leaves on blanching (Ogura et al., 1987). In fresh spinach leaves blanched for 1 min in boiling water or 0.5% NaCl and $NaHCO_3$ solution, no chlorophyllide formation was detected. On the other hand, when fresh leaves of spinach or goosefoot were dipped in hot water (60–70°C) for 5 min to 6 min, a large amount of chlorophyllide was produced, especially in goosefoot leaves, but no color change was observed. Chlorophyllase activity (mg chlorophyll hydrolyzed/hr/g fresh weight) is 1.17 in spinach and 5.18 in goosefoot leaves.

Conversion of Chlorophyll to Pyroderivatives

The development of the highly sensitive method of HPLC permitted a better understanding of the chlorophyll changes occurring during processing and storage. As discussed previously, most research limited itself to describing, calculating, and preventing the occurrence of major changes, the chlorophyll conversion into pheophytin. Only a few of the numerous reports that deal with chlorophyll degradation products in food mention products other than those previously discussed in pea puree and spinach (Buckle and Edwards, 1969; Lajolo et al., 1971). The products were not further identified.

With HPLC the chlorophylls and their derivatives in processed spinach could be resolved into ten components (Schwartz et al., 1981). In fresh spinach, only chlorophylls *a* and *b* were present. Blanching for 2.5 min at 100°C induced the formation of chlorophyll epimers, chlorophylls *a'* and *b'*. Prolonged blanching (10 min) increased the content of the C-10 epimers. Having the same chromophore as the parent chlorophylls, they do not bring about any change in the color of the spinach. Frozen spinach contained chlorophylls *a, a'*, and *b, b'*, pheophytin *a,* and trace amounts of pheophytin *b.*

In canned spinach almost all the chlorophylls were converted to pheophytins *a* and *b* accompanied by their epimers, pheophytins *a'* and *b'*, and pyropheophytins *a* and *b.* The latter are the major pigments formed by the loss of the carbomethoxy group ($-CO_2CH_3$) from the corresponding pheophytins (see Figure 2).

Pyropheophytins were not previously reported to be present in canned spinach. Failure to detect them may be attributed to lack of resolution with older methods and the identical absorption spectra of pheophytins and pyroderivatives.

Subsequently, the kinetics of chlorophyll degradation to pyropheophytins in heat-processed vegetables were investigated (Schwartz and von Elbe, 1983). The pigment alterations during heat treatment suggested the following mechanism:

$$\text{chlorophyll} \rightarrow \text{pheophytin} \rightarrow \text{pyropheophytin}$$

Both reactions follow first-order kinetics. Pheophytin *b* was found to degrade at a rate approximately 25% to 40% faster than pheophytin *a,* depending on temperature; this fact is in contrast to the difference in degradation rates usually observed for chlorophylls *a* and *b.*

Table 12 lists the concentrations of pheophytins *a* and *b* and pyropheophytins *a* and *b* found in commercially canned products of spinach, peas, asparagus, and beans.

Table 12. Pheophytins *a*, *b* and Pyropheophytins *a*, *b* in Commercially Canned Vegetables [μg (g dry wt)$^{-1}$]

	Pheophytin		*Pyropheophytin*	
Product	a	b	a	b
Spinach	830	200	4000	1400
Beans	340	120	260	95
Asparagus	180	51	110	30
Peas	34	13	33	12

Source: From Schwartz and von Elbe (1983).

Discussion. Based on laboratory studies that showed that pyropheophytins are formed approximately 4 min after heating at 121 °C, the presence of these derivatives in vegetables was expected. These findings indicate that pyropheophytins *a* and *b* are major chlorophyll derivatives responsible for the olive-green color in canned vegetables.

Chlorophyll Bleaching

Chlorophyll bleaching by oxidation plays an important part in color changes in processed and stored foods. One of the main oxidative catalysts in vegetables is lipoxygenase, which catalyzes the coupled oxidation of unsaturated lipids and chlorophyll (see chlorophyll bleaching, p. 51). A loss of chlorophyll was observed when unblanched or underblanched frozen raw peas were held for 1 year to 5 years at −18 °C. The loss was attributed to the accumulation of free fatty acids, a consequence of the progressive deterioration of the lipid material (Wagenknecht et al., 1952). The enzyme lipoxidase (earlier name of lipoxygenase) has been postulated as being the cause of chlorophyll destruction. Additional evidence was obtained when lipoxidase activity was demonstrated both in fresh and frozen peas (Wagenknecht and Lee, 1956, 1958; Buckle and Edwards, 1970a). Bleaching of chlorophyll catalyzed by lipoxidase also occurred in commercial, frozen, unblanched green bean packs stored for 12 months at −18 °C. Chlorophyll bleaching is affected by temperature. A nonenzymatic oxidation initiated by heat was also observed in green beans during frozen storage (Walker, 1964a,b).

Bleaching of chlorophyll (that is, gradual disappearance of the pigment) was observed especially in fresh vegetables during storage, the rate of chlorophyll loss being temperature-dependent.

Loss of chlorophyll as a result of storage and the effect of storage temperature were followed systematically in green asparagus (Kramer et al., 1949). The results for two sample lots are shown in Table 13.

Table 13. Effect of Storage Duration and Temperature on Chlorophyll Content in Green Raw Asparagus

Storage duration (at 0°) (days)	*Chlorophyll [μg (g fresh wt)$^{-1}$]*	
	Lot 1	*Lot 2*
1	129	88
3	119	90
5	64	79
7	48	62
3 days' storage temp.		
0°	105	84
10°	88	76
20°	91	—
30°	78	—

Source: From Kramer et al. (1949).

The chlorophyll content decreased gradually with storage time and increasing temperature.

The effect of temperature during storage on quality deterioration was also followed in freshly harvested Brussels sprouts (Lyons and Rappaport, 1959). Brussels sprouts deteriorate after harvest when held at temperatures above 10°C. Loss of quality at higher temperatures was characterized by rapid yellowing of the outer leaves, whereas at lower temperatures the green color was maintained for relatively long periods up to 72 days.

Under normal atmospheric conditions, broccoli will yellow in 2 to 3 days after harvest at temperatures of 20°C or higher. It has been reported that contact cooling is a good solution for quality maintenance in fresh broccoli. Broccoli stored at 0°C to 1°C and capped with ice retained its green color for 4 weeks or 10 weeks in iced and packed samples (Shewfelt et al., 1983), whereas bottom-iced broccoli in air-cooled units kept best overnight (Perrin and Gaye, 1986).

Attempts to explain the rapid yellowing of first-harvest vegetables appeared only recently, as was previously discussed. In parsley and spinach the peroxidase activity increased rapidly during yellowing (Yamauchi et al., 1980). In parsley the mechanism of chlorophyll degradation was suggested to be, first, an oxidation of apigenin catalyzed by peroxidase and, second, a degradation of the chlorophylls by the oxidized apigenin (Yamauchi and Minamide, 1985).

The yellowing of spinach leaves (respectively the chlorophyll changes in leaves stored at 1°C and 25°C), as well as specific enzyme activity, was also followed with the same aim (Yamauchi et al., 1985). The results are shown in Figure 7.

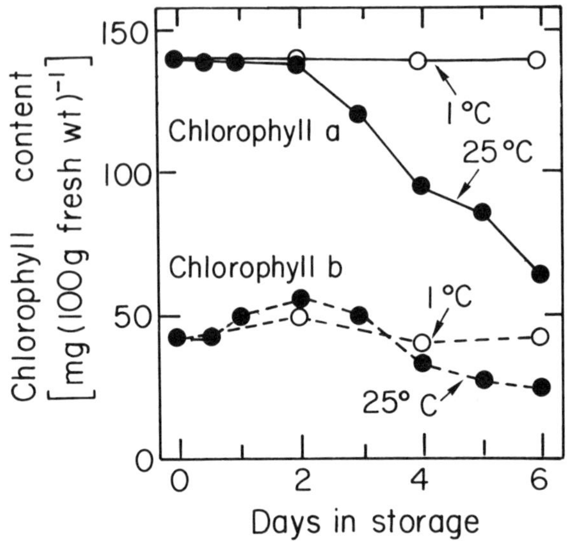

Figure 7. Changes of chlorophylls *a* and *b* contents in spinach leaves during storage. (*From Yamauchi et al., 1985.*)

Discussion. Leaves stored at 25°C kept almost harvest-fresh for the first 2 days, but deteriorated rapidly after 3 days of storage, with beginning of the yellowing. Spinach leaves stored at 1°C kept fresh for 6 days. The absorption maxima of the chlorophyll extract were not shifted after 6 days—proof that chlorophyll derivatives were not formed—but chlorophyll disappeared.

Subsequent studies of the activity and location of the enzymes involved suggested that chlorophyll bleaching in leaves is a result of the joint action of lipolytic acid hydrolase, which degrades the chloroplast lipids, and of lipoxygenase, which catalyzes the formation of freed fatty acid hydroperoxides (Yamauchi et al., 1986, 1987).

Stabilization of Chlorophylls in Processed Green Vegetables

When the causes of unattractive color during food processing were understood, many attempts were made to stabilize the pigments in processed vegetables.

pH Adjustment

The pH of the initial material and pH changes during processing and storage are determining factors in chlorophyll degradation. Controlling the pH during processing and storage is therefore the best means for color and pigment retention in heat-processed, chlorophyll-containing foods. Various methods for elevating the pH with approved alkalinizing agents were developed. Numerous patents have been obtained in which organic and inorganic salts of the alkali and alkaline earth metals were used (Blair, 1940; Gieseker, 1949; Bendix et al., 1952).

In spinach puree the pH was adjusted with magnesium carbonate, which improved chlorophyll retention, but the effect was not stable during storage (Gupte and Francis, 1964). In addition to pH adjustment, conversion to chlorophyllides was also tried, assuming that chlorophyllides are more stable than chlorophylls (Clydestale and Francis, 1968; Clydestale et al., 1970). The color data showed that all samples with an elevated pH were more acceptable, but the presence of chlorophyllide made only minor differences.

Studies on color improvement of frozen vegetables were also carried out (Sweeney and Martin, 1961). Because chlorophyll stability varies as a function of pH of the material, six different vegetables were studied. There were striking differences in the chlorophyll *a* stability of vegetables such as spinach and green peas (pH 6.7–7.0) compared with Brussels sprouts and green beans (pH 6.0–6.3). It appeared that a pH between 6 and 7 is critical. With citrate–phosphate buffers (pH 6.2–7.0), progressive increase in color was obtained in green beans, along with an increase in pH. Buffers with pH higher than 7 resulted in marked deterioration of flavor and no improvement in color. Cooking with water containing different chemical additives was tried for broccoli (Gilpin et al., 1959). For frozen broccoli, ammonium compounds were used during blanching to improve color retention (Eheart and Odland, 1973; Odland and Eheart, 1975).

High-Temperature Short-Time Processing Method

The high-temperature short-time (HTST) method, which minimizes heat exposure, produces food products of higher quality with greater chlorophyll retention and, thus, better green color than those produced by conventional heat processing (Tan and Francis, 1962). The method is generally combined with pH adjustment (Gupte and Francis, 1964; Clydestale et al., 1970).

In HTST-processed green pea puree it was demonstrated that pH control during heat processing and storage is a determining factor for pigment control in food containing chlorophyll (Buckle and Edwards, 1970b).

The effect of different color-improving methods was investigated in processed, wrinkled, and smooth green peas by Aczél (1973). The methods were

(a) short-time sterilization at high temperature (HTST);(b) the Blair method, which involves pH adjustment;(c) coloring with dosage of sodium–copper–chlorophyllin; (d) dosage of sodium glutamate; and (e) combined methods. During blanching (4–6 min at 90–95°C) the total chlorophyll content decreased by 10% to 16%. During sterilization chlorophyll was totally converted to pheophytin. The Blair method yielded an unacceptable taste, although the product was greener. The best effect was obtained with a combination of HTST and the use of sodium glutamate, the latter inducing good color and excellent taste.

HTST steam blanching has been recommended to produce high-quality frozen vegetables. The quality depends on the vegetable and the pressure and duration of the HTST steam blanch (Drake and Carmichael, 1986).

Storage in Controlled Atmosphere

Postharvest storage results in a general reduction of quality reflected by loss of water, biological oxidation, various other chemical changes and physiological breakdown. In addition, there is the deterioration caused by microorganisms.

More than 40 years ago, it was reported that reduced oxygen concentrations or a modified atmosphere of carbon dioxide and oxygen may delay ripening, thus preventing normal coloring and generally reducing the aging rate of fruits and vegetables (Biale, 1942; Platenius, 1943). Therefore, *controlled atmosphere* (CA)—that is, adjustment of the gases in the surrounding atmosphere, usually relatively high carbon dioxide and low oxygen levels, combined with some degree of refrigeration—was introduced to retard undesirable postharvest changes in fruit and vegetables. One of the most beneficial effects of CA storage of green vegetables is the decreased loss of green color. The rapid yellowing of broccoli under normal conditions was completely inhibited by storage at 24°C in an oxygen-free atmosphere, even in the absence of carbon dioxide. A modified atmosphere containing 1% to 10% oxygen reduced respiration (Lieberman and Hardenburg, 1954).

In Brussels sprouts no color changes occurred over 16-day storage at 0°C. At 5°C and 10°C, an increase in carbon dioxide and 21% oxygen delayed the loss of chlorophyll. Lowering of the oxygen content below 10% in the absence of carbon dioxide was also effective with respect to retention of the green color (Lyons and Rappaport, 1962).

Retardation of chlorophyll degradation was considered the greatest advantage of CA for green beans. The values studied were from 0% to 10% CO_2 and from 2% to 20% O_2 (Groeschel et al., 1966). At a concentration of 2% oxygen, respiration decreased to 60%. The optimum atmosphere for storage of green beans, at about 7°C, is one that has a low oxygen content (2% to 3%) and 5% to 10% carbon dioxide. The better retention of chlorophyll in CA was accompanied by a pH higher than that in the air-stored samples.

Chlorophyll degradation during CA storage was studied in excised shoots of asparagus by Wang et al. (1971). Storage in 5% CO_2 at 1.7°C and 95% relative humidity resulted in increased pH and less chlorophyll breakdown. The degradation products were exclusively pheophytins.

CA storage does not have the same beneficial effects in all vegetables. In lettuce (*Lactuca sativa*), storage in an atmosphere containing 2.5% to 10% CO_2 caused physiological injury and brown stain lesions, mainly in the outer leaf surface. These lesions usually developed within the first six leaves, but they may also be found in leaves deep within the head (Stewart and Uota, 1971). More heads developed brown stains when CO_2 was combined with 3% O_2 than when combined with 21% O_2. It was concluded that the danger of physiological injury outweighs the small improvement in color and general appearance obtained by CA storage.

The influence of CA storage on external and internal quality features of broccoli was investigated in detail by Yang and Henze (1987, 1988a,b). Storage in CA containing 5% to 10% carbon dioxide with oxygen reduced to 3%, at 0°C to 1°C, was, by visual assessment, most effective in preserving chlorophyll in comparison with other modified atmospheres used. The results were confirmed by the determination of the chlorophyll content after 4, 6, and 8 weeks of storage in various CA. The results are given in Table 14.

Table 14. Changes in Chlorophyll Content [μg (g fresh wt)$^{-1}$] in Broccoli 'Corvet' Stored in Air and Various Controlled Atmospheres

CA	*Weeks of storage*			
%CO2 : %O2	*0*	*4*	*6*	*8*
	Chlorophyll	Chlorophyll in the middle		
0.0 : 21.0		110.3	106.9	98.8
0.0 : 0.8		131.4	119.7	107.2
0.0 : 1.5	167.0	130.3	105.2	90.8
5.0 : 3.0		127.4	111.4	112.9
10.0 : 3.0		157.9	147.2	136.1

Source: From Yang and Henze (1988a).

Discussion. The highest chlorophyll retention may be observed in the CA of 10% CO_2 and 3% O_2. The broccoli stored in this CA had after 8 weeks of storage a chlorophyll content higher than that found in other samples after only 4 weeks.

Another technique to extend storage life and retain quality is the short-term application of high CO_2 levels. The usefulness of this technique must be evaluated separately for each crop because fruit and vegetables react differently. Sweet peppers, for example, are sensitive to CO_2 treatment. Exposure to 30% CO_2 for 6 days at 13°C immediately retarded wall softening and ripening, but caused subsequent development of severe calyx injury and decay at 20°C (Wang, 1977).

Broccoli was stored at 5°C for 3 days to 6 days in CA atmospheres of 20%, 30%, and 40% CO_2 and 3% and 21% O_2, respectively, and then transferred in air at the same temperature (Wang,1979). CO_2 injury was severe in samples with the highest CO_2 level. CO_2 delayed chlorophyll degradation both during and after treatment, having both an initial and residual effect. These effects of CO_2 may be related to CO_2 inhibition of ethylene production and action.

The postharvest responses of Chinese cabbage to high CO_2 treatment or low O_2 storage were also investigated (Wang, 1983). Exposure of Chinese cabbage to 30% or 40% CO_2 for 5 or 10 days was detrimental. However, a low O_2 (1%) atmosphere was very effective in extending the storage life of Chinese cabbage at 0°C. Chlorophyll loss and yellowing of the leaves was delayed, and the cabbage was still in salable condition after 5 months of storage.

Food Coloring

The quality of fresh and canned vegetables may be maintained by special measures, CA storage, or processing in milder conditions, as discussed previously.

Another means of producing a canned green vegetable with good color is to add a food colorant. Food colorants used are copper and zinc chelates of chlorophyll derivatives, pheophytins, and pheophorbides, which were first described by Willstätter and Stoll (1913). These chelates are bright green; the pheophytin chelates are liposoluble, whereas the pheophorbide chelates are water-soluble. They are very stable, with the metal firmly bound, and resistant even to concentrated hydrochloric acid, so they are safe for food coloration. The copper chelates in particular are produced commercially and used as food colorants. They are permitted in Europe and, with some restrictions, in the United States (Humphrey, 1980). These chelates were reported to be occasionally formed in canned vegetables. In canned okra, to which a weak solution of zinc chloride has been added to prevent loss of green color, a bright green color appeared due to the formation of a zinc chlorophyll chelate (Fischbach, 1943; Fischbach and Newburger, 1943). The phenomenon was observed also in canned green pea puree stored at 0°C to 37°C (Schanderl et al., 1965). In products stored at 37°C a more stable green color reappeared—a regreening due to the complexing of pheophytins with copper and zinc that are naturally present in the plant.

A water-soluble stable green pigment was formed in Brussels sprouts picked earlier in the autumn and processed at 115°C (Swirski et al., 1969). The pigment appeared to be a complex between a water-soluble protein moiety containing small quantities of copper and zinc and chlorophyll and its derivatives.

The experimental formation of zinc and copper complexes of chlorophyll derivatives in vegetable tissue by thermal processing was investigated by Jones et al. (1977b). Zinc and copper complexes of the pigments pheophytins *a* and *b* and/or pheophorbides *a* and *b* were formed in green tissues by prescribed thermal treatments. The formation of copper complexes occurred much more readily than formation of zinc complexes. Because of the spectral similarity of these complexes and the chlorophylls, their formation produces the effect of retaining the blue-green color or the previously mentioned regreening of the vegetable tissue during the following thermal treatment or storage. The analytical methods used for detection, separation, and determination of the copper and zinc complexes of chlorophyll derivatives are mentioned under Analytical Methods.

EFFECT OF PHYTOHORMONES ON CHLOROPHYLL METABOLISM IN VEGETABLES

Phytohormones or growth regulators are endogenous substances produced by plants that regulate different processes having inhibiting or promoting effects. The groups of auxins, cytokinins, and gibberellins have senescence-delaying effects, whereas ethylene and abscisic acid promote and accelerate ripening and senescence. As their mode of action is more and more understood, they may be applied to plants in pre- or postharvest treatments in order to modify the development in a certain direction.

Cytokinins

Cytokinins are derivatives of adenine. Kinetin (6-furfurylaminopurine) was first isolated from autoclaved preparations of nucleic acids, and received its name because of its activity in stimulating cell division. The group also includes synthetic compounds. Required for normal growth and differentiation, cytokinins act in conjunction with auxins to promote cell division and to retard senescence. Cytokinins stimulate chlorophyll synthesis (Fletcher and McCullagh, 1971; Dei, 1983; Arnold and Fletcher, 1986). They accelerate chloroplast differentiation and stimulate photosynthetic enzyme activity (Harvey et al., 1974). On the other hand, the inhibition of specific enzymes, such as respiratory kinase, can be another means by which cytokinins exert their antisenescence effect. Respiration inhibition was observed in several crops treated with benzyladenine (Dedolph et al., 1961; MacLean et al., 1963; Pressman and Palevitch, 1973).

A more direct effect of cytokinins concerning chlorophyll degradation is the lowering of chlorophyllase activity. This effect was reported for detached barley and oat leaves (Sabater and Rodriguez, 1978) and for *Pisum sativum* (Grossmann and Leshem, 1978). The antisenescence properties of cytokinins have been attributed to their ability to repress free-radical damage caused by the activities of lipoxygenase and superoxide dismutase in senescing foliage (Leshem, 1984).

The special effect of cytokinins—stimulation of chlorophyll synthesis and prevention of its degradation—was used commercially to ameliorate the storage of green vegetables. One of the earliest assays applied N^6-benzyladenine (BA) to various green vegetables to delay senescence during storage (Zink, 1961). The senescence inhibitor was effective in maintaining the green color and appearance in lettuce, Brussels sprouts, broccoli, parsley, and asparagus, among others. No response was observed in artichokes, snap beans, and lima beans.

The yellowing and abscission of the jacket leaves of cauliflower limits its storage after harvest. Treatment with BA prolonged the marketability of most of the heads to about 18 days (Kaufman and Ringel, 1961).

BA is effective in prolonging the life of asparagus and broccoli (Dedolph et al., 1961; MacLean et al., 1963; MacLean and Dedolph, 1964; Halevy and Wittwer, 1966) and head lettuce (Lipton and Ceponis, 1962) and for broccoli stored under high-temperature conditions (Pressman and Palevitch, 1973).

The effect of postharvest BA treatment and storage temperature (2°C and 5°C) on shelf life and quality of raw and cooked broccoli was investigated by Batal et al. (1982). The chlorophyll content was higher for samples treated with BA and stored at 2°C than it was for controls. Maintenance of good quality was also observed in cooked broccoli by BA treatment and storage at 2°C. Therefore, BA as a postharvest dip can be beneficial in maintaining greenness, tenderness, aroma, and flavor of broccoli for 3 weeks.

Preharvest spray or postharvest dip with BA was tested in broccoli stored at 0°C to 1°C for 10 weeks with two storage methods: icing or packaging (Shewfelt et al., 1983). Preharvest treatment resulted in the best color retention in iced and packed samples.

Gibberellin

Gibberellin has an effect similar to cytokinins, namely delaying senescence. Endogenous gibberellins and abscisic acid content as related to senescence were determined in detached lettuce leaves (Aharoni and Richmond, 1978). A sharp decline in gibberellins accompanied by a moderate rise in abscisic acid occurred before the onset of chlorophyll degradation. As senescence advanced, a considerable surge of abscisic acid was observed. The early depletion of gibberellins and probably cytokinins most likely represents the hormonal mechanism that triggers senescence of the leaf.

Among vegetables, the senescence-delaying effect of gibberellin was tested on tomatoes (Kader et al., 1966; Dostal and Leopold, 1967). Gibberellin was also tried on okra fruits grown in India, where the vegetable is quite popular (Singh and Dankhar, 1980). These fruits, if stored at edible stage at room temperature (32°C in India), deteriorate rapidly. Gibberellin treatment was applied as a postharvest dip at two concentrations, packaged, and stored at 32°C and 10°C. The fruits had a shelf life of 9 days and 12 days, respectively. Chlorophyll content was retained at high levels as senescence was visibly delayed. Chlorophyll *a* decreased more rapidly than chlorophyll *b*, which, curiously, exceeded the chlorophyll *a* level in some fruits.

Ethylene

Ethylene, which induces ripening in fruits, causes the transformation of the green chloroplasts to colored chromoplasts as chlorophylls break down and new pigments are synthesized. It is successfully used in horticulture to promote ripening in fruits, particularly tomatoes (Iwahori and Lyons, 1970).

Ethylene has a detrimental effect on cucumber fruits, which are consumed and processed when unripe. This effect was first observed when cucumbers were stored with other fruit species that produce substantial amounts of ethylene. Exposure of cucumbers to ethylene results in chlorophyll degradation and loss of firmness, the effect depending on ethylene concentration, storage temperature, and length of exposure (Duvekot et al., 1960). The influence of ethylene on postharvest quality deterioration on processing cucumbers was investigated by Poenicke et al. (1977). Endogenous ethylene is produced by cucumber fruits at a rate that is size-dependent, small fruits producing substantially more than large fruits. Mechanically harvested fruits produce two to three times more ethylene than do hand-harvested fruits, because ethylene production is stimulated by injury, and machine-harvested fruits are more likely to be damaged. Exogenous ethylene treatment, especially higher concentrations, decreased fruit chlorophyll content; the greatest chlorophyll loss was at the stem end of the fruit. Ambient concentration of ethylene during transportation in well-ventilated trucks does not endanger the quality of the cucumbers.

In tomatoes, ethylene stimulated the pigment changes associated with ripening, but prior treatment with gibberellin eliminated this effect (Dostal and Leopold, 1967).

In potato tubers treated with ethylene and exposed to light, chlorophyll biosynthesis was effectively inhibited (Patil et al., 1971). Ethylene is also involved in leaf senescence, as proved by investigations of Aharoni and Lieberman (1979) and Aharoni et al. (1979a,b). Chlorophyll loss accompanied by proteolysis (that is, real senescence) was promoted in oat leaves (Gepstein and Thimann, 1981). Chlorophyll loss was strongly antagonized by CO_2, with 1% CO_2 giving almost 50% chlorophyll maintenance. It was concluded that the ethylene system in leaves is more sensitive to CO_2 than it is in fruits.

This explains the effectiveness of CO_2 that is used commercially in CA storage for fruit and vegetables because it delays the ripening of fruits and the senescence of leaf vegetables.

CAROTENOIDS

Carotenoids of yellow, orange, and red color are among the most widespread and important natural pigments. They are found in higher plants, algae, fungi, and bacteria, both in nonphotosynthetic tissues and in photosynthetic tissue, accompanying the chlorophylls. Unlike chlorophylls and anthocyanins, they are also found in animals, being responsible for the color of birds, fish, insects, and some invertebrates. Since carotenoids are synthesized only in plants, plant carotenoids constitute the source of all animal carotenoids. Animal products such as milk, butter, and egg yolk contain carotenoids.

Carotenoids derive their name from the main representative of their group, β-carotene, the orange pigment first isolated from carrots (*Daucus carota*) by Wackenroder in 1831. Six years later Berzelius named the pigments of yellow leaves *xanthophylls*. In 1906 Tswett achieved the first separation of leaf pigments by inventing the chromatographic method. He separated four carotenoid pigments from two green chlorophylls and recognized that the yellow pigments formed a family. A year later the carotenoids were characterized as isoprenoid derivatives (Willstaetter and Mieg, 1907). In 1930, Karrer elucidated the structure of β-carotene and vitamin A. Owing to the work of his school, the structure of more carotenoids was established, as the symmetrical structure of many carotenoids was recognized. Zechmeister introduced the concept of polyene, and Kuhn demonstrated the relation between color and conjugation of the double bonds in the chain.

In 1934, Zechmeister published the first monograph on the carotenoids (Zechmeister, 1934). In the same year Lederer published three monographs on carotenoids, one dealing specifically with plant carotenoids (Lederer, 1934). Further work followed by Karrer and Jucker (1948), and comparative biochemistry of the carotenoids was investigated by Goodwin (1952). Meanwhile, the number of carotenoids discovered increased from 15 to 80, due to improvement in isolation methods, development of adsorption chromatography by Kuhn's school and by Strain, oxidative degradation methods, and development of synthetic methods.

Obviously in such an unsaturated substance a great many geometrical isomers of *cis* and *trans* configurations are possible. This subject was investigated by Zechmeister (1962). After 1960 the new separation of thin-layer chromatography allowed rapid progress in the field. The availability of additional spectroscopic methods, such as infrared (IR), nuclear magnetic resonance (NMR), and mass spectrometry (MS), brought a revolutionary advance. Optical rotatory dispersion (ORD) and circular dichroism (CD) spectroscopy are applied in stereochemistry. At present, the structures of 563 carotenoids are known and listed in the *Key to Carotenoids* by Straub (1987).

Publications concerned with carotenoids include the extensive monograph of Isler (1971a). Carotenoid biochemistry is covered by Goodwin (1980). Bauernfeind (1981) has emphasized the practical aspects of the use of carotenoids. Carotenoids are also dealt with in books on fruits and plant pigments by Hulme (1970/1971), Goodwin (1976), Czygan (1980), Britton (1983), and Gross (1987). Additionally, the published proceedings of the triennial International Carotenoid Symposia update the most recent aspects. They are published by the International Union of Pure and Applied Chemistry (IUPAC) (1967, 1969, 1973, 1976, 1979, 1985) and Britton and Goodwin (1982). Literature surveys on carotenoids are found in the series on terpenoids and steroids of the Chemical Society's Specialist Periodical Reports (Britton, 1976–1989).

DEFINITIONS AND NOMENCLATURE

Carotenoids are isoprenoid polyenes formed by the joining of eight C_5 isoprene units:

$$\begin{array}{l} CH_2{=}\underset{\displaystyle CH_3}{\underset{|}{C}}{-}CH{=}CH_2 \end{array} \qquad \text{Isoprene}$$

The isoprene units are linked in a regular head-to-tail manner except in the center of the molecule, where the order is inverted tail-to-tail so that the molecule is symmetrical. As a result, the two methyl groups near the center of the polyene chain are separated by six C atoms, and the other methyl groups by five C atoms, as illustrated in the prototype lycopene (Figure 8). From its basic structure almost all other carotenoids can be formally derived by hydrogenation, cyclization, oxidation, or any combination of these processes.

According to Karrer, the numbering of the molecule from the end to the center is 1 to 15, for the additional methyl groups it is 16 to 20, and for the

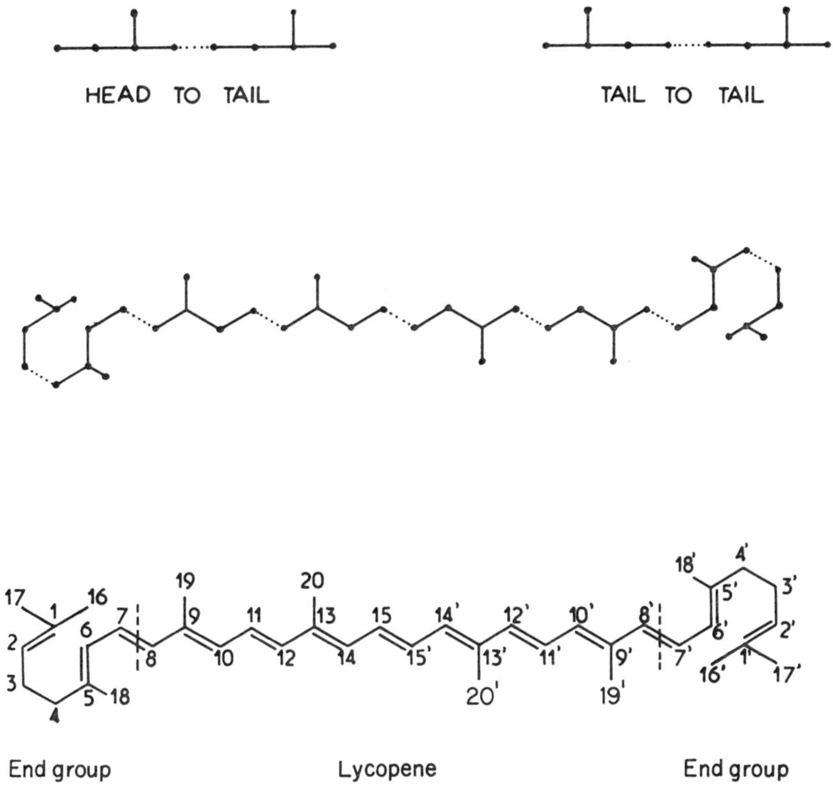

Figure 8. Joining of isoprenoid units.

symmetrical part it is 1′ to 20′. The former rules outlined by Karrer have been completed by IUPAC (Isler, 1971b; IUPAC, 1974).

The following discussion uses trivial names, which for the best-known carotenoids are well established, usually being derived from the source from which they were first isolated. These names are complemented by the more recent semisystematic names, which are based on the stem name *carotene*. The name of a specific carotenoid hydrocarbon is constructed by adding two Greek letters as prefixes to the stem name. These prefixes are characteristic of the two C_9 end groups: acyclic (ψ), cyclohexene (β, ϵ), cyclopentane (κ) (Figure 9), for example.

STRUCTURE AND CLASSIFICATION

The carotenoids can be divided into two groups: (1) carotenes that are hydrocarbons ($C_{40}H_{56}$) and (2) their oxygenated derivatives (xanthophylls). The

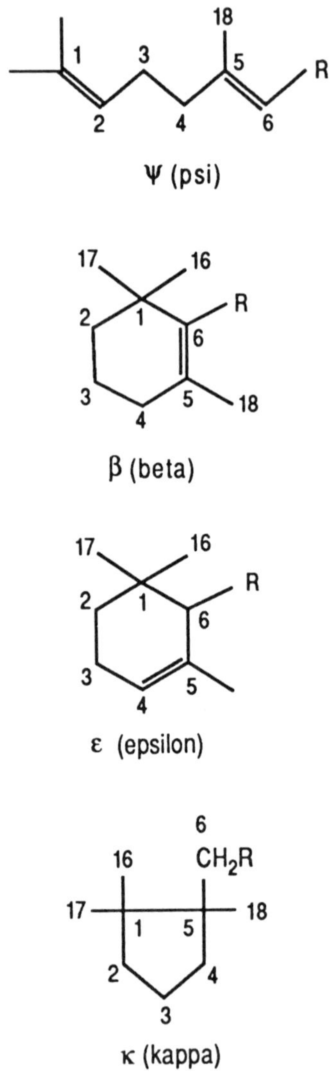

Figure 9. Some end-group-type designations of carotenes.

most frequently found functions are hydroxy- (monols, diols, and polyols), epoxy- (5,6- and 5,8-epoxides), methoxy, aldehyde, oxo, carboxy, and ester. A hydroxyl substituent occurs chiefly at C-3 in the ϵ- or β-ring. In most cyclic carotenoids the 5,6 and 5′,6′ double bonds are the most susceptible to epoxidation. The unconjugated double bond in the ϵ-ring does not undergo epoxidation.

Allenic carotenoids have a C=C=C grouping at one end of the central chain, and acetylenic carotenoids have a −C≡C−− in positions 7,8 and/or 7′,8′. Two more carotenoid types are the degraded carotenoids, with less than 40 carbon atoms in the skeleton (the *apocarotenoids*), and the higher carotenoids in which the carbon skeleton contains more than 40 carbon atoms(C_{45} and C_{50}). Carotenoids may also be classified as *acyclic* and *alicyclic,* and mono- and bicyclic. The figures in this section show the structural formulas of the most widespread carotenoids found in vegetables. The acyclic carotenes are phytoene, the first C_{40} compound in carotenoid biosynthesis with three conjugated double bonds, followed by the more unsaturated compounds phytofluene, ζ-carotene, neurosporene, and lycopene (Figure 10).

Figure 11 shows β-carotene, the most widespread bicyclic carotene and its hydroxy derivatives, the monol cryptoxanthin and the diol zeaxanthin, together with their various 5,6- and 5,8-monoepoxides and diepoxides, the most important of which are antheraxanthin and violaxanthin.

Figure 12 shows α-carotene and its hydroxy derivative, the widespread lutein and its epoxides.

Other important carotenoids are the allenic triol neoxanthin, which contains a 5′,6′-epoxide group, and neochrome, with the same formula and a 5′,8′-epoxide group (Figure 13).

Unique carotenoids found in some vegetables are discussed later.

PHYSICAL PROPERTIES

Carotenoids are lipids and are therefore soluble in other lipids—that is, they are *liposoluble*—and in fat solvents, such as acetone, alcohol, diethyl ether, and chloroform. The carotenes are soluble in apolar solvents such as petroleum ether and hexane, whereas the xanthophylls dissolve best in polar solvents such as alcohols. Except for the most unsaturated carotenes (phytoene, phytofluene, and ζ-carotene), the carotenoids are solids at room temperature and can be crystallized in various forms, colored from orange-red to dark violet, from suitable solvent mixtures.

Spectroscopic Properties

Carotenoids can specifically absorb light in the ultraviolet (UV) and visible region of the spectrum; the rest is transmitted or reflected, and they appear colored. The structural feature responsible for light absorption is the *chromophore,* which in carotenoids is the system of conjugated double bonds. Each carotenoid is characterized by an electronic absorption spectrum. In consequence, absorption spectroscopy is an important technique in carot-

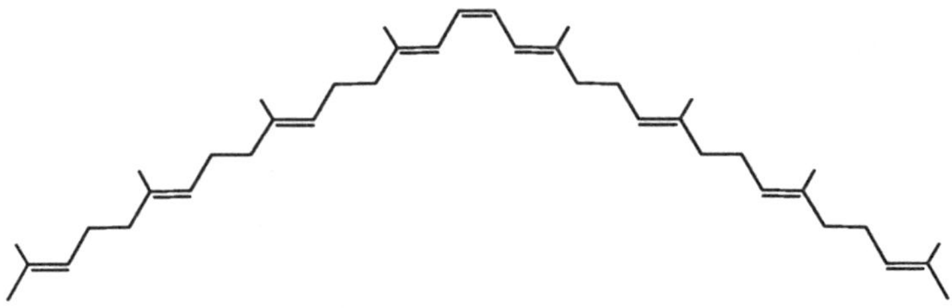

Phytoene
(15 - cis - 7,8,11,12,7',8',11',12', - Octahydro -ψ,ψ -carotene)

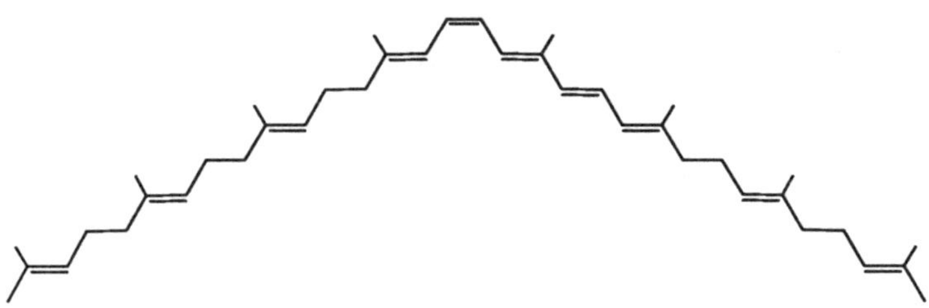

Phytofluene
(15 - cis - 7,8,11,12,7',8' -Hexahydro -ψ,ψ -carotene)

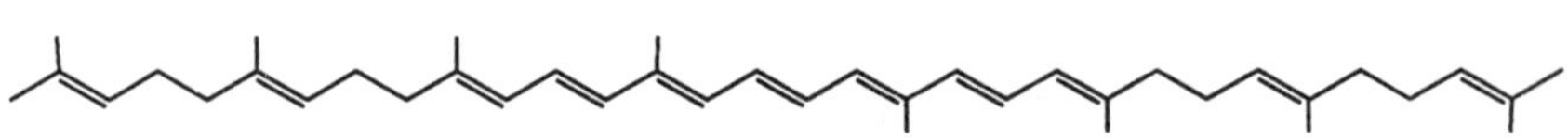

ζ – Carotene
(7,8,7',8', -Tetrahydro -ψ,ψ - carotene)

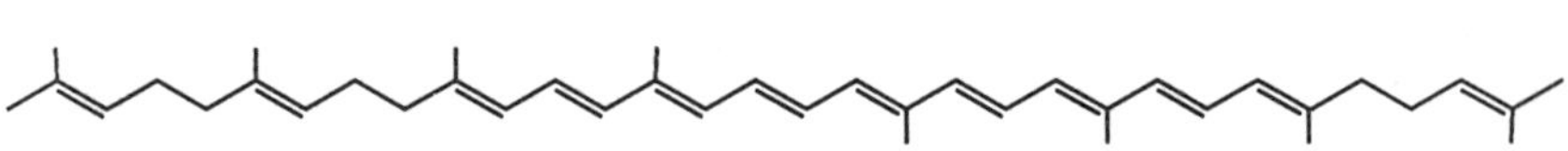

Neurosporene
(7,8 -Dihydro - ψ,ψ – carotene)

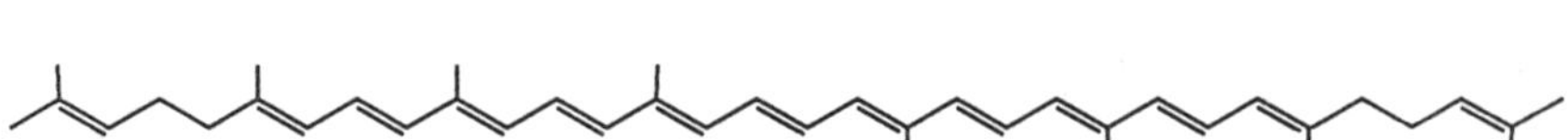

Lycopene
(ψ,ψ – carotene)

Figure 10. Acyclic carotenes.

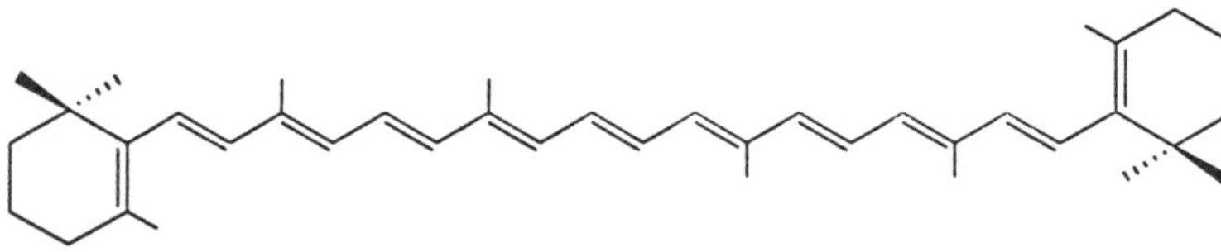

β - carotene
β, β - carotene

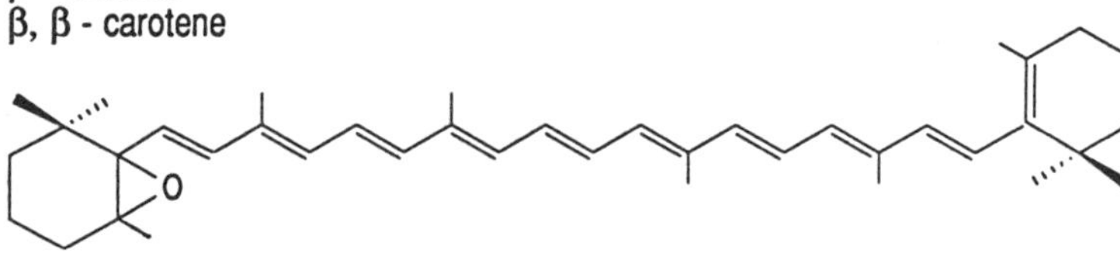

β - carotene 5,6, - epoxide
5,6, – epoxy - 5,6, dihydro - β, β - carotene

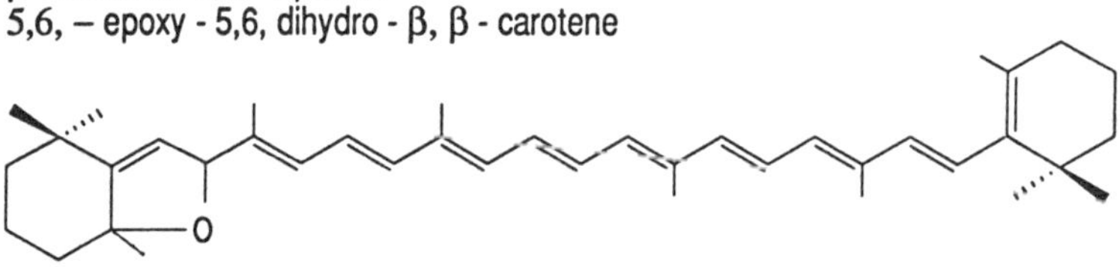

Mutatochrome
5,8 - epoxy -5,8 - dihydro - β, β - carotene

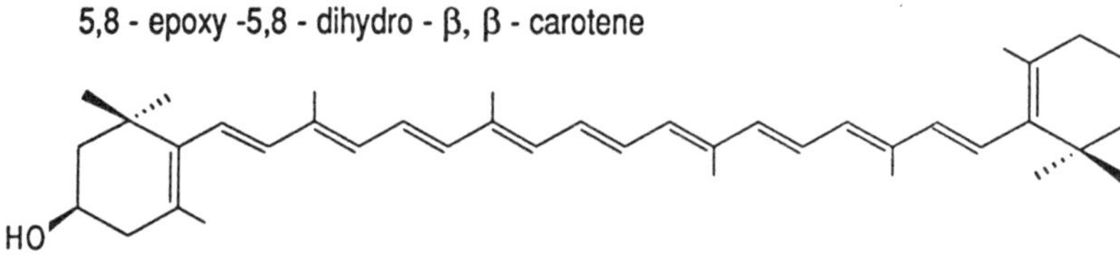

Cryptoxanthin
(3R) - β, β - caroten - 3 - ol

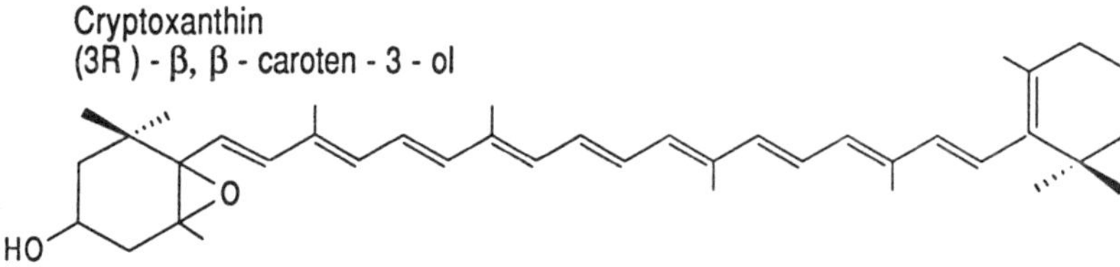

Cryptoxanthin 5,6, - epoxide
5,6, - epoxy -5,6 - dihydro - β, β - caroten - 3 - ol

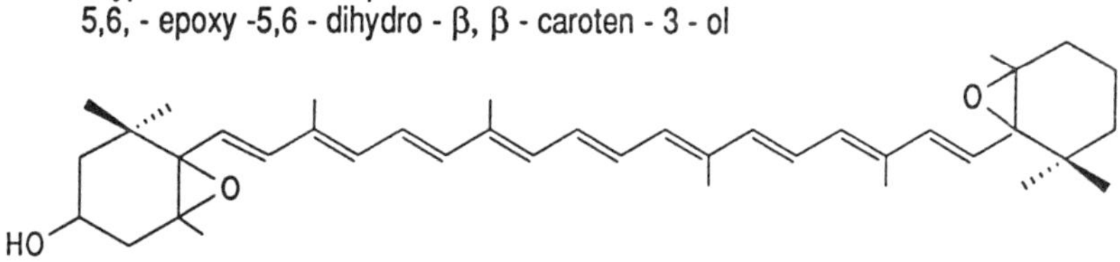

Cryptoxanthin diepoxide
5,6,5',6' - diepoxy - 5,6,5'6' - tetrahydro - β, β - caroten - 3 - ol

Figure 11. Carotenoids of the β,β series. *(continued)*

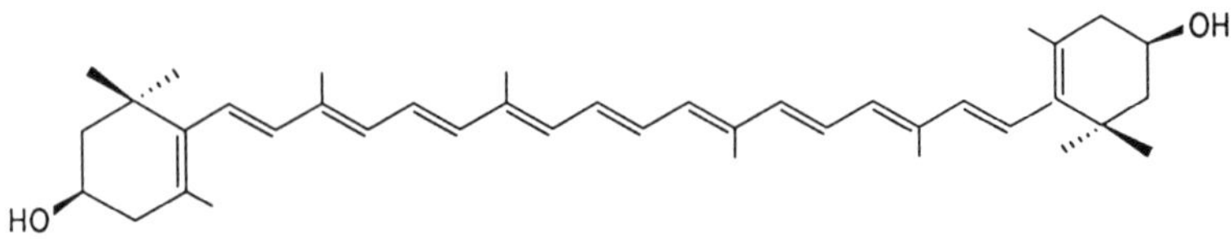

Zeaxanthin
(3R, 3'R) - β, β - carotene - 3,3' - diol

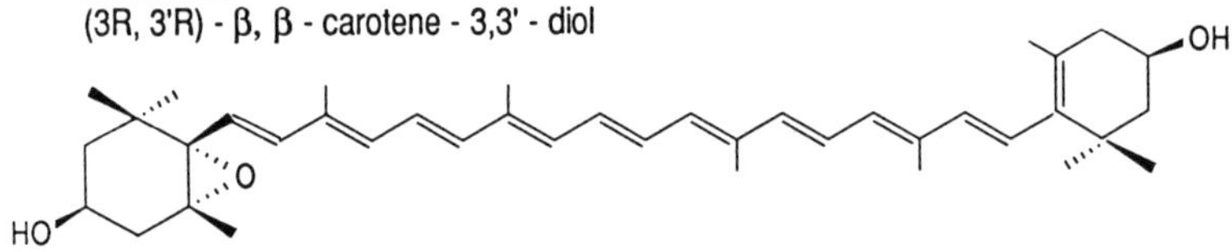

Antheraxanthin
(3S, 5R, 6S, 3'R) - 5, 6 - epoxy -5, 6 - dihydro - β, β - carotene - 3,3' - diol

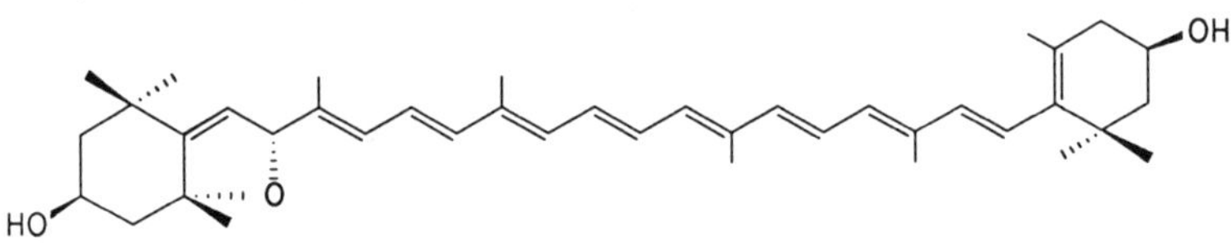

Mutatoxanthin
(3S, 5R, 8R, 3'R) - 5, 8 - epoxy -5, 8 - dihydro - β, β - carotene - 3,3' - diol

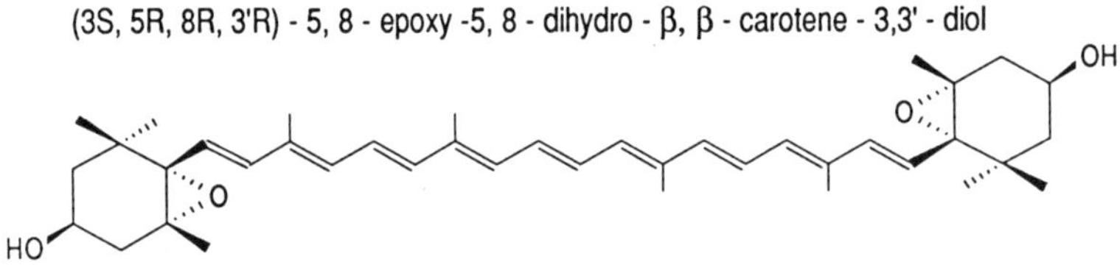

Violaxanthin
(3S, 5R, 6S, 3'S, 5'R, 6'S) - 5, 6, 5', 6' - diepoxy -5, 6, 5' 6' - tetrahydro - β, β - carotene - 3,3' - diol

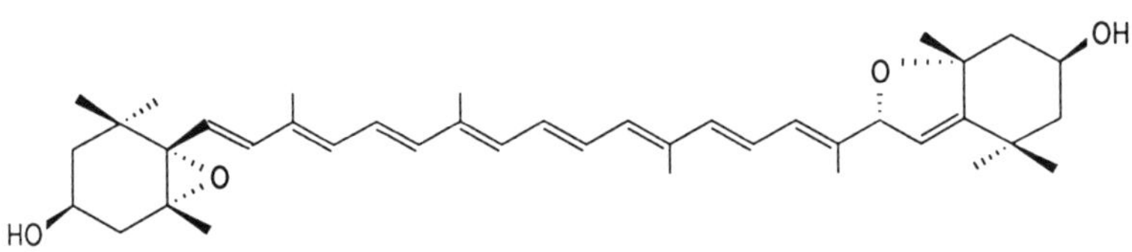

Luteoxanthin
(3S, 5R, 6S, 3'S, 5'R, 8'R) - 5, 6, 5', 8' - diepoxy -5, 6, 5' 8' - tetrahydro - β, β - carotene - diol

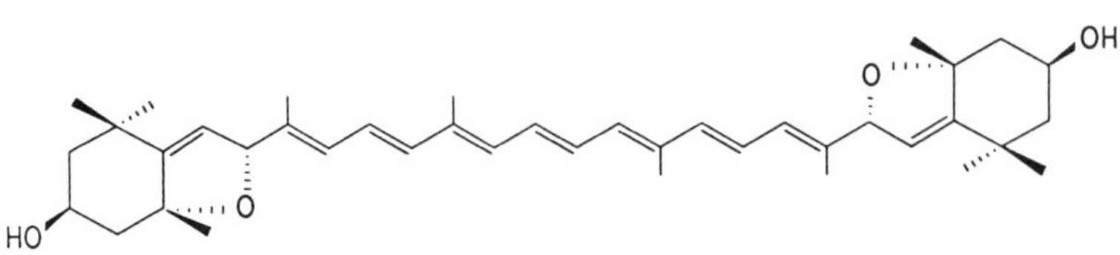

Auroxanthin
(3S, 5R, 8S, 3'S, 5'R, 8'R) - 5, 8, 5', 8' - diepoxy -5, 8, 5' 8' - tetrahydro - β, β - carotene - diol

Figure 11. *(Continued)*

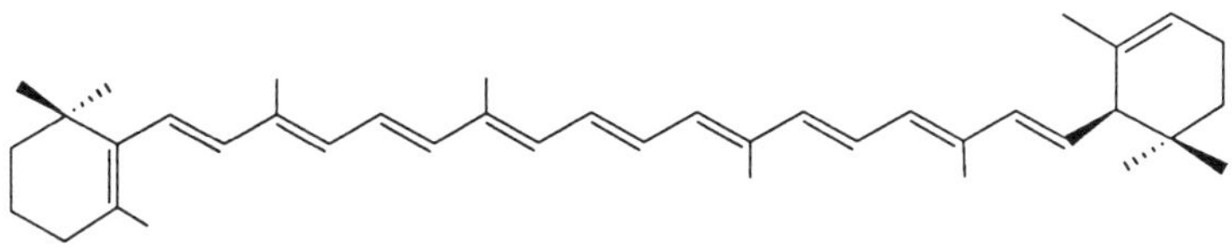

α - Carotene
(6'R) - β,ε - carotene

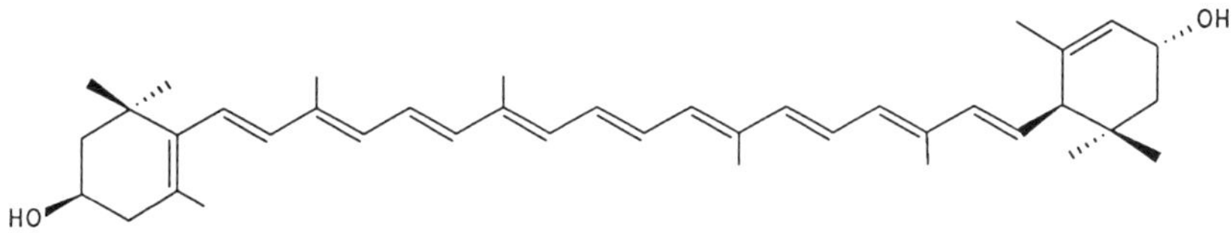

Lutein
(3R,3'R,6'R) - β,ε - carotene - 3,3' - diol

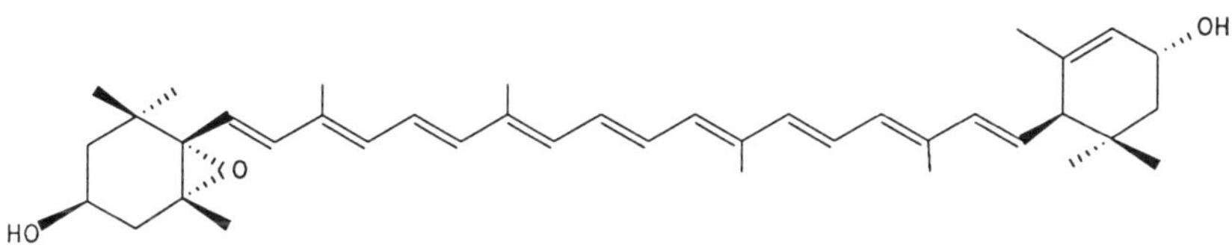

Isolutein, Lutein epoxide, Taraxanthin
(3S, 5R, 6S, 3'R, 6'R) - 5,6 - Epoxy - 5,6 - dihydro - β,ε - carotene - 3,3', - diol

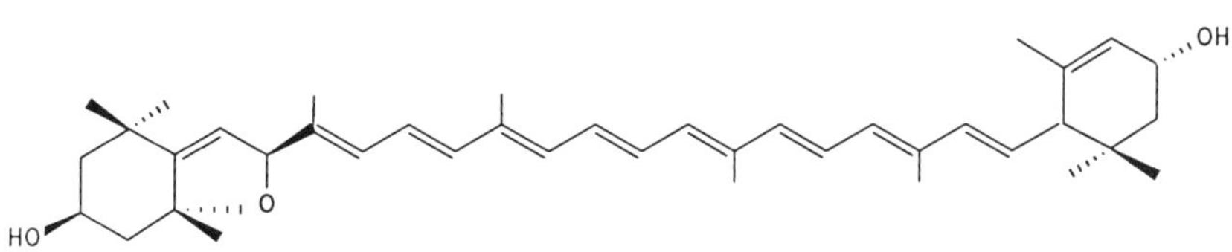

Chrysanthemaxanthin, 5,8 - epoxylutein
(3S, 5R, 8S, 3'R, 6'R) - 5,8 - epoxy - 5,8 - dihydro - β,ε - carotene - 3,3' - diol

Figure 12. Carotenoids of the β,ϵ series.

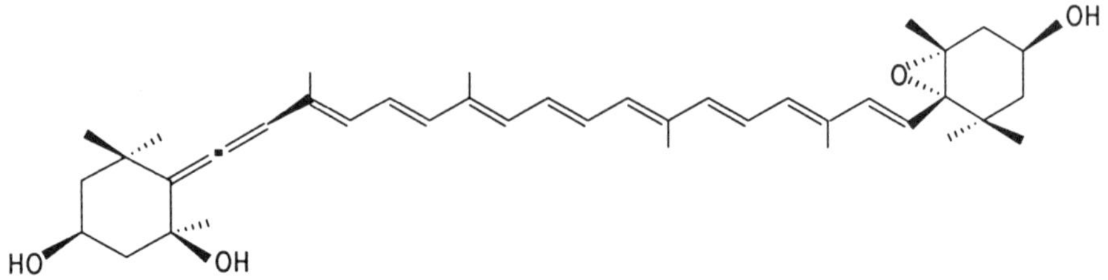

Neoxanthin
(3S, 5R, 6R, 3'S, 5'R ,6'S) - 5', 6' - epoxy - 6, 7 - didehydro - 5, 6, 5', 6' - tetrahydro -
- β, β - carotene - 3,5,3' - triol

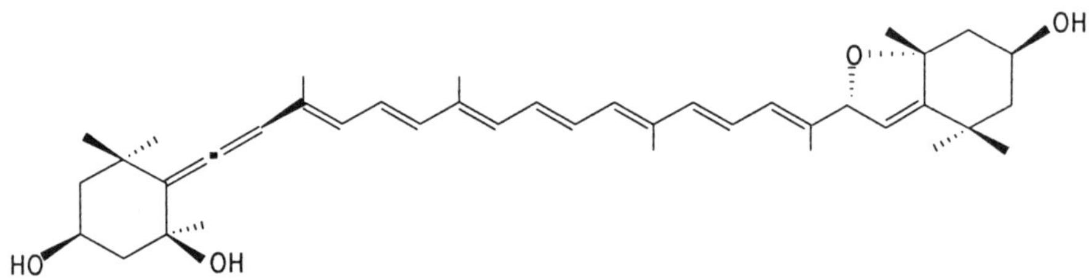

Neochrome
(3S, 5R, 6R, 3'S, 5'R ,8'R) - 5', 8' - epoxy - 6, 7 - didehydro - 5, 6, 5' 8' - tetrahydro -
- β, β - carotene - 3,5,3' - triol

Figure 13. Neoxanthin and neochrome.

enoid analysis. The position of the absorption maxima, usually 3, is a function of the number of conjugated double bonds. The positions of the absorption maxima are affected by the length of the chromophore, the position of the end double bond in the chain or ring, and the taking out of conjugation of one double bond in the ring or eliminating it through epoxidation.

The progressive bathochromic shift (movement to longer wavelength) is illustrated by the absorption spectra of the acyclic carotenoids of increasing chromophore length (Figure 14).

The chromophores of phytoene (three conjugated double bonds) and phytofluene (five conjugated double bonds) are not long enough to provide color. Both substances have their absorption spectra in the UV region, with phytofluene exhibiting a green-blue UV fluorescence. Carotenoids absorb

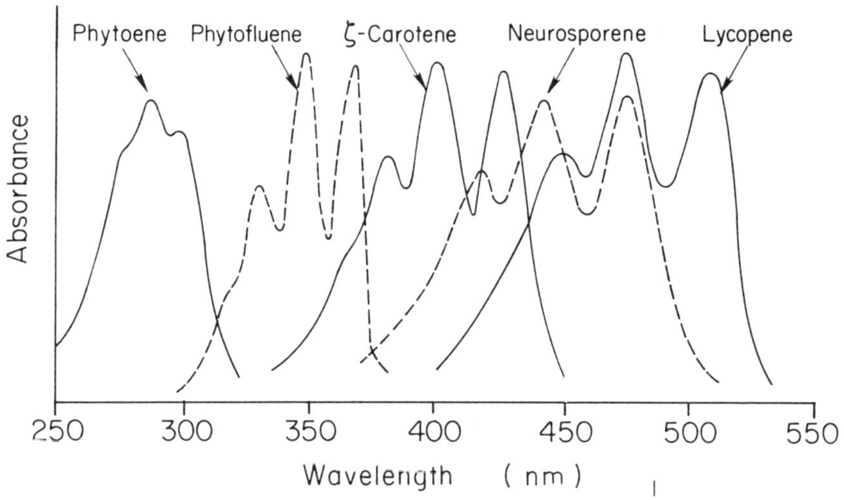

Figure 14. Absorption spectra of acyclic carotenoids of increasing chromophore length. (*From Britton, 1983.*)

mainly in the blue (430–470 nm) region, but they also absorb in the blue-green (470–500 nm) and green (500–530 nm) regions of the spectrum, and their color is determined by the light they reflect (or transmit when in solution). ζ-Carotene has a chromophore of seven conjugated double bonds with absorption bands mainly in the violet region of the spectrum and therefore exhibits a light yellow color. With the displacement of the absorption bands of a pigment toward regions of longer wavelengths, the hue increases from yellow (neurosporene) through orange to red (lycopene).

All the compounds have a typical three-peaked carotenoid absorption spectrum with well-defined maxima and minima (fine structure). A ring closure as in β-carotene produces a less defined fine structure. The introduction of a carbonyl group in conjugation with the polyene system produces a bathochromic shift and the loss of fine structure. The influence of other substituents such as OH is negligible; for example, β-carotene, cryptoxanthin, and zeaxanthin all have the same absorption spectrum.

The solvents used influence the position of the absorption maxima: petroleum ether and ethanol have little or no effect, chloroform and benzene produce a bathochromic shift of 15 nm, and carbon disulfide produces a shift of +35 nm.

Absorption maxima of some common carotenoids of vegetables are shown in Table 15.

Table 15. Absorption Maxima of Some Common Carotenoids of Vegetables

	Conjugated Double Bonds		
Carotenoid	*In Chain*	*In Ring*	*Abs. Max. (nm)*
Phytofluene	5		331, 348, 367
ζ-Carotene	7		380, 400, 424
Lycopene	11		447, 472,. 504
β-Zeacarotene	8	1	405, 428, 455
α-Carotene	9	1	423, 444, 473
β-Carotene	9	2	425, 451, 478
Zeaxanthin	9	2	425, 451, 478
Antheraxanthin	9	1	420, 444, 472
Mutatoxanthin	8	1	404, 427, 453
Violaxanthin	9		418, 440, 470
Violeoxanthin	9		415, 436, 464
Capsanthin	9 (+ 1 >C=O)	1	476

Source: From Davies (1976).
Note: Carotenes recorded in petroleum ether, xanthophylls in ethanol.

STEREOCHEMISTRY

The stereochemistry of the carotenoids involves geometrical isomerism about the carbon–carbon double bond and absolute configuration of an asymmetric carbon.

Geometrical Isomerism

Owing to the double bonds in the molecule, all carotenoids exhibit the phenomenon of *cis–trans* isomerization (stereomutation) (Zechmeister, 1962). A *cis* double bond implies a configuration with the highest-priority groups on the same side, whereas in the *trans* configuration they are on opposite sides.

The number of possible stereoisomers of a product is very high, depending on the number of double bonds in the molecule. In carotenoids the number is much more restricted because the methyl groups along the chain cause steric hindrance. Of 1056 theoretically possible *cis–trans* isomers of lycopene, only 72 sterically unhindered *cis* isomers exist, of 272 theoretically possible isomers of β-carotene, only 20 are unhindered *cis* forms. Artificial isomerization of all-*trans*-β-carotene produces 12 *cis* forms. These *cis* isomers were separated by column chromatography, and two predominant *cis* isomers were isolated. Neo-β-carotene U and neo-β-carotene B adsorbed above and below the all-*trans* isomer (Zechmeister, 1962). The structure of these two *cis* isomers was recently elucidated by applying NMR spectroscopy (Tsukida et al., 1981). Neo-β-carotene is 9-mono-*cis*-β-carotene, and neo-β-carotene B is 13-mono-*cis*-β-carotene (Figure 15).

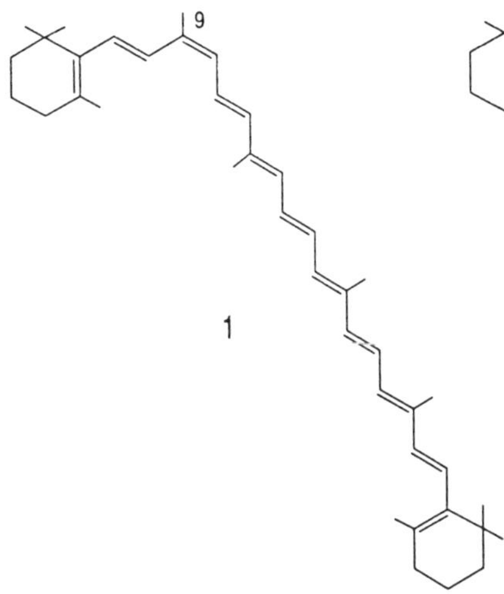

1

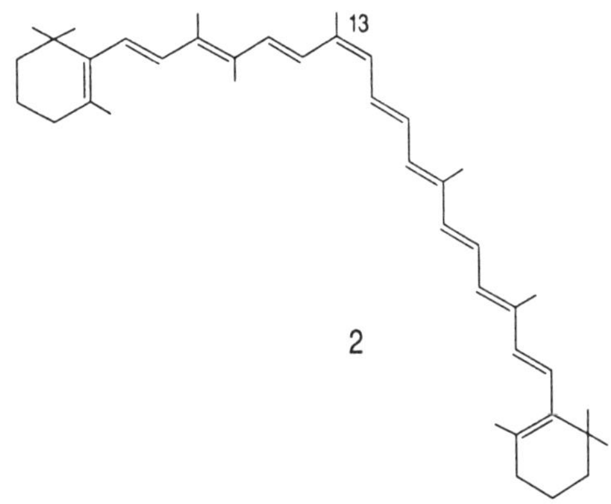

2

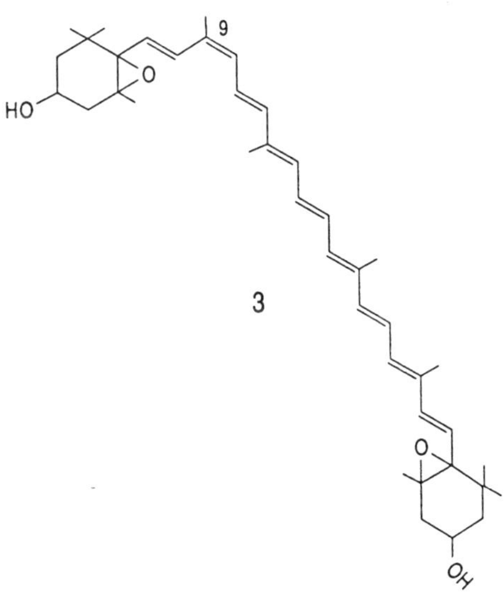

3

1. 9-mono-*cis*-β-carotene (neo-β-carotene U)
2. 13-mono-*cis*-β-carotene (neo-β-carotene B)
3. 9-mono-*cis*-violaxanthin (violeoxanthin)

Figure 15. *cis*-Carotenoids.

Naturally Occurring cis-*Carotenoids*

In nature, the carotenoids are predominantly the all-*trans* isomers, which are more stable, but naturally occurring *cis* and poly-*cis* isomers have been isolated, and their number has increased gradually as new techniques have been applied.

Phytoene, the first C_{40} compound and the precursor of carotenes, is found in higher plants as the 15-*cis* isomer (Jungalwala and Porter, 1965). In the tangerine tomato, prolycopene, a poly-*cis*-lycopene, is found as a major carotene accompanied by its congeners, which all have the poly-*cis* structure (Clough and Pattenden, 1979).

The two *cis* isomers of β-carotene, 9- and 13-mono-*cis,* were detected and quantified in various vegetables by Panalaks and Murray (1970) and Sweeney and Marsh (1971a). At uncommon high percentages they were found in cassava roots by Penteado and Almeida (1988). Eight *cis*-β-carotene isomers were separated and identified in carrots and peppers by using HPLC (Vecchi et al., 1981) (see Table 35). In various raw green vegetables every chloroplast carotenoid was a mixture of *trans* and *cis* isomers (Khachik et al., 1986) (see Table 25). 9-*cis*-Violaxanthin (violeoxanthin) is widespread in fruits and increasingly synthesized during ripening (Gross, 1987).

In red pepper, *cis*-capsorubin, 13-*cis*-capsanthin, 13′-*cis*-capsanthin, and 9-*cis*-capsanthin were identified by Tóth and Szabolcs (1981).

Other *cis* isomers isolated from natural sources may be artifacts formed during isolation procedures by stereomutation. Stereomutation occurs under differing conditions; it increases with temperature and light intensity and may be induced by the presence of iodine or acids. If a number of precautions are not respected during carotenoid analysis, *cis–trans* isomers are easily formed as artifacts. For determining the nature of the geometric isomer, a standard method of stereomutation is used (see the analytical methods section).

Spectral Properties of cis-*Carotenoids*

The absorption spectrum of a *cis* isomer presents a subsidiary peak in the near-ultraviolet, the *cis* peak. Generally it is located 143 nm from the longest wavelength maximum (e.g., 473 − 143 = 330 nm). The absorption maxima of a *cis* isomer undergo a hypsochromic shift of 2 nm to 5 nm, and the degree of fine structure and the extinction values decrease. Consequently, the *cis* isomer has a lighter hue of lower intensity than does its parent *trans* isomer (Figure 16).

Absolute Configuration

About half of the naturally occurring carotenoids possess at least one asymmetrically substituted carbon, a chiral center. Absolute configuration is the

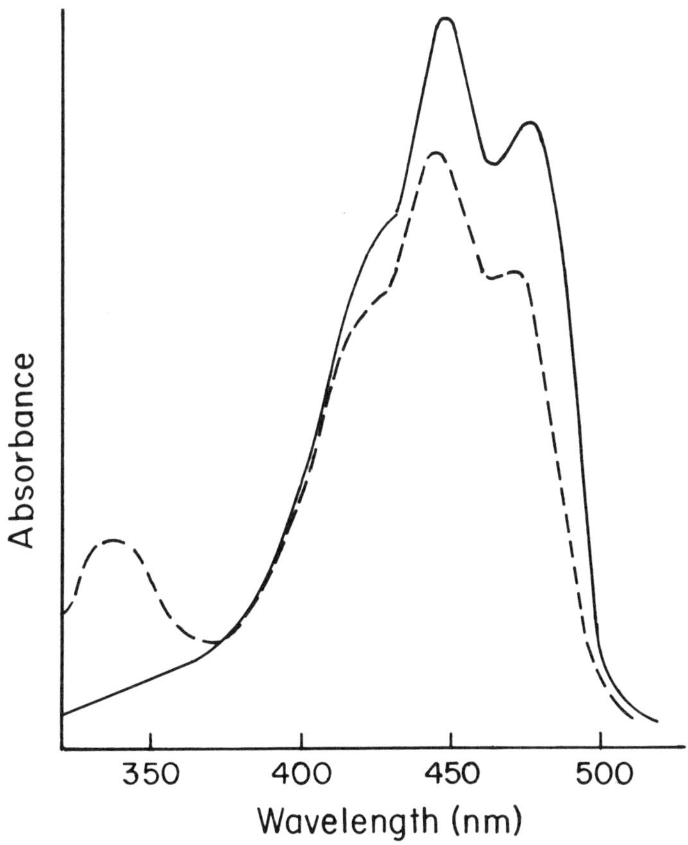

Figure 16. Absorption spectra of *trans*- (——) and *cis*-β-carotene (- - - - -).

description of the position or order of the different substituents in space. To designate the absolute configuration, one uses the general *R* and *S* convention (the sequence rule) (Weedon, 1971). The most conspicuous physical property of disymmetric substances is their optical activity—that is, their ability to rotate the plane of polarized light. Examples of chiral centers in carotenoids are C-3 of the 3-hydroxy-β-ring and C-6 of an ε-ring. (See Figures 11, 12, and 13 on carotenoid structures.)

The absolute configurations of chiral carotenoids have been determined by chiroptical methods (optical rotatory dispersion and circular dichroism) and nuclear magnetic resonance.

The absolute configuration has enormous implications in the study of biogenesis. In order to elucidate the mechanism of biosynthesis, one must know the chirality of the studied compound and the stereochemistry of its biosynthesis.

CHEMICAL PROPERTIES

Natural carotenoids consist of carbon, hydrogen, and oxygen. Because the functional groups are oxygen functions, their characterization can be carried out on a spectroscopic scale requiring only small amounts of sample, and is usually a routine analysis. Useful tests for determining carotenoid structure are discussed later. Together with spectrophotometric and chromatographic properties, they permit identification of the carotenoid. In addition, high-resolution MS, ORD, and CD measurements are now used routinely, and only small quantities of sample are necessary. The mass spectrum gives the molecular weight,and the fragmentation pattern indicates the structure. The ORD and CD spectra allow establishment of the absolute configuration of the pigment.

The chemistry of carotenoids is discussed in detail by Liaaen-Jensen (1971) and Moss and Weedon (1976).

Chemical Tests

Epoxide Test

Isomerization of the 5,6-epoxides into 5,8-epoxides occurs rapidly, even in the presence of trace amounts of dilute acids (Figure 17).

Figure 17. Isomerization of 5,6-epoxide into 5,8-epoxide.

The reaction is very fast, and the rearrangement produces a hypsochromic shift in the visible absorption spectrum as the chromophore is shortened. A shift of 20 nm indicates one epoxide group, whereas a shift of 40 nm indicates the presence of two such groups (Figure 18).

The isomerization of violaxanthin into auroxanthin involves a shift of 40 nm. The chromophore of auroxanthin is similar to that of ζ-carotene with seven conjugated double bonds. The reaction also produces a characteristic color that varies from green for 5,6-epoxides to ink blue for 5,8-epoxides. The latter group, of course, gives only the color reaction. A rapid epoxide test is to expose a thin-layer chromatogram to HCl gas for a few minutes.

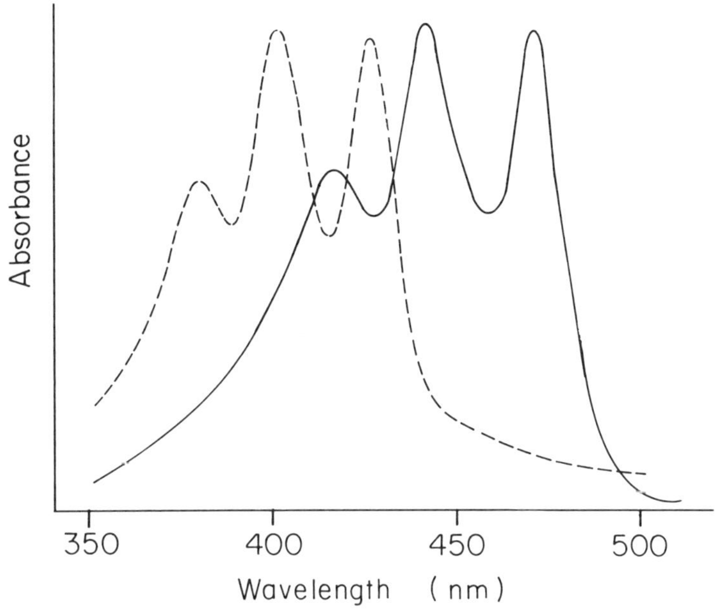

Figure 18. Conversion of violaxanthin (———) into auroxanthin (- - - - -) by HCl.

Carbonyl Reduction Test

The reduction of the carbonyl function can be carried out with metal hydrides—lithium aluminum hydride in ethyl ether or sodium borohydride in ethanol. Both the rate of reduction under standardized conditions and the magnitude of the hypsochromic shift may be of diagnostic value. A hypsochromic shift of about 6 nm to 7 nm indicates an in-ring carbonyl. A shift of 25 nm to 35 nm indicates an in-chain carbonyl. The color changes visibly to a lighter hue, and the typical fine structure reappears in the spectrum.

Hydroxyl Test

Esterification. Primary and secondary hydroxy groups are acetylated at room temperature by acetic anhydride in dry pyridine, but tertiary hydroxy groups do not react under these conditions. If the course of the reaction is followed chromatographically, the number of acetylable hydroxyls is indicated by the number of intermediary acetates observed. A monol gives a monoacetate; a diol, one intermediary acetate if the molecule is symmetrical, or a mixture of several if not; and a triol, three monoacetates and three diacetates.

Silylation. Tertiary hydroxyl groups as well as primary and secondary hydroxyls can be converted to trimethyl ethers. Sterically hindered hydroxyls cannot be silylated. Tertiary hydroxyl groups may be determined through silylation of the fully acetylated product.

Allylic Hydroxyl Test (dehydration and etherification). Allylic hydroxyl groups can be dehydrated by treatment with acidified chloroform. The introduction of an additional double bond produces a bathochromic shift due to the extension of the chromophore. Allylic hydroxyl groups undergo methylation more easily than do other alcohols. By treatment with methanol and HCl at room temperature, the methyl ether is easily formed. Because it is less polar than the alcohol, the methyl ether can be separated chromatographically.

Carotenol Esters

For investigating carotenol esters, saponification is, of course, omitted. Hydrolysis of the esters proceeds at room temperature in a weak (1–5%) alkaline solution. The kinetics of the reaction can be followed chromatographically as significant changes in polarity occur, the free xanthophylls being much more polar than the apolar esters. The mobility of the esters depends on the nature of its fatty acid constituents. The nature of the carboxylic acid liberated through saponification can be determined.

OXIDATION

Due to the system of conjugated double bonds in the molecule, the carotenoids are easily destroyed by oxidative degradation.

Chemical Oxidation

The classical methods used in carotenoid structure elucidation involved oxidative degradation (Karrer and Jucker, 1948; Liaaen-Jensen, 1971). The various oxidants are oxygen, ozone, alkaline permanganate, chromic acid, and others. The nature of degradation products depends on the location at which the attack occurs. By ozonolysis, oxidative cleavage of the carbon–carbon bonds occurs, and the formed carboxylic acids allow the characterization of the carotenoid's end group.

Stepwise degradation with permanganate or chromic acid was extensively used. The degradation products are apocarotenals and ketones (with a skeleton of less than 40 C atoms).

Oxidative fission of the central double bond of β-carotene with hydrogen peroxide yielded small amounts of vitamin A aldehyde (retinal) (Hunter and Williams, 1945). By reduction it was then transformed to vitamin A alcohol (retinol).

Oxidation of β-carotene with MnO_2 in the absence of air gave retinal (10%), β-apo-10′-carotenal (2%), and small amounts of β-apo-8′-carotenal. From lycopene, the same experimental conditions yielded small amounts of apo-8′-lycopenal (Winterstein et al., 1960).

Carotenoids containing β-rings, substituted or not at positions 3 and 3′, are readily epoxidized by monoperphthalic acid under controlled conditions (Karrer and Jucker, 1945).

Autoxidation

Autoxidation is the spontaneous combination of substances with oxygen in the air at room temperature. Both in solution and crystallized, the carotenoids undergo autoxidation in the presence of oxygen, a free-radical chain process. The process is stimulated by temperature, light, humidity, and some metals. Because carotenoid autoxidation occurs widely during food processing and storage, it has been intensely investigated by using β-carotene, the most important provitamin A, in model systems. Carotene oxidation was followed, and the decrease of the extinction at the wavelength of its maximal absorption was measured (Budowski and Bondi, 1960). Adding unsaturated fat had a prooxidant effect, which increased with the degree of unsaturation. Different antioxidants, especially butylated hydroxytoluene, were effective, thereby indicating that carotene oxidation is a chain radical reaction in which initiators and inhibitors may be involved. β-Carotene submitted to autoxidation for 15 to 20 days was oxidized to its 5,6- and 5,8- mono- and diepoxides, to cryptoxanthin, and to cryptoxanthin monoepoxide (Bodea et al., 1957).

A sample of autoxidated crystals of β-carotene kept for 5 years at 4°C in the dark was analyzed (Zinsou, 1971). The degradation products were several *cis* isomers of β-carotene and a mixture of 5,6- and 5,8-epoxides of β-carotene and some apocarotenals. A correlation between the autoxidation susceptibility and the polarity of several carotenoids was found, with the unsubstituted β-ring the most easily autoxidated (Ramakrishnan and Francis, 1980).

Temperature Effect

The rate of lycopene oxidation, in solution or crystallized, increases with temperature (Cole and Kapur, 1957a). The same effect was observed during β-carotene oxidation (El-Tinay and Chichester, 1970). As in the previous report the rate of reaction increased in the presence of cupric ions. The products of the reaction were *trans* and *cis* isomers of 5,6- and 5,8- mono- and diepoxides, showing that, as with perphthalic acid, the oxygen attack occurs at the terminal double bonds.

At higher temperatures (190–220°C) volatile compounds were formed by the thermal degradation of β-carotene, predominantly β-ionone, dihydroactinidiolide, and 5,6-epoxy-β-ionone (Schreier et al., 1979). During a simulated deodorization of palm oil, the decomposition products of β-carotene were β-13-apocarotenone, β-15-apocarotenal, and β-14′-apocarotenal (Ouyang et al.,1980). The thermal degradation products of β-carotene were characterized in a model system simulating edible oil deodorization and deep-fat frying (Onyewu et al., 1986). The process involved heating at 210°C for 5 min to 4 h. Over 70 nonvolatile compounds were observed in the study, but only 7, the most important and those formed in large amounts, were identified. Hydrocarbon-type derivatives with a much shorter polyene system with ketone, alcohol, or a combination of both of these functional groups, were obtained. By heating β-carotene at 180°C in sealed tubes and in extrusion cooking, Marty and Berset (1986) obtained the same degradation products as in autoxidation or chemical oxidation. The mixture of 5,6- and 5,8- mono- and diepoxides was separated by HPLC. These compounds were found, for the first time, to be formed during thermal degradation.

Effect of Water Content

It is known that carotenoids are more rapidly oxidized in dehydrated products, because water attached on the food surface forms a protecting film. The effect of water on carotenoid stability in low-moisture systems was studied by Chou and Breene (1972). The model system consisted of β-carotene distributed uniformly on crystalline cellulose. The decoloration reaction appeared to be first order, with the presence of water or antioxidants inhibiting the rate of decoloration.

Oxygen uptake of a model system containing β-carotene was measured by a method capable of determining headspace oxygen and entrapped oxygen, both at low concentrations (below 2%) (Teixeira Neto et al., 1981). The values of the rate constants were functions of the oxygen concentrations in the headspace and increased with increased concentrations.

In a different model system, simulating dehydrated food, the mechanism and kinetics of β-carotene oxidation were studied by Goldman et al. (1983). The presence of oxygen in the headspace is a crucial factor in β-carotene degradation. Even a low concentration of 1% to 2% leads to a significant loss of pigment. The presence of free radicals also accelerates the degradation rate. β-Carotene retention curves were sigmoidal with three regions corresponding to an induction period, a fast main period, and a retardation period, typical of an autocatalytic reaction. During the induction period the initiation of free radicals takes place. The protective role of water content can be explained by its direct effect on free radicals produced during pigment oxidation, as reported earlier by Ramakrishnan and Francis (1980).

The stability of β-carotene in a model system containing sulfite was studied by Baloch et al. (1977c). The carotene destruction decreased rapidly with increasing concentration of SO_2 up to 2000 ppm. The stability of β-carotene was also increased when nonsulfited and sulfited samples were stored at monolayer water activity (a_w).

Photooxidation (Light Effect)

Photooxidation is the oxidative bleaching produced by oxygen in the air. The rate of decoloration is enhanced by the presence of a sensitizer. In the presence of toluidine blue, degradation products of β-carotene were *cis* isomers of β-carotene epoxides and hydroxy derivatives. In the presence of Rose Bengal, diactinidiolide and small amounts of β-ionone were isolated. In the absence of sensitizers, the composition of the oxidation products was similar except that β-ionone was the main product and dehydroactinidiolide was minor (Isoe et al., 1969; Hasegawa et al., 1973). In contrast to chemical oxidation products of β-carotene, which include 5,6- and 5,8- mono- and diepoxides, the major oxidation products of β-carotene, photosensitized by hypericin, were chiefly 5,8-epoxides such as mutatochrome and aurochrome (Seely and Meyer, 1971). Photooxidation of β-carotene, whether in the absence or presence of a sensitizer, resulted generally in products that were *cis* isomers of β-carotene and its 5,6- and 5,8-epoxides. The most oxidized product was aurochrome (Zinsou and Costes, 1973).

The kinetics of carotenoid photodegradation in a vegetable juice were studied by exposing samples to 1404 lx of light (Pesek and Warthesen, 1987). The photodegradation followed first-order kinetics, and α- and β-carotene were degraded much faster than lycopene.

The dye-sensitized (Rose Bengal) photooxidation of β-carotene was investigated recently with regard to the formation of volatile noncarotenoids (Hohler et al., 1988). From the complex mixture of 56 compounds, 27 compounds, mainly carbonyl derivatives with 8 to 15 C atoms, were identified. The higher yields of C_9, C_{10}, and especially C_{13} compounds indicate that a preferential cleavage at carbon atoms -6,7, -7,8, and -9,10 of the β-carotene molecule takes place. The C_{13} compounds represent the most important group, of which β-ionone is the best known. Its 5,6-epoxide was a major compound (~ 10 mg g^{-1}).

Photosynthetic pigments were destroyed under a variety of oxidative conditions in order to find out whether the plant growth substance xanthoxin is formed (Wagner and Elstner, 1989). The photooxidation rate of the chloroplast pigments in the presence of the photosensitizer Rose Bengal was followed at 5-min intervals by reading the absorption maxima decrease of each pigment. Within 30 min, 69% of neoxanthin, 55% of violaxanthin, 42% of lutein, and 43% of β-carotene were bleached, whereas both chlorophylls remained almost intact. Xanthoxin was not formed, contrary to early reports on violaxanthin oxidation to xanthoxin occurring in the presence (or even absence) of lipoxygenase (Firn and Friend, 1972).

Enzymatic Oxidation

The in vivo oxidation of carotenoids is catalyzed by various oxidative enzymes.

Lipoxygenase

Lipoxygenase, one of the main oxidative enzymes in plants, catalyzes the oxidation by molecular oxygen of unsaturated fatty acids containing a *cis,cis*-1,4-pentadiene system to a *cis,trans*-conjugated diene hydroperoxide. This in turn oxidizes the pigments found in the vegetable tissue, chlorophylls, and carotenoids.

Lipoxygenase exists in multiple forms, the isoenzymes being classified as type 1, type 2, and type 3, based on optimum pH and substrate and product specificity.

The investigation of lipoxygenase was initiated due to the discovery of an unknown enzyme that bleached carotenoids. In 1928 it was observed that the yellow pigments of bread dough are bleached when a small amount of soybean flour was added to improve the baking quality. The procedure was subsequently patented (Haas and Bohn, 1934). The enzyme of soybean responsible for the bleaching was investigated and recognized as a specific lipid oxygenase, and was termed *lipoxidase* (André and Hou, 1932). The yellow pigments consisted mainly of carotenoids, so very soon the bleaching was associated with the enzymatic destruction of carotenoids.

The enzyme *carotene oxidase* from various beans was found to oxidize and thereby decolorize α- and β-carotene, bixin, xanthophylls, and other carotenoids. The bleaching action of the enzyme was supposed to be an instance of coupled oxidation (Sumner and Dounce, 1939; Sumner and Sumner, 1940). The mechanism of the enzyme action was further studied by Tauber (1940). It was understood that the oxidation of carotene and xanthophylls depends on the simultaneous oxidation of unsaturated fats. The unsaturated fats are probably oxidized by the enzyme named "unsaturated fat oxidase," and the oxidation product in turn oxidizes the carotenoids. The carotenoids are not attacked by the enzyme alone. In 1947 the enzyme isolated from soybean was obtained in crystalline form (Theorell et al., 1947).

The distribution of lipoxygenase in leaves and a survey of its content in a number of fruits and vegetables were reported by Holden (1970a) and Pinsky et al. (1971). Lipoxygenase was extracted and purified from various sources such as pea, alfalfa, leaves, broad bean, soybean, potato, cauliflower buds, corn germ, and eggplant (Eskin, et al., 1977). The rapid destruction of carotene (a loss of 45% to 80%) that occurs in alfalfa during the field curing process was recognized to be due to a lipoxidase (Mitchell and Hauge, 1946). Enzymatic destruction of carotenoids in chloroplasts of spinach beet leaves was also reported (Friend and Nakayama, 1959). β-Carotene disappeared more rapidly, and was followed by the disappearance of violaxanthin and lutein. The order of carotenoid disappearance was the same as that reported in autumn leaves by Goodwin (1958a). These results were confirmed by Takagi and others in studies on the carotenoid effect on spinach leaf lipoxygenase activity (Takagi et al., 1977; Takagi and Matsugami, 1977; Takagi, 1979). The degree of interaction of β-carotene and the enzyme was weaker than that of lutein and lipoxygenase, but the enzyme inhibition was similar. The differences between the two carotenoids were assumed to derive from their different polarity.

The products obtained from the coupled oxidation of β-carotene by a linoleate–lipoxidase system were also investigated (Friend, 1958). They were identified as *cis* isomers of β-carotene, β-carotene monoepoxide, aurochrome, and several conjugated ketones.

The kinetics of β-carotene oxidation coupled with the linoleate–lipoxygenase system show two phases, of which the first is temperature-sensitive (Zinsou, 1971). During the first phase, three products accumulated: 15-*cis*-β-carotene, mutatochrome, and cryptoflavin.

The volatile products that arise from bleaching with a partially purified soybean lipoxygenase were identified by Grosch et al., (1977) as β-ionone, β-ionone epoxide, and dihydroactinidiolide. The same main volatile components were obtained by thermal degradation of β-carotene in the presence of air (Schreier et al., 1979). These products are interesting flavor substances found in tomato and tea.

The products of β-carotene oxidation by an isoenzyme of lipoxygenase were the same as those reported previously (Ikediobi and Snyder, 1977). A *cis* isomer of β-carotene, mutatochrome, aurochrome, and β-apocarotenal were identified. The various lipoxygenase isoenzymes differ markedly in their carotene bleaching activity. Thus a purified isoenzyme of pea lipoxygenase (lipoxygenase-2) is a very potent oxidizing enzyme (Arens et al., 1973). The respective turnover numbers are about 1,200 moles of β-carotene per min/mole and 4,600 moles linoleic acid per min/mole. The enzyme also co-oxidizes chlorophyll, but the activity is only 1/100 of the carotene bleaching activity. Another isoenzyme, lipoxygenase-1 from seeds of soybean genotypes is devoid of carotene and chlorophyll bleaching activity (Hildebrand

and Hymowitz, 1982). However, lipoxygenase-1 is capable of bleaching carotene and chlorophyll under anaerobic conditions with the concomitant production of carbonyl compounds (Klein et al., 1984). Carotene was bleached faster than chlorophyll. The formation of an enzyme fatty acid radical complex that reacts with the plant pigments appears to be a requirement for bleaching.

The lipoxygenase-2 isomer has very little anaerobic activity (Cohen et al., 1985). Chlorophyll and carotene inhibit its aerobic activity. Increasing pigment concentrations resulted in significant inhibition of the enzymatic oxidation of linoleate as well as pigment bleaching and carbonyl formation. This suggests that the pigments may form an inactive complex with the enzyme reducing the amount of free enzyme.

Various methods were developed to measure the oxidation rate. An older procedure for lipoxygenase assay was to measure the bleaching rate of carotene in the presence of the unsaturated fatty acid substrate (Sumner and Sumner, 1940). A procedure that brings linoleate and carotene into aqueous solution with Tween 80 permits direct spectrophotometric study of carotene bleaching (Ben-Aziz et al., 1971).

Lipoxygenase affects the determination of carotene in vegetables. The carotene content of blanched vegetables is often 1.2- to 2.3-fold higher than in the raw material. This discrepancy is due to the leaching of solid matter during blanching and the destruction of carotene by lipoxygenase in raw vegetables when no antioxidant is added.

Lipoxygenase isolated from pumpkins grown in Japan was purified and characterized by Hidaka et al. (1986). According to Axelrod et al. (1981), it has the same properties as soybean lipoxygenase type 3.

The inhibitory effect of phenolic compounds in vegetables on carotene bleaching catalyzed by lipoxygenase was studied with the aid of a model system and various vegetables (spinach, broccoli, carrots, green beans, and peas) (Oszmianski and Lee, 1990). The inhibitory effect is highly correlated with the concentration and nature of the phenolic compound.

Other Enzymes with Carotenoid Bleaching Activity

Other enzymes (heme proteins) found in the vegetable cell display a similar carotenoid bleaching activity, not necessarily coupled to linoleic acid peroxidation. Oxidation of the diapocarotenoid crocin was the result of a peroxidase system in mitochondria from sugar beet leaves (Dicks and Friend, 1967). It was considered a unique example of a catalyst of carotenoid oxidation for which a lipid requirement has been precluded. Intrinsic heme proteins with carotene bleaching activity were reported in aqueous extracts of tomato and in various vegetable tissue other than seeds (Blain, 1968, 1970).

A carotene bleaching activity similar to that of lipoxygenase of the heme proteins cytochrome *c* and peroxidase was proved in a comparative study on the catalytic effect of these three enzymes (Ben-Aziz et al., 1971). Unlike lipoxygenase and cytochrome *c,* the peroxidase bleached also in the absence of added fatty acids.

A peroxidase-like protein was found to be the active carotene bleaching factor in the aqueous extracts from red pepper fruits (Kanner et al., 1977). In pumpkins the carotene bleaching activity was due primarily to lipoxygenase and, secondarily, to the heme proteins, peroxidase and catalase, which were also present (Hidaka et al., 1986).

ANALYTICAL METHODS

Analytical methods have been described by Britton and Goodwin (1971) and De Ritter and Purcell (1981), but the most detailed coverage is by Davies (1976). Updated reviews have been done by Taylor (1983) and Britton (1985).

The analysis of carotenoids is complicated because of their instability, their tendency to undergo stereomutation, their photo- and thermolability, and the readiness with which they undergo oxidation. Carotenoids are altered by acid and partially destroyed enzymatically, especially by lipoxygenase.

All analytical operations must be carried out in dim light in an inert atmosphere (under nitrogen or vacuum). Solvents must be purified, the temperature no higher than 40°C, and samples must be stored at about −20°C under nitrogen.

Isolation of carotenoids involves extraction, saponification, and separation. Identification of the separated pigment is based on spectroscopic and chromatographic properties and on chemical tests. Mass spectrometry, CD, and NMR spectroscopy are now also routinely used for structural elucidation. A series of review articles on these techniques has been published by IUPAC (Britton and Goodwin, 1982).

Extraction

Owing to the diversity of the biological material in which carotenoids are found, no standard extraction techniques can be outlined. The material has to be fresh and undamaged, and a sample of it must be representative. Extraction should be done as rapidly as possible to avoid oxidative or enzymatic degradation. Before extraction the material must be ground or cut into small pieces to facilitate complete extraction. Since the carotenoids are liposoluble, they are extracted with organic solvents. All solvents used must be pure and free of oxidizing compounds, acids, or halogens. Diethyl ether, which may contain peroxides, must be freshly prepared peroxide-free before use.

Before extraction, antioxidants [e.g., butylated hydroxytoluene (BHT)] and neutralizing agents (alkaline carbonates) are added. Vegetable tissue containing lipoxygenase may be blanched beforehand. For fresh tissue, which contains a high percentage of water, water-miscible polar organic solvents (acetone, methanol, ethanol, usually a mixture of acetone and methanol) are used. Dry material may be extracted with water-immiscible solvents such as diethyl ether or petroleum ether. Generally, the extraction is carried out in a blender, where grinding and extraction occur simultaneously. The best is the Ultra Turrax type in which the motor is mounted above the blades.

The homogenate is filtered by suction through a sintered glass funnel or through filter paper coated with filter aid in a Buchner funnel. The filter aid is diatomaceous earth (Celite). The extraction procedure is repeated until the pigment is totally extracted. From the water-containing crude extract the pigments are transferred to an appropriate water-immiscible solvent by adding enough saturated NaCl solution until two layers separate.

Saponification

Saponification is a purification procedure that removes unwanted lipids and chlorophylls. It does not affect carotenoids, which are generally alkali-stable. When alkali-labile pigments (astaxanthin, fucoxanthin) are present or carotenoid esters are to be analyzed, this step is omitted. Saponification is carried out by adding sufficient KOH in methanol or ethanol to have an overall KOH concentration between 5% and 10%. Alternative saponification methods include boiling the alcoholic lipid solution for 30 min or keeping it overnight at room temperature under N_2.

Subsequently, the saponified extract, mixed with ether in a separation funnel, is washed free of alkali, dried, and evaporated in vacuum. Water is removed by azeotropic distillation with absolute ethanol, which is preferable to drying with sodium sulfate. Unsaponifiable matter is taken up in a suitable solvent, and the total carotenoids content is determined.

Quantitative Determination

The standard method of quantitative determination is spectrophotometric and similar to the determination of chlorophylls. The absorbance (extinction or optical density) of a known volume of carotenoid solution is read at the wavelength of maximal absorption. The amount of carotenoid in micrograms per gram material is given by the expression

$$\mu\text{g carotenoid/g} = \frac{A \times V \times 10^6}{A_{1\text{cm}}^{1\%} \times 100 \times G}$$

in which V is the total volume (ml) containing G g of sample, and $A_{1cm}^{1\%}$ is the specific absorbance or extinction coefficient. The specific absorbance (extinction) coefficient is the absorbance of a 1% solution (w/v) in a 1-cm light pass cell at the appropriate wavelength. The absorbance (extinction) coefficients have been tabulated by Davies (1976). The molar absorbance (extinction) coefficient, ϵ_{mol}, is the absorbance of 1 M solution. An arbitrary $A_{1cm}^{1\%}$ of 2500 is used when the specific absorbance coefficient is unknown or, generally, for the determination of a mixture of carotenoids (total carotenoids).

Total chloroplast carotenoids found in green leaves may be determined in the unsaponified leaf extract along the chlorophylls by measuring the absorbance at the wavelengths of the chlorophylls' absorption maxima and at a wavelength where the carotenoids absorb strongly and both chlorophylls absorb little, such as 470 nm.

In the Lichtenthaler's equations for chlorophyll determination, the third equation allows one to determine total carotenoids by subtracting the contributed absorption of both chlorophylls from the absorbancy read at 470 nm (see p. 10).

Removal of Sterols

Sterols that may be present in large amounts in some extracts and may disturb further chromatographic procedure can be removed by precipitation in different solvents. After determination of total carotenoids, the extract is evaporated in vacuum and the unsaponifiable matter is dissolved in a minimum volume of methanol, petroleum ether, or acetone and kept overnight at −20°C. The sterols are removed by centrifugation in a refrigerated centrifuge.

Carotenoid Separation

Phase Separation

Separation into epiphasic and hypophasic carotenoids may be achieved by solvent partition between petroleum ether (PE) and aqueous methanol (85–95%). When the PE extract is shaken in a separation funnel with an equal volume of aqueous methanol the epiphase (PE) contains the nonpolar compounds, namely chlorophylls, carotenes and their epoxides, and the totally esterified carotenols. Monoketones and monols are evenly distributed in both phases. Polar xanthophylls, diols, and polyols are found in the hypophase methanol.

The separation of "carotene" from xanthophylls by solvent partition was used in earlier reports, "carotene" being subsequently determined in the epiphase (Halevy et al., 1957). Phase separation is still used to obtain a preliminary fractionation before chromatography. It has been applied in the carotenoid investigation of tomato seeds, white-fleshed sweetpotatoes, Japanese pumpkins, and other plants (Rymal and Nakayama, 1974; Martin, 1983, Hidaka et al., 1987).

A method based on liquid–liquid partition is carotenoid fractionation by counter current distribution with the Craig apparatus. In a 200-tube glass apparatus using a solvent system of PE and methanol 99%, the carotenoid mixture is separated into three fractions after 99 or somewhat fewer transfers (Curl, 1953). Curl applied counter current distribution preceding chromatography in his investigations on fruit carotenoids, including green and red bell peppers and tomatoes (Curl 1961, 1962, 1964). Counter current distribution was also used as a preliminary separation of the carotenoids of Centennial sweetpotatoes (Purcell and Walter, 1968).

The method was tiresome and not further used, but *droplet counter current* chromatography, which is based on the same principle of liquid–liquid partition, was recently developed. It was applied for the first time in the carotenoid field to separate the carotenoids of parsley (leaves and stem) (Francis and Isaksen, 1989). A droplet counter current chromatograph, with 72 capillaries connected in series, was used. The solvent system consists of petroleum ether (40–60°C)-acetonitrile-methanol (50:10:40), the upper layer used as the mobile phase.

Chromatography

Chromatography is the most important technique for separating and purifying carotenoids. A multitude of adsorbents are used for carotenoid chromatography, among them sucrose, cellulose, starch, $CaCO_3$, $Ca(OH)_2$, CaO, $MgCO_3$, MgO, $ZnCO_3$, Al_2O_3, silica gel, kieselguhr, which are also used as mixtures.

The adsorbents are chosen according to the types of carotenoids to be separated, because different adsorbents achieve the separation in different ways. Thus, groups of different polarity—hydrocarbons, monohydroxy and polyhydroxy carotenoids—can be separated on alumina or silica, with the most polar carotenoids being more strongly adsorbed.

For MgO and $Ca(OH)_2$, for example, separation is determined by the number and type of double bonds in the molecule. In the carotene series lycopene is the most strongly adsorbed and is followed by neurosporene and ζ-carotene. Cyclic carotenes with a double bond in the ring are adsorbed less than linear carotenoids. Because β-carotene has a longer chromophore than α-carotene, it is more strongly adsorbed. *Cis–trans* isomers can be separated on $Ca(OH)_2$ and $ZnCO_3$.

Column Chromatography (CC) or Open-Column Chromatography. The term "open-column chromatography" is now used to distinguish this method, which utilizes descending gravity columns from columns used under pressure, such as in high-pressure liquid chromatography. Open-column chromatography was first initiated by Tswett (1906) and was developed after 25 years by Kuhn and Lederer (1931) and then by Zechmeister (1934) and Strain (1941). (See also Davies, 1976 and references cited there.)

The column may be packed with adsorbents such as alumina or silica gel as a slurry (in PE). The pigments are eluted with solvents of increasing polarity (diethyl ether in PE). The polarity of the solvent is changed after the colored band is completely eluted.

Other adsorbents (MgO, $CaCO_3$, sugar, cellulose) may be packed dry. To improve the slow rate, a filter aid (diatomaceous earth) such as Hyflo Super Cel or Celite is mixed with the adsorbent in various proportions (30–60% by weight). This kind of column may be eluted by stepwise elution, but zone chromatography is usually applied. The solvent is run through the column until the colored zones are separated. The adsorbent is then extruded, and the colored zones are cut and eluted.

Open-column chromatography on MgO–Hyflo Super Cel (1:1 by weight) is the most widely applied method for investigating carotenoids of vegetables. It was used in studies on the vitamin A value and for a complete investigation of all carotenoids. A rapid method of determining "pure" carotene in plant tissue was restricted to quantify only the carotene fraction eluted from the MgO column. The widely applied official method for the determination of provitamin A was introduced in 1955 and is still in use (AOAC, 1984). The carotenoids of corn (Quackenbush et al., 1961; Neamțu et al., 1982), cucurbits (Rodriguez et al., 1976; Hidaka et al., 1987; Arima and Rodriguez-Amaya, 1988), broccoli (Gross, 1979) and peppers (Cholnoky et al., 1955, 1958b; Curl, 1962, 1964) were all separated on MgO–Hyflo Super Cel columns.

In the studies by Lee and his collaborators, carotenoid investigation of vegetables containing predominantly carotenes involved a preliminary separation of the carotenoids into three fractions: hydrocarbons, monohydroxy carotenoids, and polyhydroxy carotenoids. This separation is based on partition using a silica–methanol column (Purcell, 1958). The hydrocarbon fraction is rechromatographed on a MgO–Hyflo Super Cel column. With this method the carotenes of carrots (Ogunlesi and Lee, 1979; Lee, 1986), sweet corn (Lee et al., 1981), tomatoes (Lee and Robinson, 1980), squashes (Lee et al., 1984), and green peas (Edwards and Lee, 1986) were separated.

Of course, in the investigation of carotenoids of vegetables, other adsorbents were also used, namely $CaCO_3$, $ZnCO_3$, polyethylene, and cellulose, for example, peppers (Davies et al., 1970; Camara and Monéger, 1978). Tomato carotenoids were separated in the earlier studies on MgO, but alumina was found to be more appropriate (Britton and Goodwin, 1971; Ben-Aziz et al., 1973; Cabibel and Ferry, 1980).

The *cis* isomers of an isomerized β-carotene solution could be resolved on a hydrated-lime column for the first time (Polgár and Zechmeister, 1942). This technique was applied to various plant material, but even improved it was not suitable for routine analysis (Kemmerer and Fraps, 1945; Bickoff et al., 1949).

A so-called "high pressure liquid chromatography" method was subsequently developed for the separation of provitamin A carotene stereoisomers in fresh and cooked vegetables (Sweeney and Marsh, 1970). In the recent literature the method is considered an HPLC method, but an open column is actually used through which the eluent is forced by nitrogen under pressure. A first chromatography on MgO separates α- and β-carotene, which are rechromatographed on a column packed with a mixture of MgO and $Ca(OH)_2$ (1:6), developed with 1.5% *p*-methylanisole in petroleum ether. The method was applied in many investigations concerned with the effect of processing on the provitamin A carotenes.

Thin-Layer Chromatography. Thin-layer chromatography (TLC) has been successfully applied in carotenoid separation and purification. The methods are described in two classical handbooks (Randerath, 1963; Stahl, 1967). Thin-layer chromatography may be used alone for small quantities of sample or in combination with column chromatography. By column chromatography the carotenoids are first separated into fractions of different polarity, which are then further separated into individual carotenoids by TLC.

The method permits rapid and sharp separation and is generally applied on a micro- or semimicro scale. Substances at trace levels can be detected. Thicker layers up to 1.5 mm can be also used on a preparative scale. The most used adsorbent is silica gel; other adsorbents are basic oxides or carbonates, of which MgO is most often used. Because these adsorbents do not adhere well to the glass plate, they are mixed with bonders, such as calcium sulfate, kieselguhr, and starch.

Most methods are based on adsorption, which is determined, as previously discussed, by the polarity of the carotenoids and by the number and type of the double bonds in the molecule. Partition chromatography (r-p TLC) has also been successfully applied. It is done on layers of kieselguhr, silica gel, or cellulose impregnated with liquid paraffin or vegetable triglycerides; the mobile phase may contain water. Partition chromatography is suitable for carotenol esters that are epiphasic and weakly adsorbed. By partition chromatography the mobility of the esters is correlated with the nature of the esterifying fatty acid.

Since most carotenoids are colored, staining methods are not necessary. Colorless fluorescent carotenoids like phytofluene are detected in UV light

(360 nm). The adsorbent with the separated carotenoids is scraped from the plates directly into the eluent, filtered, and then analyzed.

Numerous methods were developed for using thin-layer chromatography on chloroplast pigments. One, which is still in use, involves separating leaf pigments, both major and minor carotenoids, on a mixture of $CaCO_3$, MgO, and $Ca(OH)_2$ by adsorption chromatography (Hager and Meyer-Bertenrath, 1966). In a recent method the complete resolution of more than 20 plastidic carotenoids from photosynthetic tissue was achieved on MgO developed with acetone in petroleum ether in different proportions (Sadowski and Wojcik, 1983).

The chloroplast pigments in autumn leaves containing carotenol esters have been analyzed by partition chromatography (Egger, 1964). The lutein esters, which predominate, were separated on cellulose layers impregnated with liquid paraffin and developed with acetone–methanol (2:1).

During investigations of some vegetable carotenoids, pigment separation was often achieved by a combination of column chromatography and TLC. Tomato carotenoids were first separated on a column of alumina, and the fractions were rechromatographed on thin layers of silica gel and MgO–kieselguhr (1:1) (Ben-Aziz et al., 1973). Broccoli carotenoids were similarly fractionated first on a MgO column, followed by a further separation on TLC both on silica gel and MgO–kieselguhr (Gross, 1979). The carotenoids of Cucurbitaceae fruits were similarly analyzed by Rodriguez-Amaya and collaborators (1975, 1976, 1989).

A combination of column chromatography and TLC in inverted order was used by Buckle and Rahman to follow pigment changes in ripening peppers (Buckle and Rahman, 1979). The pigments were first separated on cellulose layers developed with four different solvent systems suitable to each of the four ripening stages. The separated bands were subsequently separated by column chromatography on magnesium oxide or alumina.

Carotenoids of vegetables may be analyzed directly by TLC. One method allows rapid separation of carotenoids by successive thin-layer chromatographic separation on two different adsorbents (Gross, 1980). The first chromatography on silica gel G developed with the solvent system acetone–petroleum ether (30:70) gives a preliminary fractionation into groups of different polarity (hydrocarbons, monols, diols, and polyols). A further separation of each group into individual carotenoids is obtained by rechromatography on MgO–kieselguhr (1:1 w/w). The same solvent system is used, the acetone percentage being increased according to the polarity of each group (4–30%). The method was applied to separate broccoli carotenoids (Yang and Henze, 1988b).

Carotenol esters of peppers were separated by partition chromatography on cellulose layers impregnated with vegetable oil developed with acetone–methanol (3:1) (Egger, 1968).

Besides general TLC methods, special methods suitable for a certain vegetable were also developed. An analytical procedure for rapid separation of carrot and tomato carotenoids, useful in genetic studies, was developed by Umiel and Gabelman (1971). The method involves a first separation on a silica gel column of the hydrocarbon fraction, which is further separated on MgO–kieselguhr (1:1 w/w) layers on which the pigments are directly quantified by a semiquantitative technique.

Another special TLC method to separate carotenoids of both fresh and dehydrated carrots uses as adsorbent a mixture of Hyflo Super Cel, magnesium, and calcium sulfate developed with 0.3% to 1% ethanol in petroleum ether (Baloch et al., 1977b).

A special TLC method for tomato carotenoids was recently developed by Johjima and Ogura (1983). The adsorbent is MgO–Hyflo Super Cel–cellulose (10:9:1) developed with a solvent system of *n*-hexane–isopropanol–methanol (100:2:0.2). The carotenoids were directly determined with a chromato-scanner.

For Spanish red pepper carotenoids, a TLC method is used to separate the carotenoids in order to establish the relationship between total red carotenoids versus total yellow carotenoids, which appears to be a characteristic parameter for each cultivar. The method may be used to detect possible adulteration (Mínguez Mosquera and Fernández Díez, 1981; Mínguez Mosquera et al., 1984). The pepper carotenoids are separated by TLC on silica gel developed with a solvent system of hexane–ethylacetate–ethanol–acetone (95:3:2:2).

For Hungarian ground red pepper carotenoids, a new method of overpressure-layer chromatography was applied (Aczél, 1986, 1988). In this procedure the adsorbent is sealed off by a cushioning system under pressure, which forces the solvent into the layer. The adsorbent used in this method is silica gel 60 F_{254} developed with petroleum ether–benzene–acetone–acetic acid (40:10:2:2.5).

A rapid TLC method was developed for the determination of the provitamin A content of fruits and vegetables by Premachandra (1984). This simple method uses TLC plates prepared without an applicator on which the separation occurs in a short time (10 min). The adsorbents are alumina and silica gel developed with 2% ethyl ether in petroleum ether.

High-Performance (pressure) Liquid Chromatography. HPLC used at the beginning for the separation of green plant pigments is now used for carotenoids, having replaced the standard methods of open-column chromatography and TLC. With this rapid, reproducible, and highly sensitive technique, carotenoid separation, identification, and estimation are achieved simultaneously. The modern equipment contains a pump to achieve high pressure, a gradient elution capacity, and a UV–visible detector, which is a scanning spectrophotometer that determines the absorption spectra of the separated carotenoids. Pigments are identified by retention times and are quantified by peak area or peak height. Detection limits are in the nanogram range (Taylor, 1983).

The adsorbents are the same as for open-column chromatography, namely magnesium oxide for carotenes and porous silica (e.g., Zorbax SIL, Porasil, LiChrosphere) for xanthophylls. Normal-phase solvents routinely contain hexane or petroleum ether with low or increasing concentration of a polar solvent. Reversed-phase columns contain octyl (C_8)- or octadecyl (C_{18})-silylated silica gel, such as Bondapack C18, Partisil ODS, Vydac ODS, Permaphase ODS, and others. Reversed-phase solvent systems are mixtures of water or buffers in alcohol (methanol) or acetonitrile.

HPLC is widely used in the analysis of vegetable carotenoids for provitamin A determination and for a complete carotenoid investigation.

Chloroplast pigments: The first HPLC analyses of chloroplast carotenoids primarily used spinach as a characteristic sample and were previously mentioned (see the subsection on chlorophyll analytical methods). Different ameliorated methods were further developed to separate carotenoids of green vegetables.

A r-p HPLC method (column packing: Zorbax ODS; gradient solvent system: hexane–methanol, subsequently used isocratically) was developed for spinach carotenoids grown in Japan (Kakutani et al., 1983; Kakutani and Kiku, 1988). Spinach carotenoids were separated by r-p HPLC (column packing: LiChrosorb RP-18; solvent system: acetonitrile–methanol and hexane–methanol), a method that allowed separation of lutein from (added) zeaxanthin (Krinsky and Welankiwar, 1984).

Carotenoids of green leaves were separated and quantified by normal-phase HPLC [column packing: Hitachi gel; solvent system: linear gradient of acetone and *n*-hexane–benzene (1:1) (Takagi, 1985)].

All the chloroplast pigments were separated and quantified in green vegetables (raw and cooked) such as broccoli, cabbage, spinach, Brussels sprouts, and kale by r-p HPLC [column packing: Microsorb C_{18}; solvent system: methanol-acetonitrile–methylene chloride–hexane in different proportions (Khachik et al., 1986)].

Chlorophylls *a* and *b* along with α- and β-carotene of numerous vegetables grown in Japan were analyzed (Izaki et al., 1986) with a special method of Wright and Shearer (1984). (See the subsection on chlorophyll analytical methods.)

Chloroplast pigments, carotene and xanthophyll isomers, chlorophylls, and pheophytins, were separated by a r-p HPLC method (column packing: Nova-Pack C18 developed with a series of solvent systems). Besides the usual chloroplast pigments the one-step method separated α- and β-carotene, lutein epoxides, antheraxanthin, lactucaxanthin, and zeaxanthin (Siefermann-Harms, 1988).

A rapid r-p HPLC method was developed for the analysis of the action of bleaching herbicides in higher plants (Young et al., 1989). The system differentiates between those compounds that cause bleaching through inhibition of carotenoid biosynthesis and those that destroy the plant's existing pigments through photooxidative events. The biosynthetic intermediates such as phytoene, lycopene, δ- and γ-carotene, and their monohydroxy derivatives could be separated from plants treated with herbicides that inhibit carotenoid biosynthesis. In plants treated with herbicides that destroy the pigments through photooxidation, the predominating pigments were chlorophyll breakdown products and β-carotene 5,6-epoxide.

The simultaneous separation of chlorophylls and derivatives and chloroplast carotenoids was achieved by a r-p HPLC method [column packing: Chromsil C_{18}; mobile isocratic phase: acetonitrile–methanol–ethylacetate, with a stepwise increase in flow rate (Daood et al., 1989)]. The method is suitable for pigment analysis in fresh and processed green vegetables.

Provitamin A carotenoids: Much effort went into developing rapid and reproducible methods for separating the provitamin A carotenoids present in various vegetables. α-Carotene and β-carotene were separated and determined by HPLC after being rapidly extracted by a special method (Hsieh and Karel, 1983). The extract was chromatographed on a column packed with Bondapack with solvent system of chloroform–acetonitrile (8:92). The results obtained for carrots, tomatoes, Butternut squash, and sweetpotatoes differed from the former published values. The α- and β-carotene content of carrots, potatoes, and fiddlehead greens was determined by r-p HPLC [column packing: Partisil 50DS; isocratic solvent system: acetonitrile–tetrahydrofuran–water (Bushway, 1982)]. The method was further applied to determine the vitamin A value in various vegetables from different locations three times during a year (Bureau and Bushway, 1986). The results were compared with those obtained by the AOAC (1984) method.

Several other isocratic methods, both normal and r-p HPLC, have been tried by the author to separate several provitamin A carotenoids and other carotenoids as well (Bushway, 1985). One of the r-p methods was further developed and applied to the determination of α- and β-carotene in several vegetables [column packing: Vydac 218 TP 54; solvent system: acetonitrile–methanol–tetrahydrofuran (40:56:4) (Bushway, 1986)]. For the separation and quantification of carotenes in yellow/orange vegetables such as carrots, sweetpotatoes, and pumpkins, an isocratic HPLC method has been developed using nonapreno-β-carotene (C_{45}) as an internal standard (Khachik and Beecher, 1987).

The β-carotene content of various vegetables consumed in Thailand was determined by an HPLC method using a home-packed column with ODS-Hypersil 3 μm eluted isocratically with a solvent system of methanol–acetonitrile–chloroform–water (200:250:9:11) (Speek et al., 1986, 1988).

Carotenes and xanthophylls of carrots, capsicum, parsley, and tomato grown in Australia were analyzed by a r-p HPLC method (Wills et al., 1988) with column packing Novapak C_{18}, 5 μm or two columns Radialpak C_{18} connected in series and a mobile phase of various proportions of water, acetonitrile, methanol, and chloroform. Gradient elution gave a good resolution of xanthophyls and carotenes, but an analytical time of 250 min was required.

The vitamin A value of vegetables consumed in East Africa was determined by r-p HPLC on a 5 μm C18 column mobile phase, a solvent system of acetonitrile–methanol–hexane–dichloromethane (65:15:10:10) (Pepping et al., 1988).

Dietz et al. analyzed α- and β-carotene in raw and cooked leaves of lettuce, spinach, and winged beans as well as in carrot roots by r-p HPLC [stationary phase: ODS C18; mobile phase: acetonitrile–methanol–chloroform (47:47:6) (Dietz et al., 1988)].

α-Carotene and β-carotene and lutein of various vegetables consumed in Finland were also separated by r-p HPLC [column packing: Spherisorb ODS2; solvent system: acetonitrile–methanol–dichloromethane–hexane used isocratically for carotenes and as a gradient for xanthophylls (Heinonen et al., 1989)].

Separation of carotenoid isomers: Separation of β-carotene isomers by HPLC was first successfully achieved by Vecchi et al. (1981). The stationary phase is alumina (Spherisorb A 5Y), the mobile phase is hexane with controlled water content. Three mono-*cis*-, four di-*cis*-, and one tri-*cis*-β-carotene isomers could be separated. A more convenient HPLC method was developed by Tsukida, using a laboratory-packed column [column packing: $Ca(OH_2)$; mobile phase 0.1–2% acetone in hexane (Tsukida et al., 1982)]. This method was subsequently ameliorated, with higher resolution achieved by modifying the particle size of calcium hydroxide and the packing method (Koyama et al., 1988). A HPLC method that uses μ-Bondapak C18 for column packing and methanol–acetonitrile–chloroform (47:47:6) as mobile phase was developed in order to detect carotenoid isomers in fresh and canned vegetables (Schwartz and Patroni-Killam, 1985). Monitoring HPLC eluants at both 436 nm and 340 nm and calculating the peak height ratio allows detection of isomers in the analyzed plant extract. The individual isomers were further separated by two-dimensional TLC on calcium hydroxide thin layers developed with 1.5% *p*-methylanisole. Subsequently, the *cis–trans* isomers of α- and β-carotene were resolved by a rapid HPLC method [column packing: laboratory-packed $Ca(OH)_2$; mobile phase: isocratic system of acetone–hexane (3:997) (Chandler and Schwartz, 1987)]. This method was used to detect the two predominant *cis* isomers of β-carotene in various fresh and canned vegetables. It was specially adapted to follow isomerization of *trans*-β-carotene during processing of sweetpotatoes (Chandler and Schwartz, 1988).

A rapid nonaqueous r-p HPLC method (column packing: C-18 Vydac TP-201; isocratic mobile phase: methanol–chloroform) separates natural provitamin A carotenoids and also some *cis* isomers (Quackenbush and Smallidge, 1986). With this method, isomers of four carotenoids (β-carotene, α-carotene, lycopene, and lutein) could be separated (Quackenbush, 1987). It may be applied to separate β-carotene isomers in food materials, as it was successfully tried on various fresh and canned vegetables.

Special HPLC Methods for Certain Vegetables. **Carrots:** The carotenes of typical and dark orange carrots have been separated by r-p HPLC [column packing: Partisil 5ODS-3; isocratic solvent system: acetonitrile–methylene chloride–methanol (82:12:6) (Simon and Wolff, 1987)]. Six carotenes—lycopene, γ-carotene, ζ-carotene, β-zeacarotene, α-carotene, and β-carotene (in order of elution)—were separated. They were identified by cochromatography with standard pigments on thin layers of MgO–kieselguhr (4:1). In carrots, α- and β-carotene could be separated by normal-phase HPLC on alumina [column packing: LiChrosorb Alox T; mobile phase: dichloromethane–hexane (5:95) (Zonta and Stancher, 1986)].

Corn: The main carotenoids of corn grain have been separated by normal-phase HPLC [column packing: Ultrasphere–Si, developed with a step gradient of isopropanol (3.1%) in hexane increased to 4% (Weber, 1987)]. The method did not separate the carotenes, but it did separate well the major corn xanthophylls (cryptoxanthin, lutein, and zeaxanthin).

Peppers: The carotenoids of red paprika products have also been separated by r-p HPLC [column packing: Nucleosil 10 C_{18}; mobile phase: a gradient of acetone–water, from 100:40 to 100:5 (Baranyai et al., 1982)]. Another r-p HPLC method for ground paprika separated the carotenoids without saponification [column packing: 10 μm Chromsil C_{18}; isocratic solvent system of acetone–water (9:1) (Pavisa et al., 1987)]. The carotenoids in paprika extract (oleoresin), saponified and unsaponified, were analyzed by r-p HPLC with column packing of Zorbax C-18 and solvent system consisting of a gradient of acetone–water (75:25) and acetone–methanol (75:25) (Fisher and Kocis, 1987). The carotenoids in the unsaponified extract were eluted in the following order: xanthophylls, carotenoid monoesters, monohydroxy carotenoids, carotenes, and carotenoid diesters.

A quantitative rapid analysis of carotenoids and carotenoid esters in red bell peppers was carried out by r-p HPLC that used octadecylsilane as column packing and a linear gradient of methanol–ethyl–acetate as mobile phase (Gregory et al., 1987). Of the total of 38 peaks, 13 were not identified.

Later, the adsorption and partition properties of several fatty acid esters of capsanthin and capsorubin isolated from red peppers and other carotenoid esters were studied by normal-phase (silica) and reversed-phase (octadecylsilane) HPLC (Philip and Chen, 1988). Normal-phase separation was based on the functional group of the carotenoids, and individual esters of the same carotenoid did not resolve. Reversed-phase separation was more dependent on the number of acyl carbons than the functional group, so individual esters of the same carotenoids differing only by two acyl carbons were well separated. These results confirmed earlier findings of Egger (1964, 1968), who separated plant carotenoid esters by thin-layer partition chromatography.

Four provitamin A carotenoids (α-, β-, and γ-carotene, and cryptoxanthin) in five different immature Mexican pepper cultivars were determined by r-p HPLC [column packing: Novapak C_{18}; mobile phase: acetonitrile–methanol–tetrahydrofuran (58:35:37) (Mejía et al., 1988)].

Another r-p HPLC method, which separates free and esterified paprika carotenoids, was recently developed by Biacs et al. (1989) [column packing: Chromsil C_{18}; mobile phase: isocratic solvent system; acetonitrile–2-propanol–water (39:57:4)].

Potato (yellow and white-fleshed): Potato carotenoids, free and esterified in all reports of Tevini and his group were separated by a r-p HPLC method with RP8 5 μm column packing and a mobile phase: gradient of 85% methanol in water (Tevini et al., 1984a).

Squash: The main carotenoids, free and esterified, of three varieties of squash have also been separated by r-p HPLC [column packing: Microsorb C-18, 5 μm; mobile phase methanol–acetonitrile–methylene chloride–hexane used isocratically and as a gradient (Khachik and Beecher, 1988a)]. The method may be generally applied to separate a number of straight-chain fatty acid esters (decanoate, laurate, myristate, palmitate) of various carotenoids, such as violaxanthin, auroxanthin, lutein, zeaxanthin, isozeaxanthin, and β-cryptoxanthin (Khachik and Beecher, 1988b).

A mixture of free and esterified carotenoids was also separated by a r-p HPLC method that used Supelcosil LC 18-S as column packing and a gradient of 1,2-dichloroethane–2-propanol (1:2) and acetonitrile as mobile phase (Zonta et al., 1987).

Tomato: One of the earliest and best-known r-p HPLC methods is that developed for determining provitamin A carotenes in tomatoes [column packing: Partisil-PXS-5/ODS, 5 μm; mobile phase: 8% chloroform in acetonitrile (Zakaria et al., 1979)]. β-Carotene and lycopene were separated and quantified in six tomato samples. The method separates α-carotene from β-carotene, lycopene, and other tomato carotenes; however, they were not identified. Cabibel et al. (1981) developed another HPLC method for tomato carotenes. It is a normal-phase method based on the first HPLC method applied to carotenoid analysis (Stewart and Wheaton,1971). It used a laboratory-packed MgO column developed with a linear gradient of hexane–acetone containing 10% benzene. With this method eight carotenes could be separated. Tomato pigments including chlorophyll were separated by r-p HPLC [column packing: Chromsil C_{18}; mobile phase: acetonitrile–isopropanol–water (200:288:13) (Daood et al., 1987)].

IDENTIFICATION

Carotenoids can be identified on the basis of their chromatographic and spectrophotometric behavior. Adsorption affinity is determined by the length of the chromophore and the polarity of the functional groups, as discussed previously. According to Davies (1976), the numerous factors that influence the mobility of the pigments on TLC make the R_f value per se of little value.

The absorption spectrum of each chromatographically pure pigment is recorded and compared with values from the literature (Davies, 1976). The slightly different values that may appear in the literature are inherent to the spectrophotometer's accuracy. For quantitative determination of each pigment, the extinction is read at the maximal absorption, and the reported extinction coefficient is used in calculations. For functional groups the described chemical tests are carried out on a spectroscopic scale.

For determining the nature of the geometric isomer, stereomutation is induced by iodine-catalyzed photoisomerization. To a PE solution of the pigment, a few drops of an iodine solution in PE are added and the mixture is exposed to light. If the pigment is an all-*trans* isomer, a hypsochromic shift occurs, whereas a *cis* isomer gives a bathochromic shift. The isomers may be separated chromatographically. A *cis* isomer differs from the all-*trans* isomer in its adsorption affinity and electronic spectrum.

When the pigment is finally identified, the ultimate test to confirm its identification is the cochromatography test. The unknown pigment is mixed with an authentic carotenoid and chromatographed. It is a clear indication that they are identical if the two pigments do not separate on different adsorbents when developed with different solvents.

At present, for identifying carotenoids, MS is used to determine molecular weight; NMR spectroscopy is used for double-bond stereochemistry, and IR spectroscopy for identifying functional groups and polyene chain characteristics.

BIOSYNTHESIS

Carotenoid biosynthesis is discussed in previously mentioned books and in the IUPAC symposia proceedings (Goodwin, 1952, 1980; Hulme, 1970/71; Isler, 1971a; Czygan, 1980; Britton, 1983). A comprehensive discussion on carotenoid biosynthesis (Spurgeon and Porter, 1983) was followed by a review of carotene biosynthesis in higher plants (Jones and Porter, 1986). The in vitro biosynthesis of carotenoids was discussed by Bramley (1985).

The general characteristics of carotenogenesis are similar in higher plants, algae, fungi, and bacteria. This section discusses carotenoid biosynthesis in higher plants only.

Carotenoid biosynthesis involves six stages:

1. Formation of mevalonic acid
2. Formation of geranylgeranyl pyrophosphate
3. Formation of phytoene
4. Desaturation of phytoene
5. Cyclization
6. Formation of xanthophylls

The early biosynthetic pathway of carotenoids from mevalonate is common to other isoprenoids, such as rubber, steroids, and terpenes found in essential oil, which are important natural products.

Genetic studies, inhibition studies, and in vivo labeling techniques, replaced gradually by in vitro incubation with enzyme systems isolated from plant tissue, especially from tomato and pepper fruits, have all been used to elucidate the overall pathways and their stereochemistry. Multiple reports in the field have proved the existence of all the precursors involved and the supposed mechanisms in different organisms.

Formation of Mevalonic Acid

The biological isoprene precursor of carotenoids and all terpenoids is isopentenyl pyrophosphate (IPP). It is formed from mevalonic acid (MVA), which is successively phosphorylated into mevalonic acid-5-phosphate and 5-pyrophosphate by kinase enzymes and ATP. MVA-5-pyrophosphate is then decarboxylated to give the isoprene unit IPP (Figure 19).

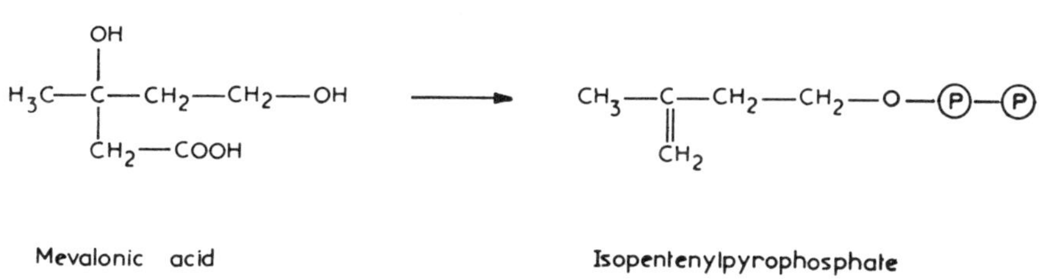

Figure 19. Mevalonic acid and isopentenyl pyrophosphate.

Mevalonic acid itself is formed through the condensation of three molecules of acetyl CoA via acetoacetyl CoA and β-hydroxy-β-methylglutaryl CoA, the latter being reduced to MVA.

Formation of Geranylgeranyl Pyrophosphate

The chain elongation from IPP (C_5) to geranylgeranyl pyrophosphate (C_{20}), which is catalyzed by the prenyl transferase enzyme, is shown in Figure 20.

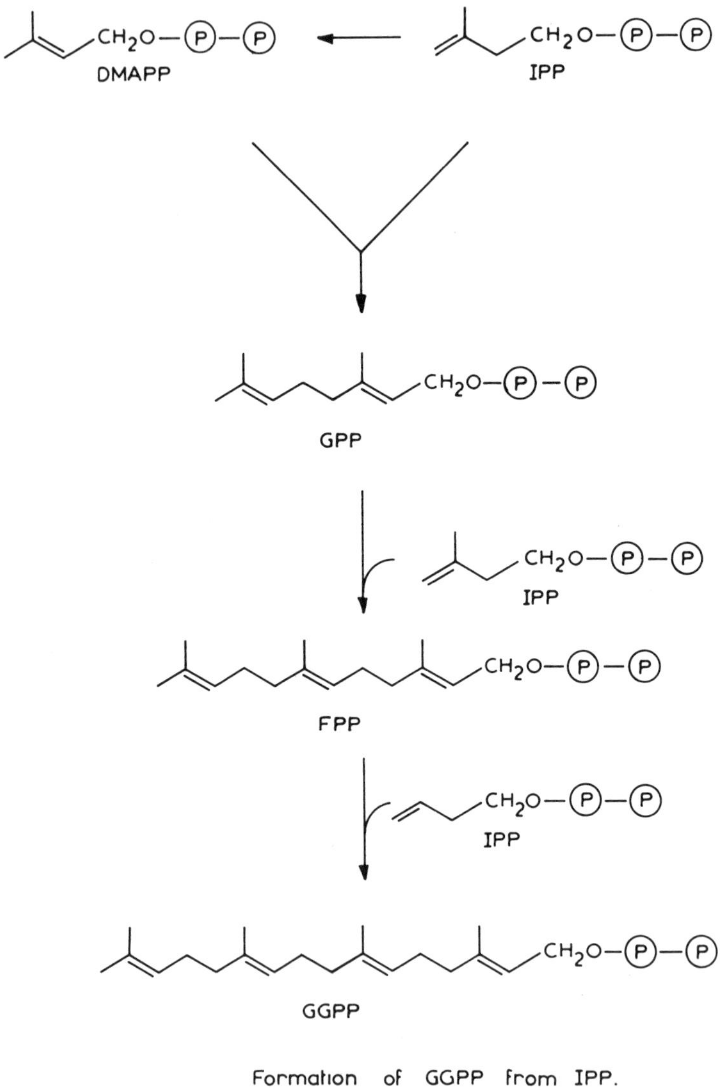

Figure 20. Conversion of isopentenyl pyrophosphate into geranylgeranyl pyrophosphate.

IPP is first isomerized into dimethylallyl pyrophosphate (DMAPP). DMAPP undergoes condensation with IPP to give the C_{10} intermediate geranyl pyrophosphate (GPP). Successive addition of 2 molecules of IPP gives the C_{15} intermediate farnesyl pyrophosphate (FPP) and the C_{20} geranylgeranyl pyrophosphate (GGPP). GGPP may be used to form diterpenes such as phytol, the side chain of chlorophyll, or C_{40} carotenoids.

Incorporation of labeled acetate and CO_2 into carotenoids by tomato slices was demonstrated by Braithwaite and Goodwin (1960a). Biosynthesis of phytoene from isopentenyl and farnesyl pyrophosphates succeeded with a partially purified tomato enzyme system (Jungalwala and Porter, 1967). The development of the acetone-extracted tomato plastid system was a major breakthrough that permitted further investigations of the overall biosynthesis pathway of carotenes. A soluble extract of the acetone powder of tomato plastids catalyzed all the reactions occurring in the conversion of IPP into phytoene (Porter, 1969). The enzyme named phytoene synthetase complex was further purified and partially characterized (Maudinas et al., 1977; Porter and Spurgeon, 1979).

Isopentenyl pyrophosphate isomerase, which catalyzes the isomerization of IPP into DMAPP, and prenyl transferase, which catalyzes the successive condensation of 3 molecules of IPP with DMAPP, GPP, and FPP, both present in tomato plastids, have been partially purified and characterized (Spurgeon et al., 1984).

The in vitro method for the enzymatic biosynthesis of phytoene with a soluble tomato plastid system, both from IPP and GGPP, is described by Jones and Porter (1985). Carrot root slices and gratings have effectively incorporated [2-^{14}C]acetate and DL-[2-^{14}C]mevalonate into carotene, but mevalonate is more efficiently incorporated (Braithwaite and Goodwin, 1960b).

Chloroplasts isolated from French bean leaves incorporated labeled mevalonate into phytoene (Charlton et al., 1967; Buggy et al., 1969, 1974). Mevalonate kinases, the enzymes that catalyze the synthesis of phospho- and pyrophosphomevalonate, were found in French bean chloroplasts (Rogers et al., 1966; Gray and Kekwick, 1973). The conversion of IPP-4-^{14}C to acyclic carotenes was catalyzed by an ammonia sulfate spinach leaf enzyme (Subbarayan et al., 1970).

Chloroplasts are capable of synthesizing chloroplast carotenoids by using photosynthetically fixed CO_2. $^{14}CO_2$ was rapidly incorporated into β-carotene by illuminated etiolated maize seedlings, since the formed chloroplasts utilize the photosynthetically fixed CO_2 (Goodwin, 1958b). Intact spinach chloroplasts converted ^{14}C-labeled CO_2, phosphoglycerate, acetate, and mevalonate into carotene and xanthophylls, a proof of their autonomy in terpenoid biosynthesis (Bickel and Schulz, 1976; Grumbach and Forn, 1980). Isolated intact spinach chloroplasts effectively form β-carotene from $NaH^{14}CO_3$ under photosynthetic conditions. The results obtained indicate a tight linkage of photosynthetic carbon fixation and plastid isoprenoid synthesis (Schulze-Siebert and Schultz, 1987).

Isolated *Capsicum annuum* chromoplasts incorporated labeled IPP into phytoene (Camara et al., 1982). The biosynthetic capabilities and localization of enzymatic activities in carotenoid metabolism of these chromoplasts were then studied (Camara and Monéger, 1982). The stroma and membrane fractions of orange pepper fruit chromoplasts have been separated. The stroma alone is able to incorporate IPP into phytoene, whereas the following steps in carotenoid biosynthesis occur only in the membrane fraction.

The incorporation of β-hydroxy-β-methylglutaric acid (HMG) and HMG CoA into carotenoids by a soluble system of maize seedlings was reported by Berry et al. (1972). HMG CoA reductase, which catalyzes the formation of MVA, was found in plastids of pea seedlings (Brooker and Russell, 1975). In sweetpotato root tissue the enzyme is active only in wounded roots (Ito et al., 1979). Mevalonate kinase activity has been demonstrated in several plant species, the best source being pumpkin seedlings (Loomis and Battaile, 1963).

Formation of Phytoene

The first step unique to carotenoid biosynthesis is the condensation of 2 molecules of GGPP to form phytoene. Phytoene is formed via the C_{40} cyclopropane intermediate, prephytoene pyrophosphate (PPPP) (Altman et al., 1972). In higher plants 15-*cis*-phytoene is generally found, the *trans* isomer occurring in other organisms. The formation of one isomer or another depends on the stereochemistry of the hydrogen removal involved in the condensation of GGPP.

Tritiated PPPP was incubated with etioplasts isolated from young *Triticum sativum,* chloroplasts and chromoplasts from *Capsicum annuum* fruits (Camara et al., 1985a). PPPP was efficiently incorporated into carotenoids, *cis*-phytoene being the major labeled product.

Then, with PPPP and IPP as substrates, the synthesis of phytoene was demonstrated in chloroplasts, etioplasts, and amyloplasts obtained from different plant materials such as *Capsicum* fruits, peas, spinach, and maize seedlings (Dogbo et al., 1987). It was found that the stromal fraction, not specifically from chromoplasts but from plastids at different stages, is the sole site of phytoene synthase. For this synthesis, the presence of plastid membranes, envelopes, or thylakoids is not required. In a former report phytoene synthetase associated with spinach chloroplast envelopes was demonstrated (Lütke-Brinkhaus et al., 1982). In a recent report, the same authors demonstrated that the phytoene synthase complex in etioplasts and etiochloroplasts is present in a soluble form in the stroma (Lütke-Brinkhaus and Kleinig, 1987).

Desaturation of Phytoene

From phytoene, according to the Porter–Lincoln pathway, lycopene is formed through stepwise dehydrogenation. Through each dehydrogenation sequence the chromophore is extended by two double bonds alternately from either side. Phytoene is converted into lycopene, via phytofluene, ζ-carotene, and neurosporene (see Figure 10). The outline of the pathway was based on the pattern of pigment distribution in various tomato mutants at a time when the structure of all the pigment involved had not yet been elucidated (Porter and Lincoln, 1950). Since all the colored carotenes are *trans* isomers, a *cis–trans* isomerization must occur at the phytoene or phytofluene level.

In tomato, all steps involving isomerization of *cis*-phytofluene into *trans*-phytofluene and four dehydrogenations were demonstrated (Qureshi et al., 1974a). In *Capsicum* fruits the labeling pattern obtained after 1-^{14}C IPP incorporation also argues for a *cis–trans* isomerization at the phytofluene level.

Following the stereochemical assignment of poly-*cis*-carotenes in tangerine tomato fruits, it could be concluded that phytoene is the clear branch point for poly-*cis*-carotene formation (Clough and Pattenden, 1983). Additionally it was established that during desaturation of phytoene to phytofluene the introduction of the C-11′ *trans* double bond is associated with a simultaneous *trans–cis* isomerization at the C-9′ double bond. These simultaneous changes further continue during chromophore elongation in prolycopene biosynthesis. The enzymatic conversion of *cis*-phytofluene into poly-*cis*-acyclic carotene was demonstrated with a cell-free preparation of tangerine tomato fruit plastids (Qureshi et al., 1974b).

Labeled phytoene was converted into acyclic carotenes by a soluble spinach enzyme system (Subbarayan et al., 1970).

Formation of Cyclic Carotenes

Biosynthesis of cyclic carotenes occurs through cyclization of one or both end groups of lycopene, as suggested by Porter and Lincoln. In higher plants two ring types are found: the β-ring and the ε-ring. The biosynthesis of the two rings occurs separately, not by interconversion of the β-ring to the ε-ring, as postulated earlier (Figure 21). The separate formation of the two ring systems was suggested by inheritance studies on tomatoes (Tomes, 1967) and established by elucidation of the cyclization mechanism using labeled mevalonate (Williams et al., 1967).

The same intermediate "carbonium ion" gives rise to both ε- and β-rings by elimination of protons either from C-6 or C-4, showing that these rings are formed independently and not by isomerization of one to another.

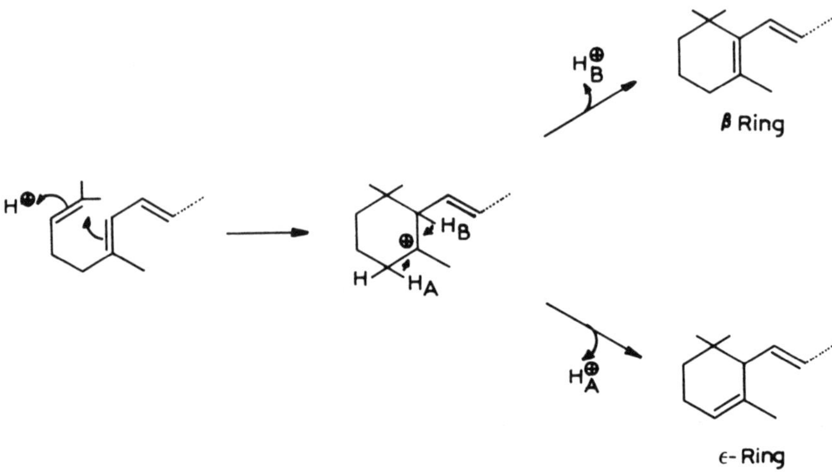

Figure 21. Cyclization mechanism of β- and ϵ-ring formation.

The general scheme for the biosynthesis of the common β- and ϵ-ring cyclic carotenes is given in Figure 22.

The monocyclic intermediates of α- and β-carotene are δ-carotene and γ-carotene, respectively. Cyclization may also occur at the neurosporene level, as indicated by the existence of α-zeacarotene and β-zeacarotene, which are found in maize grains. Cyclization occurs in carotenes that have at least a double bond at C-7,8.

Conversion of lycopene into β-carotene was demonstrated with plastids from French bean seedlings and tomato fruit (Hill et al., 1971). Conversion of labeled lycopene to mono- and dicyclic carotenes was demonstrated by soluble enzymes from tomato fruit plastids (Kushwaha et al., 1970; Papastephanou et al., 1973).

Labeled neurosporene and lycopene were transformed into β-carotene by incubation with *Capsicum* fruit chromoplasts. At any concentration used, lycopene was the preferred substrate (Camara and Monéger, 1982). Subsequently the cyclization of lycopene to β-carotene was studied with chromoplast membranes prepared from *Capsicum* fruits (Camara et al., 1985b). The enzymic activity is tightly bound to the membranes, and under optimal conditions lycopene is cyclized to β-carotene without accumulation of the intermediate γ-carotene. Lycopene cyclase from *Capsicum* chromoplast membranes has been demonstrated and solubilized (Camara and Dogbo, 1986).

Additional evidence that lycopene is the major cyclization substrate came from in vitro studies with cyclization inhibitors such as 2-(4-chlorophenylthio) triethylamine hydrochloride (CPTA) and nicotine, both of which cause accumulation of lycopene as cyclization is inhibited. The effect

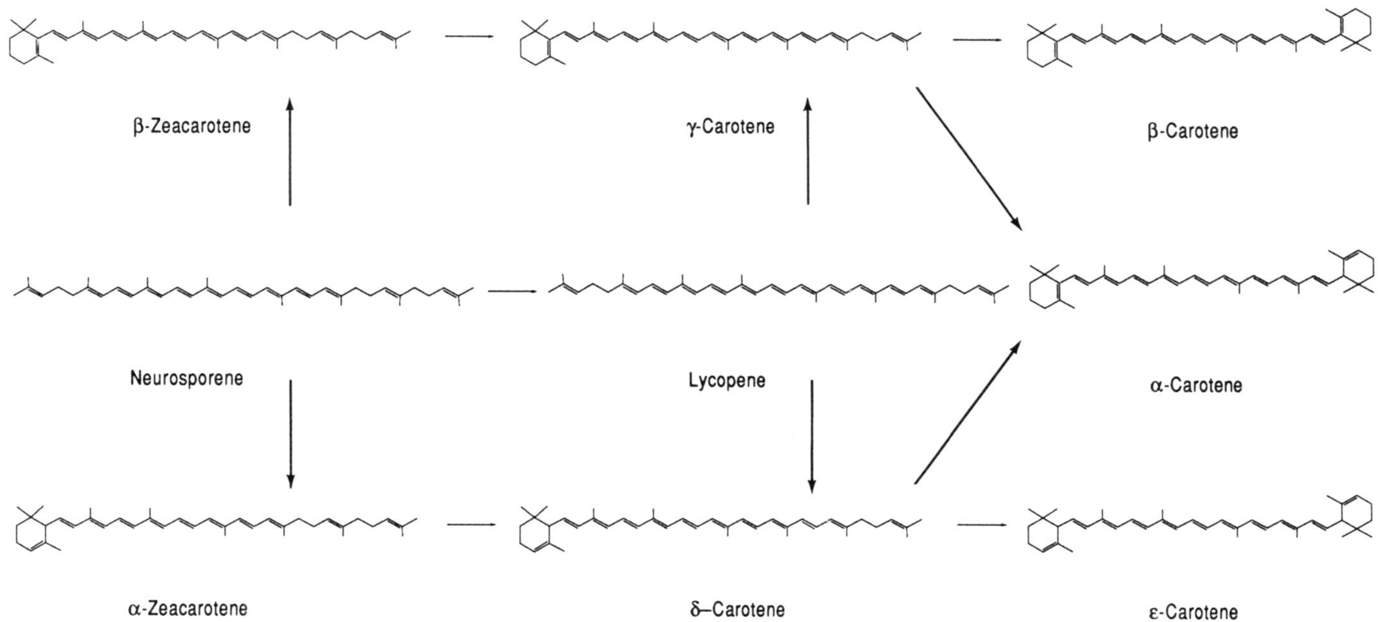

Figure 22. General scheme for the biosynthesis of the common β- and ϵ-ring cyclic carotenes. (*From Britton, 1983.*)

of CPTA was studied on citrus fruits, which after treatment developed an intense red color like that of a ripe red tomato (Yokoyama et al., 1971, 1972). The effect of nicotine was also studied in higher plants, *Cucurbita* species, and banana leaves (Howes, 1974; Gross and Costes, 1976). The latter study indicated that different enzymes may be involved in the synthesis of δ- and γ-carotene as well as in the second cyclization step, since they react differently to inhibition.

Formation of Xanthophylls

Insertion of the oxygen function, hydroxy and epoxy groups occurs after cyclization as a late step in biosynthesis when corresponding carotenes are oxidized. Hydroxylation occurs mainly at C-3 at one or both rings when the most widespread xanthophylls are formed, but it may also occur at other positions of the ring or in chain.

Most of the studies on xanthophyll biosynthesis have been done with algae and bacteria. In higher plants, as in bacteria, zeaxanthin is formed by hydroxylation of β-carotene. The oxygen is derived from molecular oxygen (Yamamoto et al., 1962), the reaction being catalyzed by a mixed-function oxidase (Goodwin, 1980) with retention of the absolute configuration (Milborrow, 1982).

Epoxy carotenoids are formed by the oxidation of the corresponding carotene or xanthophyll; for example, violaxanthin is formed from epoxidation of zeaxanthin via antheraxanthin, the monoepoxide. The mechanism is not known, and the source of oxygen in violaxanthin seems to be different, depending on the site where biosynthesis occurs in the chloroplast (Costes et al., 1979). Natural 5,8-epoxides are formed through rearrangement of 5,6-epoxides in a slightly acid medium. Allenic carotenoids are formed through a rearrangement of the epoxide group according to a mechanism proposed for neoxanthin from violaxanthin (Davies, et al., 1970).

The biosynthesis of the *Capsicum* cyclopentyl ketocarotenoids occurs via xanthophyll epoxides. The suggested mechanism involves a pinacolic rearrangement of the 3-hydroxy-5,6-epoxy end group (Cholnoky and Szabolcs, 1960; Weedon, 1967). Thus, cryptoxanthin 5,6-epoxide is the precursor of cryptocapsin, antheraxanthin is the precursor of capsanthin, and violaxanthin is the precursor of capsorubin, with capsanthin 5,6-epoxide as intermediate. Besides structural consideration, this pathway corresponds to the carotenoid distribution in different-colored varieties of *C. annuum* (Davies et al., 1970). Additional confirmation is obtained by following carotenogenesis in ripening pepper, which reveals precursor–product relationships (Valadon and Mummery, 1977; Rahman and Buckle, 1980).

The in vitro conversion of violaxanthin into capsorubin and of antheraxanthin into capsanthin was demonstrated with a chromoplast-enriched fraction of *Capsicum* fruits (Camara, 1980; Camara and Monéger, 1981).

FUNCTIONS

Photofunctions

The photofunctions of carotenoids have been reviewed comprehensively by Krinsky (Isler, 1971a) and updated by Mathis and Schenck (Britton and Goodwin, 1982). The universal presence of carotenoids in photosynthetic organisms, as chlorophyll–carotenoid–protein complexes located in the photosynthetic membranes, has a justification. From the variety of functions attributed to these pigments, two important photofunctions, essential for the plants' survival, have been clearly established: (1) accessory pigments in photosynthesis and (2) protective agents of the photosynthetic apparatus against potential damage from visible light.

Role in Photosynthesis

Chlorophylls play the major role in photosynthesis. For maximum light utilization they are aided by accessory pigments that absorb at different wavelengths from the chlorophylls and transfer the absorbed light energy to the chlorophylls. Evidence that carotenoids can transfer energy to chlorophylls was found by measuring the enhancement of chlorophyll fluorescence on illumination with light of the wavelengths absorbed by carotenoids. The yield of transfer varies greatly in different organisms and from one carotenoid to another.

The mechanism of energy transfer from carotenoids is still not well understood, but the transfer is efficient only when both molecules are close together in the pigment–protein complex (Goedherr, 1979). It has been proved that both β-carotene and xanthophylls have the ability to transfer absorbed energy to chlorophylls (Siefermann-Harms, 1981). For the major light-harvesting chlorophyll *a/b*-protein complex of lettuce chloroplasts, the efficiency of the energy transfer from carotenoids to chlorophyll *a* has been determined to be 100% (Siefermann-Harms and Ninnemann, 1982). Since this pigment–protein complex mainly binds xanthophylls, this proves that xanthophylls are involved in energy transfer.

Photoprotection

The photoprotection exerted by the carotenoids was reviewed by Krinsky (1979, 1989). The damage caused by visible radiation to cells can be altered by the presence of endogenous carotenoids. This fact was first observed in a purple photosynthetic bacterium when the wild-type strain was compared with a blue-green mutant that lacked carotenoids. In the presence of air and light the mutant stopped growing, its chlorophyll was destroyed, and the cells were killed, but the wild type continued to grow. It could be proved that the chlorophyll was the endogenous photosensitizer responsible for the damage (Sistrom et al., 1956).

Photosensitization involves harmful photochemical reactions initiated by the excited sensitizer in the triplet state. In photosynthetic plants the sensitizer is the chlorophyll, which in the excited state can either initiate redox reactions, with formation of free radicals, or transfer the energy to oxygen, thus forming the highly reactive singlet oxygen $^1O_2$. Carotenoids protect the organisms by directly quenching the excess energy of excited chlorophyll or quenching the singlet oxygen. These mechanisms, whereby carotenoids exert the photoprotection, were outlined by Krinsky (1971, 1979, 1989).

Photoprotection is related to the length of the chromophore in the carotenoid molecule, a minimum of nine conjugated double bonds being necessary (Mathews-Roth et al., 1974). Photosensitive mutants from maize and sunflower (*Helianthus annuus*) have also been described in higher plants, in addition to bacteria and algae (Walles, 1972; Mustardy et al., 1976).

The action of some herbicides is based on the photoprotective function of carotenoids. When carotenoid biosynthesis is blocked by herbicides, the protection is no longer provided and the plant is killed in high light intensities (Britton, 1979).

Carotenoids are also presumed to be implicated in photosynthetic electron transport. By reason of their close association with the reaction center of PS II and their extended chromophore, carotenoids may mediate redox reactions. Mathis and others have reported that light excitation of chloroplasts produces absorption changes due to formation of a carotenoid cation radical at the reaction center of PS II (Mathis and Schenck, 1982; Mathis and Rutherford, 1984).

Generation of carotenoid cation radicals via electrochemical oxidation is now under study to gain a better understanding of the natural processes (Grant et al., 1988).

The Violaxanthin Cycle

Reversible changes in the violaxanthin content of leaves held under light and dark conditions was first observed by Sapozhnikov (Sapozhnikov et al., 1957). The light-dependent conversion of xanthophylls is due to the *violaxanthin cycle*, which involves de-epoxidation of violaxanthin in the light and re-epoxidation of the formed zeaxanthin in the dark via antheraxanthin.

$$\text{violaxanthin} \underset{\text{dark}}{\overset{\text{light}}{\rightleftharpoons}} \text{antheraxanthin} \underset{\text{dark}}{\overset{\text{light}}{\rightleftharpoons}} \text{zeaxanthin}$$

The biochemistry of the reactions was investigated in detail, mainly by Hager (1975) and Yamamoto (IUPAC, 1979). Two enzymes are involved: the violaxanthin de-epoxidase situated on the inner side of the thylakoid membrane, and the epoxidase situated on the stroma side of the membrane. Thus, the cycle arrangement is transmembraneous, which also implies the transport of the substrates across the membrane.

The enzyme violaxanthin de-epoxidase has been isolated from spinach (Hager and Perz, 1970) and lettuce chloroplasts (Siefermann and Yamamoto, 1974). Violaxanthin de-epoxidase was active against all naturally occurring 3-hydroxy-5,6-epoxy carotenoids, such as violaxanthin, antheraxanthin, neoxanthin, cryptoxanthin epoxide, and lutein epoxide, with the absolute configuration 3*S*,5*R*,6*S* (Yamamoto, 1979). Only part of the violaxanthin is available for the de-epoxidase (Siefermann and Yamamoto, 1974). The distribution of available and unavailable violaxanthin in spinach thylakoids was examined, and a heterogeneous organization was found (Siefermann-Harms, 1984). Violaxanthin epoxidasc, tightly membrane bound, has not been isolated. It is a mixed-function oxidase requiring the presence of oxygen and NADPH, and its activity is independent of light.

Although numerous links between the violaxanthin cycle and photosynthetic processes have been established, a firm conclusion regarding its function has not been drawn. A link between violaxanthin de-epoxidation and photosynthetic electron transport was established by Grumbach (1983b). In isolated intact spinach chloroplasts preincubated in the dark and then reilluminated, the de-epoxidation of violaxanthin occurred simultaneously with the oxidation of plastohydroquinone-9 to plastoquinone-9. When illumination was discontinued, zeaxanthin was transformed into violaxanthin, but no significant change in the redox state of the plastoquinone pool occurred.

The occurrence of the enzymatic violaxanthin de-epoxidation in etiolated leaves has been demonstrated (Pfündel and Strasser, 1988). The occurrence of violaxanthin de-epoxidase well before the formation of chlorophyll-containing thylakoids and the onset of photosynthetic activity extend the scope in which the function of the xanthophyll cycle may be searched.

Antioxidants

In addition to quenching reactive species formed by photochemical reactions (triplet sensitizers, $^1O_2$), carotenoids can act as antioxidants and thereby protect cells and organisms from oxidative damage (Krinsky, 1979, 1989).

The protection is due to the ability of the carotenoids to quench free-radical species. The direct quenching or inhibition of radical species by β-carotene was reported (Packer et al., 1981). In another in vitro study of the ability of β-carotene to deactivate free radicals that initiate lipid peroxidation, it was found that β-carotene is an unusual type of lipid antioxidant (Burton and Ingold, 1984). Specifically, β-carotene exhibits good radical-trapping antioxidant behavior only at a low partial pressure of 3.10^{-2} atm (2% oxygen). Such low oxygen partial pressures are found in most tissues under physiological conditions. At higher oxygen pressures, β-carotene loses its antioxidant activity and shows an autocatalytic prooxidant effect. These results explain the effectiveness of β-carotene in inhibiting lipid peroxidation (Krinsky and Deneke, 1982).

It was also shown that several naturally occurring carotenes and xanthophylls supersede β-carotene in their $^1O_2$ quenching efficiency, with lycopene being the most effective (Di Mascio et al., 1989).

Biological Function: Role of Carotenoids in Human Nutrition

Provitamin A Carotenoids

One of the most important physiological functions of carotenoids is to act as vitamin A precursors in animal organisms. Almost all animal species are able to enzymatically convert plant carotenoids of a special structure into vitamin A. Vitamin A is the primary alcohol retinol:

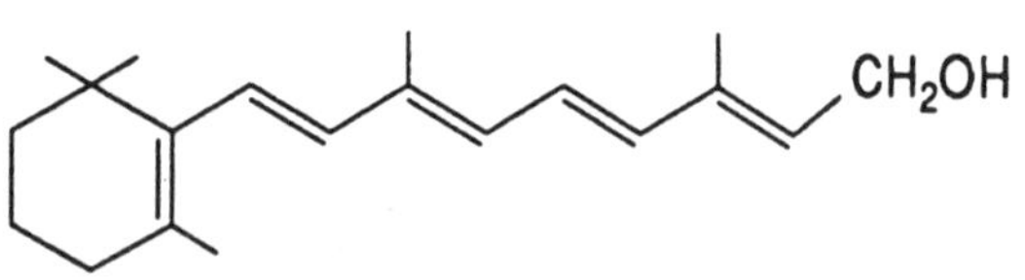

A condition sine qua non for provitamin A activity of a carotenoid precursor is at least one unsubstituted β-ring and the unchanged side chain. β-Carotene with two β-rings is the provitamin A with highest activity. All other provitamin A carotenoids have the intact β-ring at only one end of the molecule (e.g., α-carotene, γ-carotene, 5,6- and 5,8-monoepoxides of β-carotene, cryptoxanthin and its monoepoxides, and the β-apocarotenals). *Cis* isomers have a lower potency than *trans* isomers.

Vitamin A_2 is 3-dehydroretinol and has about half of the vitamin activity of retinol. Carotenoids possessing a 3,4-dehydro-β-ring act as provitamin A_2, as found by Gross (1967). (See also Budowski et al., 1963; Budowski and Gross, 1965.) An interesting exception is the finding of provitamin A activity of astaxanthin and cantaxanthin in fish (Gross and Budowski, 1966).

Forty carotenoids are listed as vitamin A precursors by Bauernfeind (1981). Of these the following ten are found in vegetables: β-carotene, α-carotene, γ-carotene, β-zeacarotene, β-carotene 5,6-epoxide, β-carotene 5,8-epoxide, β-cryptoxanthin, cryptoxanthin 5,6-epoxide, 3′-hydroxy-α-carotene, and cryptocapsin. β-Carotene is the most widespread provitamin A, it being present in almost all vegetables. In green vegetables it is the only provitamin A. It is found at high levels in yellow vegetables such as carrots, corn, and yellow sweetpotatoes. α-Carotene accompanies β-carotene in carrots and is found at low levels in some leaf vegetables and in pumpkins and squashes. γ-Carotene is found in carrots and tomatoes; β-carotene 5,6-epoxide in white sweetpotatoes and yellow peppers; cryptoxanthin in corn, pumpkins, squashes, and peppers; β-zeacarotene in corn and squashes; cryptoxanthin epoxide in peppers; and cryptocapsin in peppers.

The vitamin A activity of some carotenoids found in vegetables is listed in Table 16 (in terms of β-carotene valued 100%).

Theoretically, β-carotene must give rise through central cleavage to 2 molecules of vitamin A. The conversion of β-carotene occurs mainly in the intestinal mucosa from which a soluble enzyme has been isolated. The enzyme that catalyzes the central cleavageof β-carotene is 15,15′-dioxygenase, which occurs in different species of animals (Lakshmanan et al., 1972). This enzyme yields 2 molecules of retinal, which is subsequcntly reduced to retinol. The newly formed retinol is then esterified with long fatty acids, transported in the lymph, and stored in the liver. The in vitro central cleavage of β-carotene catalyzed by the dioxygenase enzyme could not be duplicated (Hansen and Maret, 1988). But subsequently the enzymatic conversion of β-carotene to retinal by a cytosolic enzyme from rabbit and rat intestinal mucosa was demonstrated unequivocally both in vivo and in vitro (Lakshmanan et al., 1989).

Although 1 molecule of β-carotene may be converted into 2 molecules of vitamin A, bioassays show that only half of the molecule is active. This fact was explained inter alia as a result of incomplete resorption.

Another acceptable explanation (Glover and Redfearn, 1954; Glover,

Table 16. Relative Vitamin A Activity of Some Carotenoids

Carotenoid	*Activity (%)*
all-*trans*-β-Carotene	100
9-*cis*-β-Carotene (U)	38
13-*cis*-β-Carotene (B)	53
all-*trans*-α-Carotene	53
9-*cis*-α-Carotene (U)	13
13-*cis*-α-Carotene (B)	16
all-*trans*-Cryptoxanthin	57
9-*cis*-Cryptoxanthin (U)	27
15-*cis*-Cryptoxanthin	42
β-Carotene 5,6-epoxide	21
β-Carotene 5,8-epoxide (mutatochrome)	50
γ-Carotene	42–50
β-Zeacarotene	20–40

Sources: From Zechmeister (1962) and Bauernfeind (1972).

1960) postulates that β-carotene is degraded through oxidation at one end or the other of the polyene chain. The intermediates are the homologous apocarotenals, which can be further metabolized into retinal or the corresponding apocarotenoic acids. The metabolism scheme is given in Figure 23.

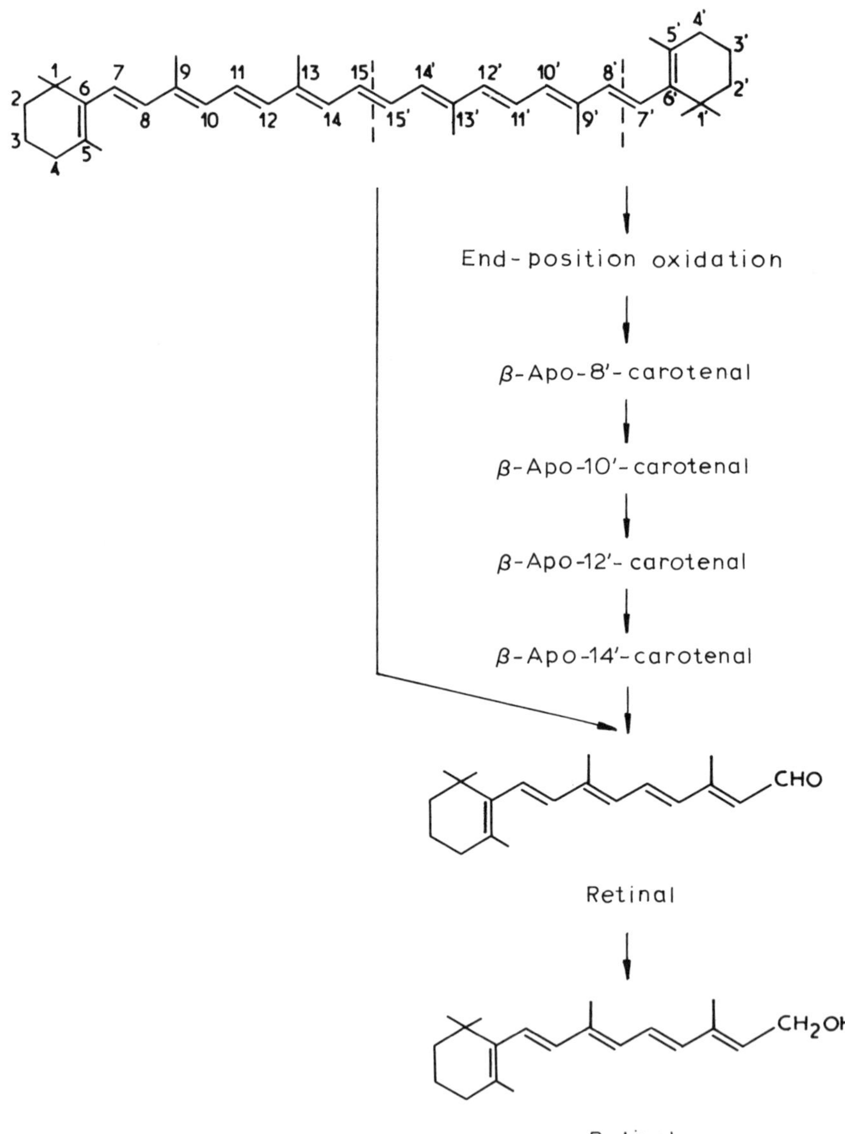

Figure 23. Metabolism schemes of β-carotene to vitamin A. (*From Glover, 1960.*)

The vitamin A activity of the intermediate apocarotenals was tested in curative growth tests with vitamin A-deficient animals and found to be 72% to 78% of β-carotene (Sharma et al., 1976, 1977). According to the results, it has been concluded that the dioxygenase nonspecifically attacks any of the double bonds of β-carotene. The weight of evidence available so far overwhelmingly favors random cleavage and is against any specific attack at the central double bond of the carotene molecule (Ganguly and Sastry, 1985).

The sources of man's vitamin A are dietary. Animal food provides half of the total vitamin A as preformed vitamin A, and plant food (fruit and vegetables) provides the other half as provitamin A carotenoids. The vitamin A activity of the carotenoids depends on the nature and amounts of biologically active carotenoids, their state of isomerization, their stability in the gastrointestinal tract, and their digestibility. The FAO/WHO recommendation of 1967 was that one sixth of dietary carotenoids is metabolically available as vitamin A, assuming an intestinal absorption of one third of dietary β-carotene and a conversion efficiency of 50% (FAO/WHO, 1967).

In 1974 the NAS-Recommended Dietary Allowances Panel introduced the term *retinol equivalent* to express vitamin A content in addition to international unit activity (IU $= 0.3\ \mu$g retinol) (NAS-NRC, 1974). By definition,

$$\begin{aligned}\text{retinol equivalent} &= 1\ \mu\text{g retinol}\\ &= 6\ \mu\text{g } \beta\text{-carotene}\\ &= 12\ \mu\text{g other provitamin A carotenoid}\\ &= 3.33\ \text{IU vitamin A activity from retinol}\\ &= 10\ \text{IU vitamin A activity from } \beta\text{-carotene}\\ &= 20\ \text{IU vitamin A activity from other}\\ &\qquad \text{provitamin A carotenoids}\end{aligned}$$

For the calculation of total retinol equivalents corresponding to β-carotene and other provitamins in micrograms, the following equation is used:

$$\text{total retinol equivalents} = \frac{\mu\text{g } \beta\text{-carotene}}{6} + \frac{\mu\text{g other provitamin A carotenoids}}{12}$$

The recommended daily allowance is 1000 retinol equivalents. Most data on β-carotene content in food, reported in food consumption tables, are too high because they are based on methods that determine the total carotenes without separation of β-carotene from other provitamin A carotenoids that have lower activity. This error was pointed out repeatedly by Simpson (Simpson and Chichester, 1981; Simpson et al., 1985; Lee et al., 1989). It is essential to develop a state-of-the-art analytical technique for carotenoid separation. Due to the diversity of the material, it is difficult to develop a single method to be used for all vegetables. In a critical review of provitamin A determination, all the problems associated with this analysis are discussed by Rodriguez-Amaya (1989).

The carotenoids ingested from food by man are not selectively absorbed (Bauernfeind, 1981). β-Carotene can be absorbed intact or converted to vitamin A. Carotenoids are usually present in fat and blood, but also in milk, retina, heart, liver, pancreas, nerves, bone marrow, adrenals, seminal vesicles, corpora lutea, and placenta.

The levels in blood (serum or plasma) indicate the dietary intake. Factors affecting carotenoid availability, such as fiber, protein, and fat content of diets, were discussed in Bauernfeind (1981) and Lee et al. (1989).

Only three reports have been published on carotenoid levels in human adipose tissue, the last in 1988 (Parker, 1988). Total carotenoid concentration varied 40-fold between individuals, from 0.34 to 13.5 $\mu g\ g^{-1}$ adipose tissue. β-Carotene and lycopene were predominant (20.2 and 18.5%, respectively).

If high-carotene food is consumed excessively, a high carotenoid content is found in blood (*carotenemia*) and in skin, which becomes yellow (*carotenodermia*), but is not considered harmful because it does not produce hypervitaminosis A (Bauernfeind, 1981).

USES

Colorants

The use of carotenoids as colorants has been comprehensively discussed by Bauernfeind (1981). Some natural plant pigments used as food colors are annatto (bixin, a C_{25} diapocarotenoid being the main pigment), paprika extracts (capsanthin and capsorubin, see p. 140), alfalfa and tagetes extracts (xanthophylls, mainly lutein), tomato extracts, and carrot extracts. These natural extracts are gradually being replaced by synthetic carotenoids, namely β-carotene, β-apo-8′-carotenal, and canthaxanthin (4,4′-diketo-β-carotene). β-Carotene, which has vitamin A value, is the most recommended, producing yellow to orange colors; the other two colorants produce a red hue. These carotenoids are added directly to food for human consumption. Other carotenoids, called *pigmenters,* when added to animal feed will color either the body tissue (skin and fat) or animal products such as milk, eggs, butter, and cheese.

Medicinal Applications

The medical application of carotenoids have been reviewed by Mathews-Roth (Bauernfeind, 1981; Britton and Goodwin, 1982). The primary use of food carotenoids is the prevention or correction of vitamin A deficiency in man. β-Carotene is used in the therapy of erythropoietic protophorphyria, in which sensitivity to visible light is a cardinal symptom (Mathews-Roth, 1987). Observations that the carotenoid pigments protect green plants and photosynthetic bacteria against photosensitization suggested that these pigments may be used in the treatment of photosensitivity diseases (Mathews, 1964). The first results led to further investigations, and β-carotene was found to be effective in treatment of photosensitivity diseases. Porphyria is related to overproduction in vivo of endogenously synthesized porphyrin molecules. These molecules, which have a structure similar to chlorophyll, act as endogenous photosensitizers producing sensitivity to visible light. The therapeutic effect of β-carotene is based on the same mechanism as photoprotection in plants. Quenching of singlet oxygen $^1O_2$ inhibits peroxide formation, which damages cellular components and produces itching, burning, crythema, and edema in patients.

Carotenoids are being considered as potential cancer prevention agents (Mathews-Roth, 1985). This ensued from experimental studies with specific animal model systems and retrospective and prospective epidemiological studies. In animal models it was investigated if carotenoids have the same well-known antitumor effect as vitamin A. Both β-carotene and canthaxanthin, irrespective of their vitamin A activity, can prevent or slow down the growth of skin tumors induced by UV-B (290–320 nm) radiation (Mathews-Roth, 1982). It was established that the carotenoids interfered with the progression phase of UV-B carcinogenesis (Mathews-Roth and Krinsky, 1987). In other animals, treatment with β-carotene, canthaxanthin, and an algal extract produced regression of carcinoma (Shklar and Schwartz, 1988). The prevention of indirect and direct chemical carcinogenesis by carotenoids (β-carotene and canthaxanthin) is attributed to their ability to react as powerful antioxidants (Santamaria et al., 1988).

Epidemiological evidence has indicated a possible association of increased dietary consumption of certain vegetables with lower rates of cancer, the effect being attributed to a specific preventive action of β-carotene (Peto et al., 1981). This hypothesis incited the most interest and intensive investigations.

Human prospective and retrospective studies have strongly indicated that β-carotene protects against cancer of the lung and other organs (Temple and Basu, 1988). The possible mechanisms of β-carotene action are multiple: (a) acts following its conversion into vitamin A, (b) alters carcinogen metabolism, (c) acts as an antioxidant, and (d) enhances immune defenses.

In various studies β-carotene has been quantified in blood, where its level is considered a reliable index of dietary intake. Some recent studies on the effect of vegetable ingestion upon blood carotenoid levels illustrate the complexity of this metabolic process (Kim et al., 1988; Brown et al., 1989).

An investigation of serum carotene concentration and serum lipid peroxides level in healthy man revealed that these levels were inversely correlated (Ito et al., 1988). This correlation could be a proof that α- and β-carotene play a role in cancer prevention through reducing lipid peroxidation by quenching active oxygen species and trapping free radicals. Lycopene was found to be the most efficient carotenoid singlet oxygen quencher and, being present in human blood at about the same levels as β-carotene, may give the same protective action against cancer (Di Mascio et al., 1989).

LOCALIZATION

Carotenoids are found in nongreen plastids, in chloroplasts, and especially in chromoplasts. Chloroplasts are ubiquitous in photosynthetic tissue, especially in leaves and in unripe fruits, whereas chromoplasts are found in flowers, ripe fruits, and some roots and tubers.

Chloroplasts

In chloroplasts the carotenoids are masked by the presence of chlorophylls. The pattern of chloroplast carotenoids is universally uniform and contains four basic "chloroplast carotenoids," namely one carotene and three xanthophylls, a possible indication that all higher plants have evolved from a common ancestor (Strain, 1966). Additional carotenoids are found as minor pigments: α-carotene, α- and β-cryptoxanthin, isolutein (lutein 5,6-epoxide), zeaxanthin, and antheraxanthin (zeaxanthin 5,6-epoxide). An unusual carotenoid, lactucaxanthin, is present in lettuce (see p. 138, unique carotenoids). The carotenoid distribution in leaf vegetables and other green vegetables is discussed in the respective section.

Although the four chloroplast carotenoids are qualitatively the same, various carotenoid patterns are found in different species as a result of the innumerable variations in both the relative percentages of the four main pigments and the nature and relative percentages of the minor pigments. The approximate levels of chloroplast carotenoids are as follows: lutein, which predominates, 40–57%; β-carotene, 25–40%; violaxanthin, 9–20%; and neoxanthin, 5–15%.

In all photosynthetic organisms, practically all the carotenoids located in photosynthetic membranes are present in the form of chlorophyll–carotenoid–protein complexes. Likewise, practically all the complexes contain at least one molecule of carotenoid (Mathis and Schenck, 1982). Chloroplasts from higher plants contain for an average photosynthetic unit of 400 chlorophyll molecules about 83 molecules of carotenoids (24 β-carotene, 38 lutein, 10 violaxanthin, and 10 neoxanthin). About half of the β-carotene molecules are associated with the reaction centers, whereas all xanthophylls belong to the light-harvesting complex.

The carotenoid and chlorophyll composition of light-harvesting and reaction-center proteins of the spinach thylakoids was analyzed by HPLC (Braumann et al., 1982). On average, reaction-center I proteins may bind about 100 Chl *a* and 10–15 β-carotene per 1–2 P_{700} (Siefermann-Harms, 1985). The β-carotene fraction extracted from spinach thylakoids and analyzed by HPLC contained 80% all-*trans* isomer and 20% different kinds of *cis* isomers such as 15-, 13-, 9-*cis,* 9,13-, and 9,9′-di-*cis* isomers (Ashikawa et al., 1986). Most of the *cis* isomers originated from PS I.

The light-harvesting Chl *a/b*–lutein-proteins from higher plants have been studied extensively. The pigment content of Chl *a/b*–lutein-protein complexes of some vegetables, investigated by Siefermann-Harms (1985), is shown in Table 17.

Table 17. Pigment Content of Light-Harvesting Chl *a/b*–Lutein-Proteins from Higher Plants (molar ratio)

Species	*Chl* a	*Chl* b	*Total Carotenoids*	*β-Carotene*	*Lutein*	*Neoxanthin*	*Violaxanthin*
Lactuca sativa	100	67.0	49.7	1.7	26.7*	9.5	9.8
Phaseolus vulgaris	100	62.5	43.3	0.6	30.7	10.0	2.0
Spinacia oleracea	100	64.9	53.7	5.0	28.2	14.2	3.9

Source: From Siefermann-Harms (1985).
*26.7 = 18.8 lutein and 7.9 lactucaxanthin.

The carotenoid distribution in the yellow envelope membrane of the chloroplast is different from that of the green thylakoid membrane. Whereas lutein is the predominant pigment in the thylakoids, in envelopes the violaxanthin is almost three times higher than that of lutein (Douce et al., 1973; Siefermann-Harms, 1981).

Carotenoids of envelope membranes from nongreen plastids are further discussed in the distribution section (see cauliflower and potatoes). The presence of carotenoids in nongreen plastid envelope membranes suggests that the plastid envelope membranes play a major role in their biogenesis (Alban et al., 1988). The synthesis of phytoene was demonstrated in etioplasts and amyloplasts, the stromal fraction being the sole site of phytoene synthesis (Dogbo et al., 1987).

The biogenesis of carotenoids takes place in the chloroplasts and is an integral part of their development. It is closely linked to biosynthesis of other chloroplast components. The pigment–protein complexes and other material lipids, and so on, are used in the construction of the thylakoid membranes. If one component is not available, the entire chloroplast construction process is disrupted (Britton, 1982).

During leaf senescence, at the ultrastructural level, chloroplasts are transformed into gerontoplasts that contain plastoglobules (see p. 50). The yellowing of autumn leaves is due to the degradation of the chlorophylls and the unmasking of the carotenoids.

The carotenoids of autumn leaves were called *autumn xanthophylls* because they were entirely epiphasic but could be separated from the carotenes (Tswett, 1911). The yellow autumn leaf pigments were esterified xanthophylls; the esterification was thought to be the only change occurring in autumn (Kuhn and Brockmann, 1932).

During leaf senescence the chlorophyll degradation is paralleled by carotenoid degradation, which occurs at a much slower rate and depends on species and climatic conditions. The investigation of carotenoid changes in the leaves of deciduous trees during autumn necrosis revealed that in all cases β-carotene and zeaxanthin disappeared more rapidly. In one species, lutein and violaxanthin esters were formed (Goodwin, 1958a). In a similar study it was found that violaxanthin disappeared most rapidly, followed closely by neoxanthin. Lutein and β-carotene were the most stable carotenoids (Sanger, 1971). The lutein esters of autumn leaves were found to be a mixture of mono- and diesters, simple and mixed, the esterifying acids being palmitic and linolenic acid (Egger and Schwenker, 1966). The other xanthophylls are esterified with the same acids.

The composition of whole leaves and isolated plastoglobules of beech (*Fagus silvatica*) has been studied during four natural autumnal senescent stages (Tevini and Steinmüller, 1985). The total carotenoid content decreased only slightly during leaf senescence. At the last two stages, the leaves were yellow, with carotenoids predominating. During senescence the carotenoids liberated from thylakoids were esterified and increasingly deposited in plastoglobules. At the last stage carotenoid esters were the major component of the yellow plastoglobules, in which prenylquinones and fatty acids are also deposited. Of the total carotenoids found in plastoglobules, 76% were in esterified form.

Chromoplasts

Chromoplasts have been reviewed by Sitte et al. (1980). Chromoplasts are the principal carotenoid-bearing organelles in higher plants. They are the site of a wide range of chromoplast carotenoids, as they possess a remarkable metabolic activity. They contain all the carotenogenic enzymes and are therefore autonomous from this point of view (see earlier on carotenoid biosynthesis). They serve an important ecological function by mediating pollination and seed dispersal.

Chromoplast Types

Chromoplasts present a remarkable structure diversity both in form and size. Chromoplasts have a double-membrane envelope and a stroma matrix containing ribosomes and filaments of DNA. The internal structure is very different, and four types of chromoplasts occurring in plants have been characterized: globulous, tubulous, membranous, and crystallous. Within these structural elements the carotenoids are accumulated. A list of the occurrence of the various chromoplast types found in plant organs is given (Sitte et al., 1980). According to these authors, variations in the type and amounts of carotenoids, polar lipids, and proteins may be the key to the development of each type of chromoplast.

The most common type of chromoplast is the globulous, which is considered to be the oldest in terms of evolution. Globulous chromoplasts are lens-shaped or spheroidal and contain carotenoid-carrying lipid droplets (*plastoglobules*) of different sizes. Globulous chromoplasts occur in pumpkins and peppers. Tubulous chromoplasts contain tubules that may or may not be associated with plastoglobules. They occur in peppers and pumpkins. Membranous chromoplasts contain numerous concentric membranes and occur in *Capsicum* fruits, tomato, and summer squashes. Crystal-containing chromoplasts contain carotenoid crystals, which may determine the chromoplasts' shape. The crystals consist of pure pigment, and because they are large and colored they were first observed 100 years ago with a light microscope. In some cases the pigments may crystallize as small crystallites, up to 100 being formed per chromoplast. Chromoplasts containing crystals occur in carrots, tomatoes, and pumpkins. Some fruits may contain two types of chromoplasts: one in the inner region, another in the outer region of the pericarp, as further exemplified.

The chromoplasts can be isolated and the different internal structures separated by applying modern techniques. The chemical composition of the chromoplast substructure was determined by Sitte et al. (1980). With the exception of carotene crystals all the substructures, globules, tubules, and membranes contain lipids and proteins, the carotenoid constituting but a minor component.

Ultrastructural Changes in Ripening Fruits and Developing Roots

Fruit ripening involves carotenoid synthesis and ultrastructural changes. The plastid ultrastructural changes in ripening fruits have been studied independently and simultaneously following pigment changes, because these changes are closely correlated.

In carotenogenic fruits that synthesize large amounts of carotenoids during ripening, such as tomato and pepper, chloroplasts change into chromoplasts, thereby determining the parallel change of the carotenoid pattern. The gradual disintegration of the chloroplasts involves the disappearance of the chloroplast carotenoids. After disaggregation of the grana thylakoids, an enhanced carotenogenesis starts. At this stage the biosynthesis of new chromoplast carotenoids, such as capsanthin and capsorubin in pepper or lycopene in tomato, begins. Plastoglobules increase in size and number until they become the predominant structural elements in which the lipophilic carotenoids accumulate. In chromoplasts of other types they may crystallize out or accumulate in the respective structural elements.

Although the beginning of the chromoplast carotenoid synthesis follows the breakdown of the thylakoid membranes, it does not mean that the carotenoid-carrying structures of chromoplasts are derived from decomposing or transforming thylakoids (Sitte et al., 1980). In pepper fruit plastids it was observed that the beginning of an active biogenesis of ketocarotenoids coincides with the development of achlorophyllous membranes derived in part from the inner envelope membranes (Camara and Brangeon, 1981). These membranes are considered to be a framework for enzyme binding in carotenoid synthesis.

Asparagus Fruits. In asparagus (*Asparagus officinalis* L.) fruit pigment and plastid ultrastructural changes were followed in parallel (Simpson et al., 1977). Ripening commenced in interior of the fruit and progressed toward the epidermal cells. Chromoplasts of the epidermis and subepidermis were round and full of plastoglobules. Chromoplasts in the pericarp inner regions were tubulous (fibrillar). Spindle-shaped chromoplasts accumulated large-sized plastoglobules, most of which transformed into fibrils of exceptionally large diameter. The microfibrillar ultrastructure of these fibrils was different from that found in *Capsicum* fibrils.

Pepper Fruits. In various *Capsicum* fruit cultivars, tubulous or globulous chromoplasts may be found. During chloroplast–chromoplast transformation in red peppers, the bundle of paralleled fibrils characteristic of the mature chromoplasts begins to appear at the same time as the disintegration of the thylakoids becomes evident. The tubules may originate within the plastoglobules and may often be arranged in closely packed array. They are composed of smaller microfibrils (Simpson and Lee, 1976; Kirk and Tilney-Basset, 1978).

In other *Capsicum* fruits the early ripening stage was characterized by a typical grana–intergranal structure and a highly developed peripheral reticulum. At a later stage the thylakoid system disintegrates and is replaced by achlorophyllous single thylakoids derived in part from the inner envelope membrane (Camara and Brangeon, 1981).

Pumpkins. The ultrastructural changes of plastids in ripening *Cucurbita pepo* var. *pyriformis* were followed (Ljubešić, 1972). A mature *C. pepo* fruit has two colored parts, the proximal part being yellow throughout the entire development of the fruit. In this part the proplastids develop directly into chromoplasts. The distal part of the fruit is an intense green color. The yellowing is the direct consequence of the chloroplast–chromoplast transformation. The chloroplasts of the outer cell layers have highly stacked grana. The thylakoid system disintegrates according to two patterns: direct disintegration, and an unusual and more complex one, both of which lead to globulous chromoplasts.

In *Cucurbita maxima* Duch. *turbaniformis* fruits, tubulous chromoplasts were formed (Ljubešić, 1977). Tubules may be formed through reorganization of degraded thylakoids, or they may develop inside the plastoglobules. Both types of tubular formation occur in the fruit.

In the inner endocarp of *Cucurbita pepo* L. cv. Small Sugar pumpkins, the chromoplasts were developed by modification of already developed amyloplasts that are present in large amounts (Grilli, 1965).

Tomato. Chloroplast–chromoplast transformation was described in normal red tomatoes and in various strains (Kirk and Tilney-Bassett, 1978). Red tomato fruits produce two types of chromoplasts, as observed in cherry tomatoes (Laval-Martin, 1974). In the jelly part of the pericarp are globulous chromoplasts containing mainly β-carotene, whereas chromoplasts in the outer part contain voluminous sheets (crystalloids) of lycopene. The development and ultrastructure of the lycopene crystalloid body was described by Ben-Shaul and Naftali (1969). In high-color red tomatoes the pigment body development is the same as in normal red tomatoes, namely loss of granal membranes, increase in size and number of globules, and crystal deposition along an extended thylakoid system (Mohr, 1979). The globules are believed to be β-carotene dissolved in lipid material derived from membrane lysis.

Carrot Roots. In carrot roots the chromoplasts evolve directly from proplastids (Frey-Wyssling and Schwegler, 1965). The chromoplasts are of the crystal-containing type, the carotene crystals being formed when a critical pigment concentration is reached. The development of the carotene body was followed in growing carrots and reported by Ben-Shaul et al. (1968). The thylakoids of the proplastids became thickened and elongated, and were transformed into a specialized membrane system, which develops into the carotene body proper. The crystalline inner core is surrounded by lipoprotein layers.

It may be that the carrot chromoplasts serve no useful purpose—that they represent an abnormal situation produced perhaps as a result of a mutation in one of the genes concerned with carotenoid formation, which has been selected out and propagated by man. This idea is supported by the fact that the root of the wild carrot is white and does not contain chromoplasts (Kirk and Tilney-Basset, 1978).

CAROTENOIDS IN VEGETABLES

Carotenoids in edible vegetables may be concentrated in different parts of the plants, such as roots, stems, leaves, flowers, fruits, or seeds. Since chlorophylls are invariably accompanied by the four chloroplast carotenoids, carotenoids are to be found in all leaf vegetables, in unripe fruits, or immature inflorescence vegetables. They predominate in yellow–orange or red vegetables, such as carrots, tomatoes, and cucurbits.

Reports on carotenoid distribution are very uneven: some include extensive analysis, some include partial analysis, and a great many have not been analyzed at all. Many modern techniques are now available and should be applied to fill the gaps found in distribution tables and to update the old data.

Because of the importance of the nutritive value of the food and because β-carotene is the ubiquitous provitamin A carotenoid, investigation in vegetables has been limited mainly to the determination of β-carotene. This determination was first carried out as soon as the provitamin activity of β-carotene was established, and continues even now. Since external color is an important quality attribute, it was analyzed, for the most part, colorimetrically by nondestructive methods.

On the other hand, carotenoids of highly carotenogenic vegetables (carrots, tomatoes, and peppers) were thoroughly investigated, they being quite important for elucidation of carotenoid biosynthesis. The data described here on carotenoid distribution also include old reports, which have to be considered since they sometimes proved to be correct or else they have historic value. The terms used within are generally reproduced "as is."

The vegetables are presented in alphabetical order according to common name. Distribution in the plant other than the principal edible part is also reported. Factors affecting the carotenoid distribution, such as intraplant and intervarietal variations, environmental factors, and maturity, complete the description.

Characterization of the carotenoid pattern found in each vegetable is difficult due to the heterogeneity of the plant parts in which they are found. The general pattern of distribution established for fruit carotenoids has limited application.

Changes During Storage, Cooking, and Processing

In harvested vegetables, which are living organisms, life processes continue. Changes that occur in harvested vegetables include water loss, conversion of starch to sugars, flavor and color changes, sprouting, softening, and decay. To maintain fresh vegetables in the living state,it is usually necessary to slow down these processes by controlling the storage conditions (humidity, oxygen, and temperature as discussed previously; see chlorophylls).

Carotenoid changes occurring during storage depend on the physiological date of the harvested vegetables; that is, there is an increase with continuing maturation or a decrease with senescence.

Most vegetables are processed rather than eaten raw. Processing includes household cooking and industrial canning, freezing, dehydration, and pickling. These processes involve thermal treatment and exposure to oxidation, which is enhanced by light or the presence of enzymes. Highly unsaturated carotenoids are very susceptible to such treatments (see "Chemical Properties").

During heating, naturally occurring structures are disrupted. The carotenoprotein complexes, which are more stable than the free carotenoids, are destroyed, and the carotenoids are exposed to degradation. In the absence of oxygen, the only effect of heating is to induce conversion of the natural *trans* isomers to *cis* isomers, which are less stable and have reduced vitamin A activity. During processing, shortening the thermal treatment time, reducing the presence of oxygen to a minimum, and adding antioxidants results in minimal carotenoid oxidation. Blanching inactivates the enzyme activity of oxygenases and lipoxygenase. Freezing, after blanching, does not cause carotenoid degradation. Dehydrated vegetables are exposed to the most destruction. Freeze drying is preferable to air drying, which implies a relatively long exposure to oxygen and heat. Dehydrated products, powdered or lyophilized, with a great surface area susceptible to destruction are less stable. They have to be stored at the optimal water activities, in inert atmosphere, and with added antioxidants.

Investigations of the effects of storage and processing on carotenoids are concerned principally with the nutritive value of the food and therefore focus on the retention of β-carotene, the main and ubiquitous provitamin A.

Surprisingly, after canning, carotene retention was generally found to increase. Gradually it was found that the cause of this apparent increase was the leaching of soluble solids during processing. The method of calculation applied, which gives spuriously high values, does not allow for loss of solids (Murphy et al., 1975; Baloch et al., 1977a; Elkins, 1979). This method is based on the determination of the nutrient on a dry basis of the raw and processed product, which is the *apparent retention*. This method must be replaced by calculation of the nutrient in a known weight of food before and after cooking, whose result is the *true retention*.

Although many kinds of vegetables can be stored and processed, there are marked differences within each species in adaptability to a given method. Therefore, changes during storage, cooking, and processing are discussed for each vegetable.

Unique Carotenoids

Some vegetables have a special biosynthetic capacity and produce carotenoids found exclusively in the respective family or genus or even species, although rarely they may be found in other genera as well.

Lettuce

In green lettuce an unusual xanthophyll, named *lactucaxanthin*, is present along the chloroplast carotenoids (Siefermann-Harms et al., 1981). It belongs to the stereoisomers of ϵ,ϵ-carotene-3,3′-diol, of which ten isomers are theoretically possible. Seven were found in the animal kingdom (e.g., chiriquixanthin in frogs and tunaxanthins in marine fish). Lactucaxanthin is the only isomer found in plants, mainly in the *Lactuca* genus and in a few related genera of Cichorieae. Its structure is shown in Figure 24.

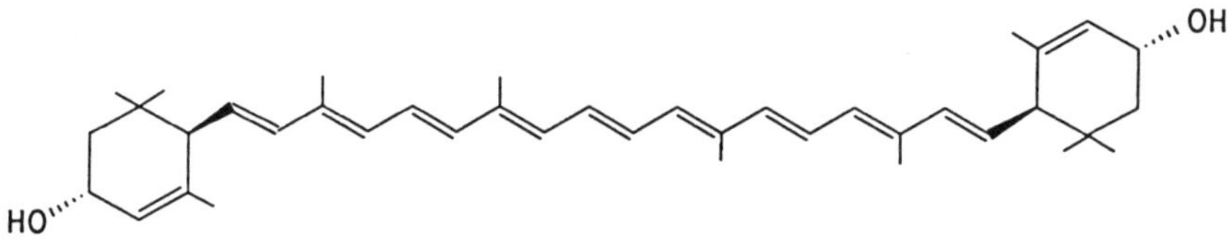

Figure 24. Lactucaxanthin (3*R*, 6*R*, 3′*R*, 6′*R*) - ϵ,ϵ - carotene - 3,3′ - diol.

One of the fish tunaxanthins (3*S*,6*S*,3′*S*,6′*S*) is the enantiomer of lactucaxanthin (Ikuno et al., 1985).

Maize

α-Cryptoxanthin or zeinoxanthin is the monohydroxy derivative of α-carotene. It may have the hydroxyl at the β- or ϵ-ring, the second compound being a provitamin A. The structure of the pigment was controversial for a long time, but it appears that the most widespread derivative is that devoid of provitamin A—3-hydroxy-α-carotene (Figure 25).

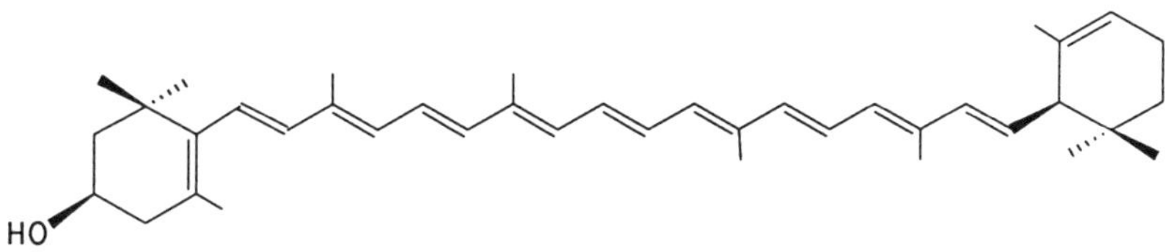

Figure 25. α-Cryptoxanthin or zeinoxanthin (3*R*, 6′*R*,) - β,ε - caroten - 3 - ol.

The pigment was detected for the first time in yellow corn grains and later in corn seedlings but in amounts too small to be characterized (White et al., 1942; Moster et al., 1952). Isolated in substantial amounts from corn gluten, it was obtained in crystalline form (Petzold and Quackenbush, 1960). Evidence indicated that it is widely distributed in corn inbreds, so it was named *zeinoxanthin.* On the basis of its physical properties and lack of provitamin A activity, it was suggested to be 3-hydroxy-α-carotene.

Meanwhile, α-cryptoxanthin was isolated from other plant sources: citrus fruits, yellow pepper, and winter cherries, to which one or another structure was assigned (Curl, 1956; Bodea and Nicoară, 1957; Cholnoky et al., 1958a). The structure proposed for the citrus α-cryptoxanthin was correct. Subsequently, it was proved that α-cryptoxanthin isolated from yellow peppers (*Capsicum annuum* var. *lycopersiciforme flavum*) and zeinoxanthin are identical, both being 3-hydroxy-α-carotene (Szabolcs and Ronai, 1969).

Recently, 3′-hydroxy-α-carotene was detected for the first time in nature, in red algae (Bjørnland et al., 1984). It is suggested that β,ε-caroten-3-ol should be named zeinoxanthin, and β,ε-caroten-3′-ol, α-cryptoxanthin.

In Brazilian pumpkins and squashes both monohydroxy-α-carotenes were present (Arima and Rodriguez-Amaya, 1988). The trivial names as suggested in the preceding report were used. α-Cryptoxanthin (a provitamin A) was found in almost all samples, and in *C. pepo* both α-cryptoxanthin and zeinoxanthin were found. The confusing use of the two trivial names has to be settled.

Pepper

The carotenoids that impart the beautiful red color to *Capsicum* fruits, both the sweet bell pepper and the hot-type paprika, are unique, having the keto group at the central chain and a hydroxylated cyclopentane ring (K) at one or both ends. Their structures are given in Figure 26.

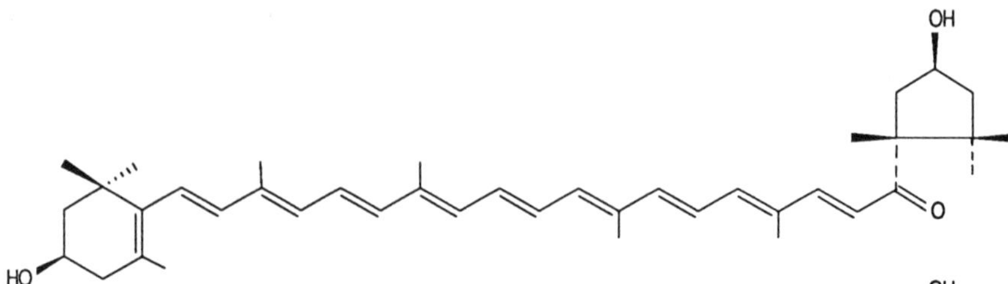

1

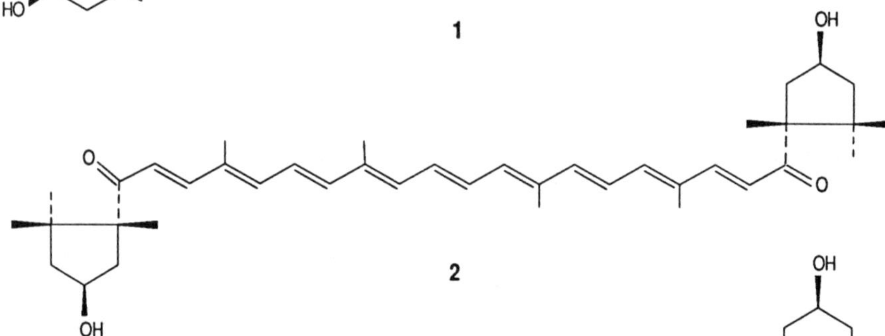

2

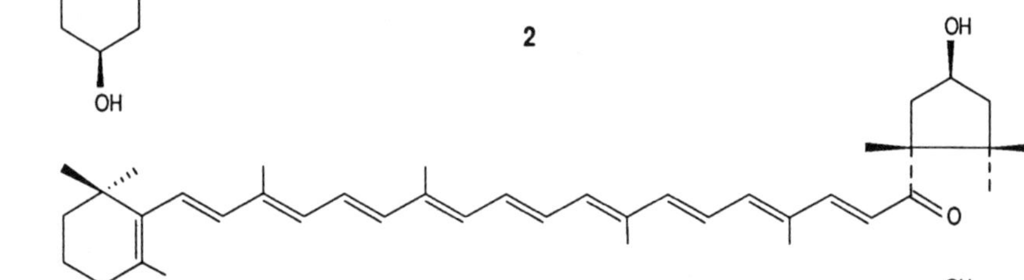

3

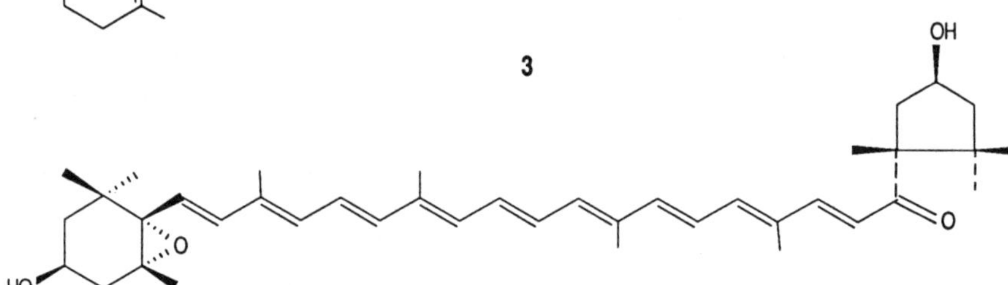

4

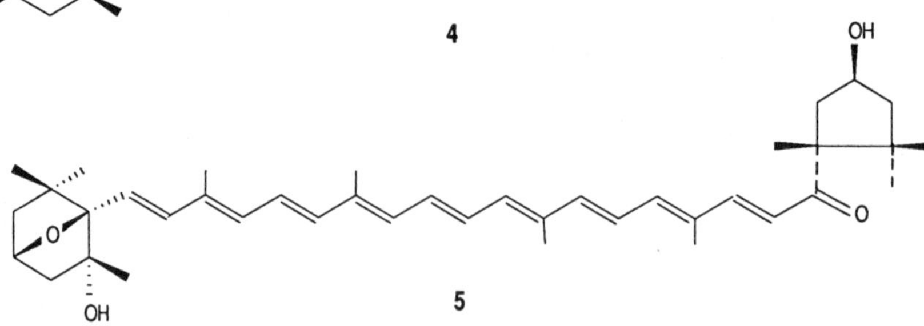

5

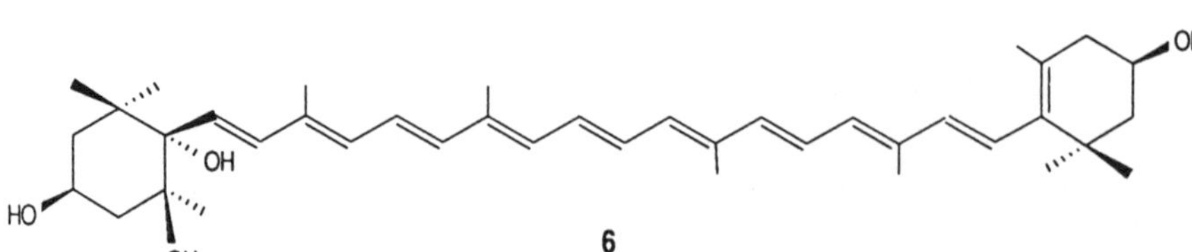

6

1. Capsanthin (3*R*, 3′*S*, 5′*R*) - 3,3′ - dihydroxy - β,κ - caroten - 6′ - one
2. Capsorubin (3*S*, 5*R*, 3′*S*, 5′*R*) - 3,3′ - dihydroxy - κκ, - carotene - 6,6′ - dione
3. Cryptocapsin (3′*S*,5′*R*)-3-hydroxy- β,κ-caroten-6′-one
4. Capsanthin 5,6-epoxide (3*S*, 5*R*, 6*S*, 3′*S*, 5′*R*) - 5,6 - epoxy - 3,3′ -dihydroxy - 5,6 - dihydro - β,κ - caroten - 6′ - one
5. Capsanthin 3,6-epoxide (3*S*, 5*R*, 6*R*, 3′*S*, 5′*R*) - 3,6 - epoxy - 5,3′ - dihydroxy - 5,6 - dihydro - β,κ - caroten - 6′ - one
6. Tetrol (Shamoutixanthin) (3*S*, 5*S*, 6*S*, 3′*R*) - 5,6 - dihydro - β,β - carotene - 3,3′,5′,6′ - tetrol

Figure 26 *(at left).* Unique carotenoids found in *Capsicum* fruits.

The most abundant pigment is capsanthin. The second major pigment is capsorubin, and next is cryptocapsin. Capsanthin was first obtained in crystalline form in 1927, capsorubin in 1934. In 1955, cryptocapsin was isolated (Zechmeister and Cholnoky, 1927, 1934; Cholnoky et al., 1955). A fourth ketocarotenoid, capsanthin 5,6-epoxide, was found as a minor pigment in *Capsicum* (Curl, 1962; Davies et al., 1970). The elucidation of their structure expanded for about 30 years due to the unusual end group—the cyclopentanol ring, which is not generally found in carotenoids—and to the special position of the keto group conjugated to the central polyene chain as a part of the molecule's chromophore (Zechmeister and Cholnoky, 1935).

The structure of the *Capsicum* ketones, respectively of the novel end group, was elucidated at the beginning of the 1960s (Barber et al., 1960; Cholnoky et al., 1963; Faigle et al., 1964). Unequivocal proof of the structure was provided by the total synthesis of optically inactive capsorubin (Cooper et al., 1962). The synthesis of optically active *Capsicum* carotenoids was reported by Ruettimann (1982).

Capsanthin 5,6-epoxide and its isomer *capsochrome,* capsanthin 5,8-epoxide, have been obtained in crystalline form and their chirality established (Parkes et al., 1986), thereby confirming one of their structures that were tentatively assigned earlier (Baranyai et al., 1977).

Novel 3,6-epoxides were also isolated. They contain the interesting ring 7-oxabicycloheptanol. One of the pigments is cucurbitaxanthin A, which was isolated from pumpkins (see further on), and the second is capsanthin 3,6-epoxide. Another interesting carotenoid described in the same report is the tetrol shamoutixanthin, isolated for the first time from Shamouti oranges, whose constitution was first established by Gross (Gross et al., 1975; Gross, 1987). The chirality assigned to the pepper tetrol is different from that assigned to the carotenoid with the same constitution isolated from *Rosa pomifera* fruits (Maerki-Fischer and Eugster, 1985). Pepper carotenoids are rarely found in other genera. They were isolated from some fruits of *Berberis* species, lily flowers, and asparagus fruits (Simpson et al., 1977; Valadon and Mummery, 1977; Goodwin, 1980).

Sweetpotatoes

Although having a low total carotenoid content of about 5 μg (g fresh wt)$^{-1}$, the pattern of Brazilian white sweetpotatoes is uncommon (Almeida and Penteado, 1988). It is dominated by two diepoxides of β-carotene: luteochrome (47%) and β-carotene 5,6,5′,6′-diepoxide 26%. The structures of these epoxides were established and reported by Almeida et al. (1986, 1988). Luteochrome consists of a mixture of two chiral isomers (Figure 27).

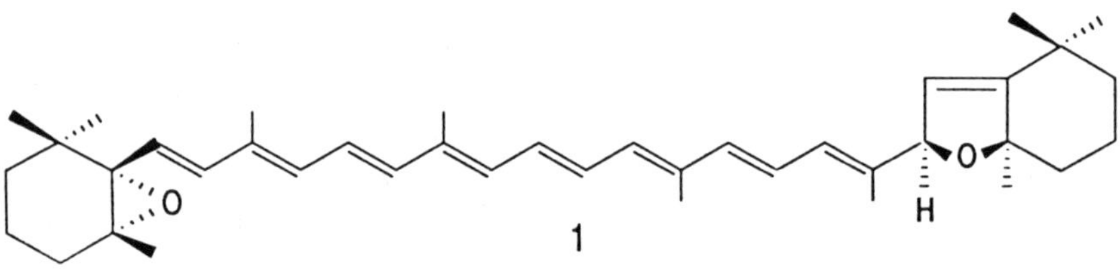

1

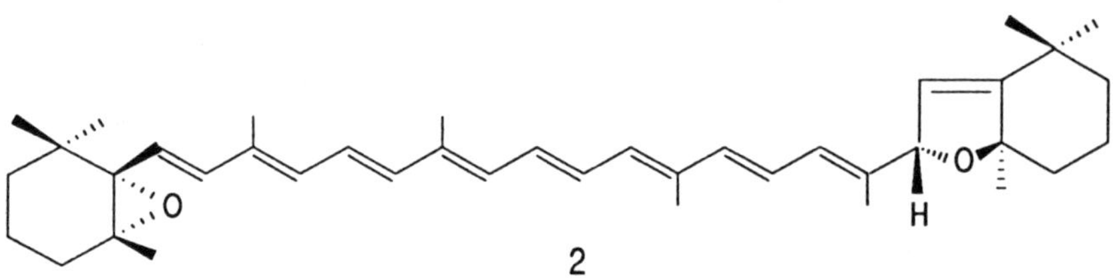

2

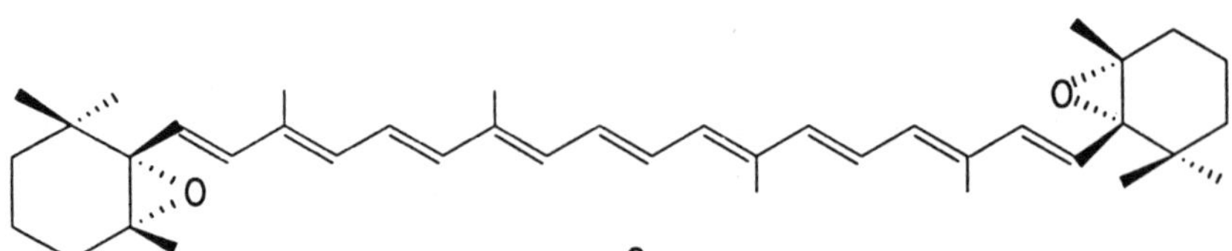

3

1, 2. Luteochrome (5*R*, 6*S*, 5′*R*, 8′*R*) and (5*R*, 6*S*, 5′*R*, 8′*S*) - 5,6,5,8 - diepoxy - 5,6,5′,8′ - tetrahydro - β,β - carotene

3. β,β-Carotene diepoxide (5*R*, 6*S*, 5′*R*, 6′*S*) - 5,6,6′,5′ - diepoxide - 5,6,5′,6′ - tetrahydro - β,β - carotene

Figure 27. β-Carotene diepoxides of Brazilian white-fleshed sweetpotatoes.

For the first time, carotenoids with an unhydroxylated 5,6-epoxy-5,6-dihydro β-ring in optically active form were isolated. The chirality of C-5 and C-6 is the same as in violaxanthin.

Luteochrome was detected as a minor pigment in the orange Centennial sweetpotato (0.2%) (Purcell, 1968). As a minor pigment it was present also in fruits, in banana pulp (Gross et al., 1976). β-Carotene 5,6,5′,6′-diepoxide was detected as a minor pigment in apples and yellow cherries (Valadon and Mummery, 1967; Gross, 1985).

Tomato

Lycopene is the major carotene of tomato, accompanied by its more saturated precursors, but other, minor carotenoids were isolated and identified. Two hydroxy derivatives of lycopene—the monohydroxy lycopene *lycoxanthin* (ψ,ψ-caroten-16-ol) and the dihydroxy lycopene *lycophyll* (ψ,ψ-carotene-16,16′-diol)—were isolated from tomato in 1936 by Zechmeister and Cholnoky. Lycoxanthin, which is found in greater amounts than lycophyll, was reported several times in tomato (Went et al., 1942; Curl, 1961; Ben-Aziz et al., 1973), and both lycopene xanthophylls were followed in ripening Marzano tomatoes and cherry tomatoes (Edwards and Reuter, 1967; Laval-Martin et al., 1975). Their structure as primary alcohols was established (Markham and Liaaen-Jensen, 1968) (Figure 28).

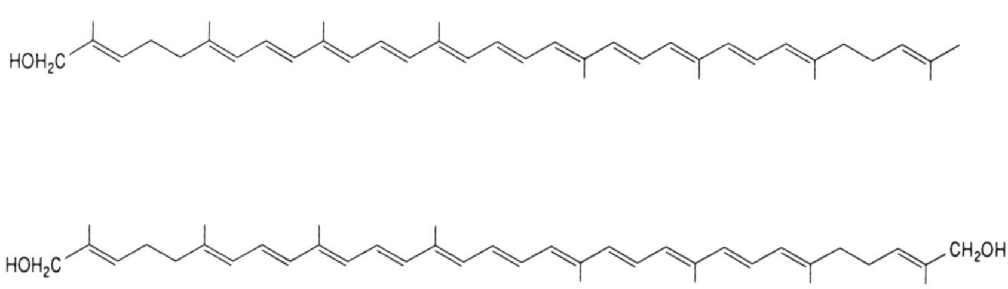

Figure 28. Lycoxanthin (ψ,ψ-caroten-16-ol) and lycophyll (ψ,ψ-carotene-16,16′-diol).

A series of epoxides in very small amounts were isolated from red tomatoes and identified as epoxides of phytoene, phytofluene, ζ-carotene, and lycopene (Britton and Goodwin, 1969; Ben-Aziz et al., 1973). They all were 1,2-epoxy derivatives. The 5,6-epoxide of lycopene was also detected, as were two apoderivatives of lycopene, namely apo-6′-lycopenal and apo-8′-lycopenal (Figure 29).

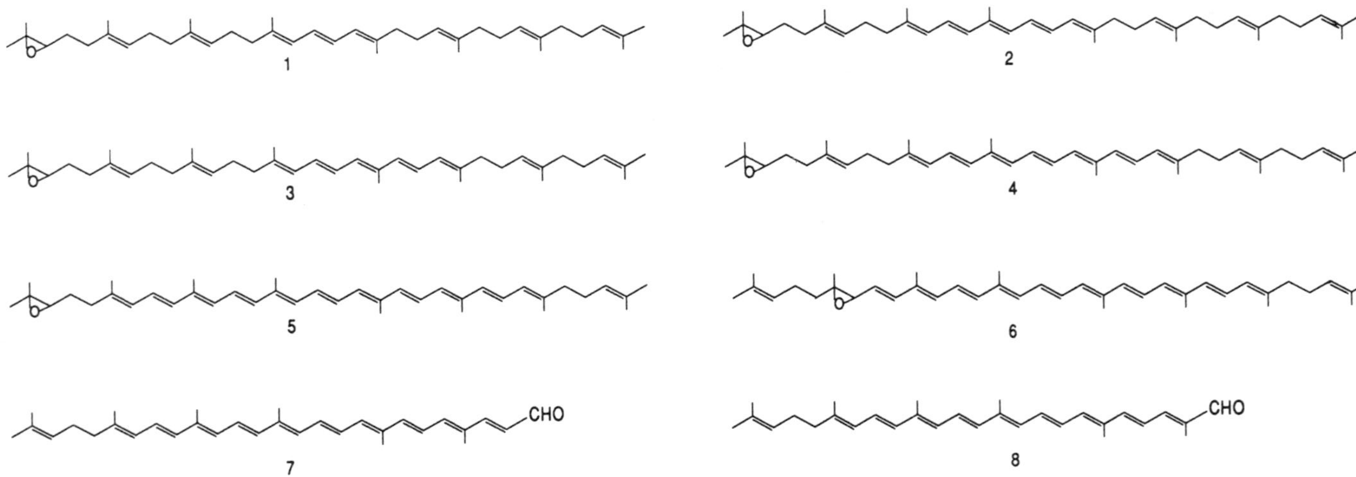

1. Phytoene epoxide (1,2-epoxy-1,2,7,8,11,12,7′8′,11′,12′-decahydro-ψ,ψ-carotene)

2, 3. Phytofluene epoxide (1,2-epoxy-1,2,7,8,7′,8′,11′,12′-octahydro-ψ,ψ-carotene) and (1,2-epoxy-1,2,7,8,11,12,7′,8′-octahydro-ψ,ψ-carotene)

4. ζ-Carotene epoxide (1,2-epoxy-1,2,7,8,7′,8′-hexahydro-ψ,ψ-carotene)

5. Lycopene 1,2-epoxide (1,2-epoxy-1,2-dihydro-ψ,ψ-carotene)

6. Lycopene 5,6-epoxide (5,6-epoxy-5,6-dihydro-ψ,ψ-carotene)

7. Apo-6′-lycopenal (6′-apo-ψ-caroten-6′-al)

8. Apo-8′-lycopenal (8′-apo-ψ-caroten-8′-al)

Figure 29. Minor pigments in red tomato.

Tomato Strains

From delta tomatoes in which δ-carotene is the major carotene, the 1,2-epoxides of δ- and γ-carotene were isolated (Britton and Goodwin, 1975) (Figure 30).

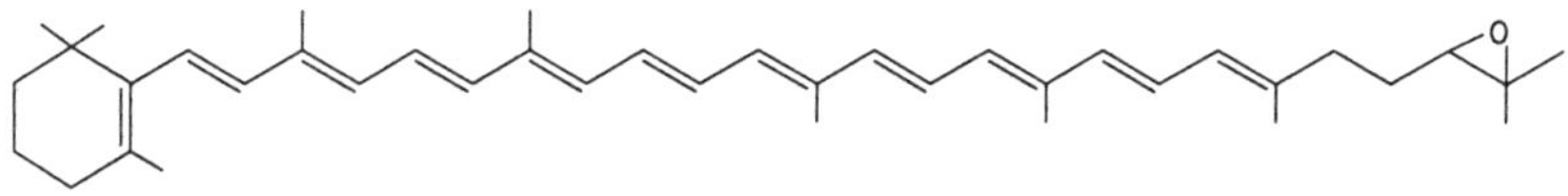

1

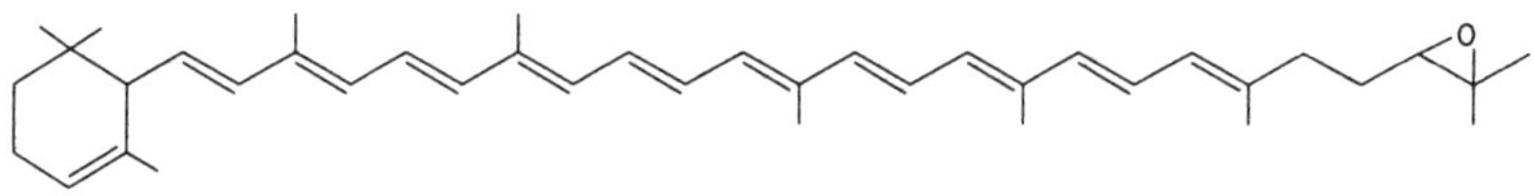

2

1. γ-Carotene 1′,2′-epoxide (1′,2′-epoxy-1′,2′-dihydro-β,ψ-carotene)
2. δ-Carotene 1′,2′-epoxide (1′,2′-epoxy-1′,2′-dihydro-ε,ψ-carotene)

Figure 30. γ- and δ-carotene 1′,2′-epoxides.

Tangerine-type tomatoes have the capacity to synthesize *cis* isomers of lycopene. Their major carotene is prolycopene, a poly-*cis* isomer responsible for the orange color of these fruits. It was first isolated from the tangerine tomato Tangella (Zechmeister et al., 1941). Prolycopene shows two spectroscopic features: its absorption maximum in hexane at 438 nm is at about 35 nm shorter wavelength than that of lycopene; the spectrum shows decreased extinction values and is almost devoid of fine structure (Zechmeister et al., 1943). These characteristics are for a poly-*cis* isomer possessing one or, perhaps, two "hindered" *cis* double bonds (Zechmeister, 1962). Although the stereochemistry was investigated extensively for 20 years, the structure of prolycopene was not elucidated. Elucidation became possible only with the advent of modern methods, which established that prolycopene (7,9,7′,9′-tetra-*cis*-ψ,ψ-carotene) is a tetra-*cis*-lycopene with two hindered *cis* double bonds (which contradicts the theory) (Clough and Pattenden, 1979; Englert et al., 1979).

The stereochemistry of prolycopene and its congeners was definitively established by Clough and Pattenden (1983). All the precursors except phytoene (i.e., the 15-*cis*-isomer) have a poly-*cis*-structure (Figure 31).

Poly-*cis*-ζ-carotene and poly-*cis*-γ-carotene were isolated earlier from the Tangerine Golden Jubilee Tomato (Raymundo and Simpson, 1972; Glass and Simpson, 1976). Several poly-*cis*-lycopenes were detected in roots of *Brassica rutabaga* and *Brassica napus* (Joyce, 1954, 1959).

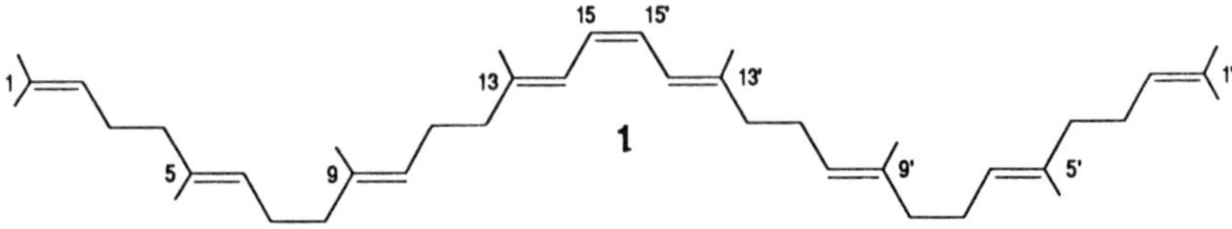

1

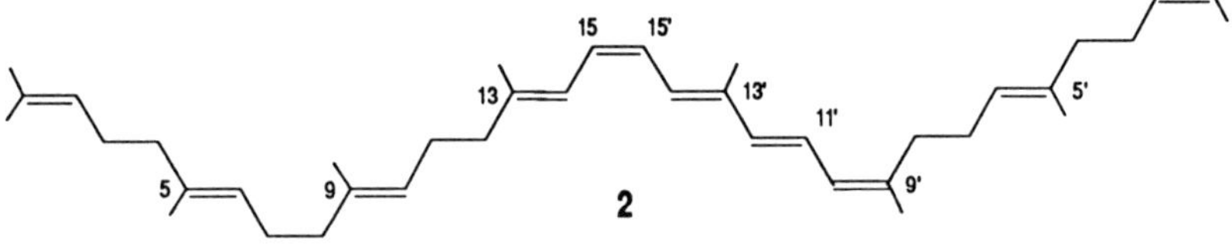

2

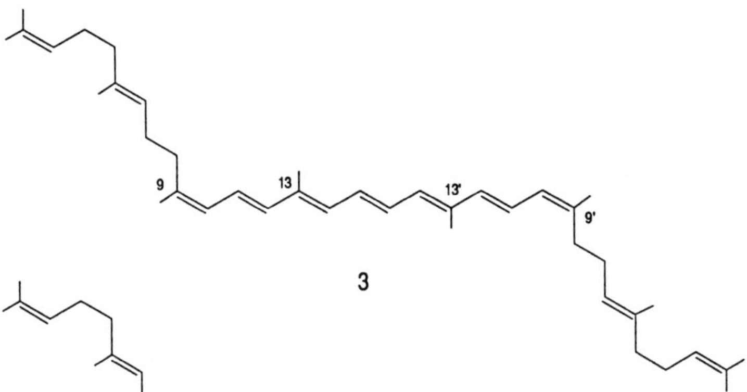

3

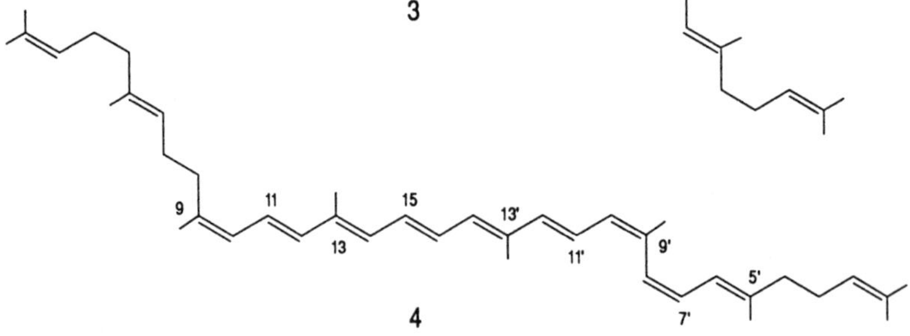

4

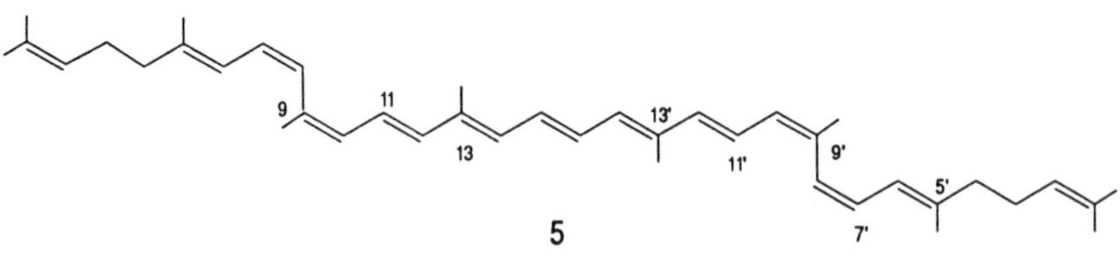

5

1. 15-*cis*-Phytoene
2. 15,9′-di-*cis*-Phytofluene
3. 9,9′-di-*cis*-ζ-Carotene
4. 9,7′,9′-tri-*cis*-Neurosporene
5. 7,9,7′,9′-tetra-*cis*-Lycopene

Figure 31 *(at left).* Carotenes of the Tangella tomato. (*From Clough and Pattenden, 1983.*)

Cucurbita, spp. Fruits

The structures of two new carotenoids isolated from the pulp of the Japanese pumpkin (*Cucurbita maxima*) were elucidated and reported by Matsuno et al. (1986). The new carotenoids were named cucurbitaxanthins A and B (Figure 32).

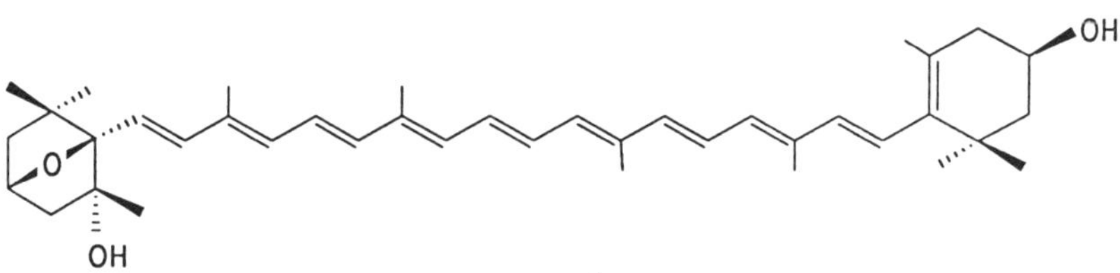

1

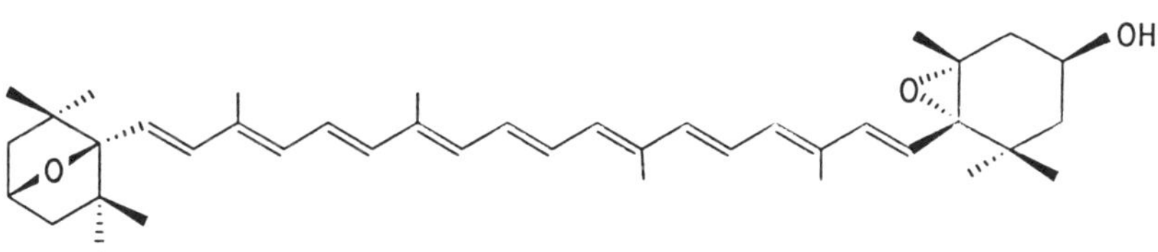

2

1. Cucurbitaxanthin A (3*S*, 5*R*, 6*R*, 3′*R*) - 3,6 - epoxy - 5,6 - dihydro - β,β - carotene - 5,3′ - diol
2. Cucurbitaxanthin B (3*S*, 5*R*, 6*R*, 3′*S*, 5′*R* , 6′*S*) - 3,6,5′,6′ - diepoxy - 5,6,5′,6′ - tetrahydro - β,β - carotene - 5,3′ - diol

Figure 32. Novel carotenoids in pumpkin flesh.

The pigments have an unusual structure, the novel 5-hydroxy-3,6-oxabicycloheptane ring. Naturally occurring carotenoids with such a ring system were isolated for the first time from marine algae (*Euglenophyceae*) by Liaaen-Jensen (1985).

CAROTENOID DISTRIBUTION IN VEGETABLES

Asparagus (*Asparagus officinalis* L.) Liliaceae

A complete analysis of the carotenoids found in asparagus was never carried out. Only the β-carotene content in the stem, the edible part of the plant, was determined. In green asparagus, chloroplast carotenoids should be present, and in white asparagus one expects to find the carotenoids typical for etiolated tissue. The presence of vitamin A in asparagus was proved very early (Christ and Dye, 1929). A daily dose of 0.1 g green asparagus, whether fresh, cooked, or canned, was sufficient to promote health and growth in rats. Fed with even higher doses of bleached asparagus, the animals died. The data supported the presumption that vitamin A content is associated with "greenness."

In green asparagus the β-carotene content in μg (g fresh wt)$^{-1}$ ranges from 4.3 to 7 (Fellers et al., 1934; Zimmerman et al., 1941; Gleim et al., 1944). In asparagus containing 8.5 μg (g fresh wt)$^{-1}$ carotene, all-*trans*-β-carotene constituted 83.8% of the fraction (Beadle and Zscheile, 1942). The β-carotene content of white, violet, and green asparagus grown in Romania was reported in 1983 (Dobreanu et al., 1983). β-Carotene content was highest in green asparagus (4.8 μg g^{-1}); violet and white asparagus, respectively, had 1.4 μg g^{-1} and 1.0 μg g^{-1}.

Distribution in Other Parts of the Plant

The fruits of asparagus are red due to a high carotenoid content (Simpson et al., 1977). They were investigated in detail at three ripening stages. The results are given in Table 18.

Discussion. The total carotenoid content in ripe fruits is very high [2520 μg (g fresh wt)$^{-1}$]. Unripe (green) asparagus fruit contained chloroplast carotenoids. During ripening, lutein and neoxanthin disappeared. The most remarkable feature of the ripe fruit was the carotenoid pattern (6) containing the unique ketocarotenoids usually found in *Capsicum,* capsanthin, capsorubin, and cryptocapsin.

Table 18. Carotenoid and Chlorophyll Pigment Composition of Green, Breaker, and Ripe *Asparagus officinalis* Fruit

	Percentage of Total Carotenoids		
Carotenoid Pattern	*Green*	*Breaker (brown)*	*Ripe (red)*
Phytofluene	—	tr	1.7
β-Carotene	32.7	49.4	33.9
Lutein	58.6	9.9	—
Cryptoxanthin	—	tr	2.1
Zeaxanthin	—	tr	12.4
Antheraxanthin	tr	6.9	1.4
cis-Antheraxanthin	—	5.5	—
Violaxanthin	6.8	6.9	3.6
Neoxanthin	1.4	—	—
Cryptocapsin	—	—	0.5
Capsanthin	—	14.0	37.5
Capsanthin isomer I	—	7.2	1.0
Capsanthin isomer II	—	—	2.0
Capsanthin isomer III	—	—	2.4
Capsorubin	—	—	1.5
Total carotenoids [μg (g fresh wt)$^{-1}$]	57.2	184.0	2520.0
Chlorophyll *a* [μg (g fresh wt)$^{-1}$]	3.9	1.6	—
Chlorophyll *b* [μg (g fresh wt)$^{-1}$]	1.8	0.3	—

Source: From Simpson et al. (1977).
Note: tr = trace.

Changes During Storage, Cooking, and Freezing

Changes of the carotene content in fresh asparagus during storage and cooking were followed (Gleim et al., 1944). During storage for 24 h at 19°C to 25°C asparagus lost 9% of its carotene content. Storage for 1 week at 0°C to 4°C resulted in a loss of 14%. During storage for 10 days at 10°C, carotene content of 4.8 μg (g fresh wt)$^{-1}$ increased to 7.4 μg (g fresh wt)$^{-1}$, probably due to the loss of water (Dobreanu et al., 1983).

During cooking in small or large amounts of water, fresh or stored asparagus retains all its carotene content, the quantity of water not influencing the carotene retention. On the other hand, losses of carotene during the cooking of frozen asparagus were 38% in the larger amount of water and 18% in the smaller amount.

During freezing, asparagus lost little carotene (Zimmerman et al., 1941).

Balsam Pear
(*Momordica charantia* L.)
Cucurbitaceae

The tropical fruit balsam pear is also called Chinese bitter melon or bitter gourd. The orange–yellow, warty, spindle-shaped fruit has brown seeds. In Asian countries the orange pulp is a popular vegetable preferred for its bitterness.

The carotenoid composition of the pericarp of *Momordica charantia* was investigated at four maturity levels (Rodriguez et al., 1976). The results are shown in Table 19.

Table 19. Qualitative and Quantitative Distribution in Developing *Momordica charantia* Fruit

	Immature-green		*Mature-green*		*Partly Ripe*		*Fully Ripe*	
Carotenoid	*(μg g)$^{-1*}$*	*%*	*(μg g)$^{-1*}$*	*%*	*(μg g)$^{-1*}$*	*%*	*(μg g)$^{-1*}$*	*%*
Phytofluene	tr		tr		tr		tr	
α-Carotene	2.2 ± 1.1	34.9	1.5 ± 0.3	28.3	0.3 ± 0.1	3.5	0.4 ± 0.1	1.8
Flavochrome	—		—		tr		tr	
β-Carotene	1.5 ± 0.8	23.8	2.3 ± 0.8	43.4	2.5 ± 0.3	29.4	5.3 ± 2.3	24.1
5,6-Monoepoxy-β-carotene	—		—		tr		tr	
Mutatochrome	—		—		tr		tr	
ζ-Carotene	—		—		tr		tr	
δ-Carotene	—		—		tr		tr	
Zeinoxanthin	0.2 ± 0.1	3.2	0.5 ± 0.3	9.4	0.3 ± 0.1	3.5	0.6 ± 0.2	2.7
γ-Carotene	—		—		tr		tr	
Cryptoxanthin	—		—		4.7 ± 1.7	55.3	13.7 ± 5.4	62.3
Lycopene	—		tr		0.2 ± 0.1	2.4	0.5 ± 0.1	2.3
Rubixanthin	—		—		tr		tr	
Lutein	2.4 ± 1.2	38.1	1.0 ± 0.4	18.9	0.1	1.2	tr	
Zeaxanthin	—		—		0.4	4.7	1.5 ± 0.1	6.8
Total	6.3		5.3		8.5		22.0	

Source: From Rodriguez et al. (1976).
Note: tr = trace.
*Each value is the mean of five samples. The average fresh weight of the fruits was 38.2 g at the immature-green stage, 77.1 g at the mature-green stage, 61.5 g at the partially ripe stage, and 64.4 g at the ripe stage.

Discussion. The number of carotenoids increased from 4 to 5 in the immature and mature-green fruit and to 6 or 7 in the partly and fully ripe fruit. An additional 8–9 pigments in trace amounts, characterized by their electronic spectra, produced a complex carotenoid pattern.

The major carotenoid pattern in the ripe fruit is 4, according to the classification of fruit carotenoids established by Goodwin (1976). That means large amounts of β-carotene and its hydroxy derivatives cryptoxanthin and zeaxanthin. Prior to color break, α-carotene and its mono- and dihydroxy derivatives, zeinoxanthin (3-hydroxy-α-carotene), lutein, and β-carotene, were the major carotenoids. At color break, mono- and dihydroxy derivatives of β-carotene (cryptoxanthin and zeaxanthin) began to accumulate. Cryptoxanthin accumulated rapidly to become the major pigment in the ripe fruit (62%) and was followed by β-carotene (24%) and zeaxanthin (6.8%). The hydroxy carotenoids of the ripe fruit were almost entirely esterified. During fruit ripening the change of the carotenoid pattern dominated by pigments of the β,ϵ-series to a pattern dominated by pigments of the β,β-series was evident.

Distribution in Other Parts of the Plant

The carotenoids of the balsam or seeds, which have a brilliant red color, were also investigated at four ripening stages (Rodriguez et al., 1975). In fact, it is probable that the seed coating and not the seeds themselves were analyzed (Shrivastava, 1972). The qualitative and quantitative carotenoid distribution is shown in Table 20.

Table 20. Qualitative and Quantitative Carotenoid Distribution in *Momordica charantia* Seed (coating)

	Immature-green		*Mature-green*		*Partly Ripe*		*Fully Ripe*	
Carotenoid	*(μg g)$^{-1}$**	*%*	*(μg g)$^{-1}$**	*%*	*(μg g)$^{-1}$**	*%*	*(μg g)$^{-1}$**	*%*
Phytofluene	tr		tr		tr		tr	
β-Carotene	1.1 ± 0.4	39.3	1.2 ± 0.4	8.3	tr		tr	
ζ-Carotene	—		—		tr		tr	
Lycopene	tr		11.8 ± 13.8	81.4	244.9 ± 54.3	97.1	261.0 ± 60.2	96.3
Zeinoxanthin	—		—		0.5 ± 0.4	0.2	0.4 ± 0.2	0.1
Lutein	1.7 ± 1	60.7	1.5 ± 0.6	10.3	1.1 ± 0.7	0.4	1.1 ± 1	0.4
Cryptoxanthin	—		tr		0.8 ± 0.4	0.3	1.9 ± 1.5	0.7
Rubixanthin	—		tr		2.3 ± 0.3	0.9	2.9 ± 0.7	1.1
Lycoxanthin	—		—		2.7 ± 1.3	1.1	3.6 ± 1.5	1.3
Total xanthophylls	1.7		1.5		7.4		9.9	
Total	2.8		14.5		252.3		270.9	

Source: From Rodriguez et al. (1975), by permission of the copyright owner, ASPP.
Note: tr = trace.
*Each value is the mean of five samples. The average fresh weight of the seeds was 6.1 g at the immature-green stage, 12 g at the mature-green stage, 9.8 g at the partly ripe stage, and 11.3 g at the fully ripe stage.

Discussion. The carotenoid content showed a 100-fold increase from 28 $\mu g\ g^{-1}$ at the immature stage to 270 $\mu g\ g^{-1}$ at the fully ripe stage. This massive increase was almost exclusively attributable to lycopene, which accounted for 96% of the carotenoids in the ripe seeds.

Precursors of lycopene were not detected except for phytofluene in trace amounts; lycoxanthin (16-hydroxylycopene), the hydroxylated lycopene derivative, was formed in detectable amounts already at the partly ripe stage. Four additional minor pigments: zeinoxanthin (3-hydroxy-α-carotene), β-cryptoxanthin, rubixanthin, and lutein, were also present in detectable amounts. This carotenoid pattern, although dominated by lycopene, cannot be characterized as pattern 3, which is characterized by large amounts of lycopene and the related, partly saturated, acyclic polyenes.

Green Beans
(*Phaseolus vulgaris* L.)
Leguminosae

Legume carotenoids were reviewed by Sri Kantha and Erdman (1987). The pods of the green beans (French or snap bean) have a β-carotene content in μg (g fresh wt)$^{-1}$ between 2 and 4 (Zimmerman et al., 1940; Flynn et al., 1946; Hayden et al., 1948; Weits et al., 1970; Elkins, 1979; Böttcher, 1986). The "crude carotene" content [23.7 μg (g dry wt)$^{-1}$] of green beans grown in Canada was chromatographically separated into its constituents: α-carotene (3.2 μg), neo-β-carotene B (3.7 μg), neo-β-carotene U (2.5 μg), and all-*trans*-β-carotene (14.3 μg). The biopotency was calculated accordingly (Panalaks and Murray, 1970). A close average value of 21 μg (g dry wt)$^{-1}$ was found for 11 varieties of green beans grown in New York State (van Buren et al., 1982). The qualitative and quantitative distribution of the major carotenoid constituents of green beans was reported by Khachik et al. (1989). (See Table 21.)

The vitamin A activity of various vegetables grown in the United States was determined after separating α- and β-carotene and β-cryptoxanthin by HPLC three times a year (November, March, July) (Bureau and Bushway, 1986). In green beans the activity was 77.3 μg RE (100 g)$^{-1}$ in March and 49.3 μg RE (100 g)$^{-1}$ in July. The average carotene content was 64 μg α-carotene (100 g)$^{-1}$ and 364 μg β-carotene (100 g)$^{-1}$.

In seeds of 10 bean varieties the carotene content was 0.23 to 3.34 μg (100 g)$^{-1}$, thus showing a considerable variation among varieties (Fordham et al., 1975). In the respective sprouts the carotene content was 0.23 to 4.05 μg (100 g)$^{-1}$.

Distribution in Other Parts of the Plant

For a study on carotenogenesis, the carotenoids of etiolated French bean seedlings were analyzed (Goodwin and Phagpolngarm, 1960). A different carotenoid distribution was found in cotyledons and true leaves of etiolated French bean seedings. The cotyledons had a total carotenoid content of 13.5 μg (g dry wt)$^{-1}$, with the oxygenated xanthophylls chrysanthemaxanthin and auroxanthin as major pigments. The true leaves had a total content of 107 μg (g dry wt)$^{-1}$, the constituents being β-carotene lutein and neoxanthin (50%). Violaxanthin was absent.

Lima Beans
(*Phaseolus lunatus* L.)
Leguminosae

Green baby lima beans have a β-carotene content of 12.2 μg (10 g)$^{-1}$ (Cook et al., 1961). In dry lima bean seeds of 11 U.S. commercial cultivars, the carotene content ranged from 0.17 to 2.43 μg (10 g)$^{-1}$ (Abdul-Baki, 1971). In cultivars originating from South America the carotene content was higher. Cultivars from Argentina, Bolivia, and Colombia contained 9.6, 12.2, and 18.6 μg (10 g dry wt)$^{-1}$, respectively. During seed maturation of the cultivars G2 and Green Fordhook 861, the carotene content was 5.0 and 3.6 μg (g dry wt)$^{-1}$ in the mature green seeds. It decreased to 0.27 and 0.23 μg (g dry wt)$^{-1}$ in dry, nonbleached seeds and almost disappeared in bleached seeds.

Mung Beans
(*Phaseolus aureus* Roxb.)
Leguminosae

The nutritive value of sprouted legumes and grains, which are largely consumed in oriental countries, was investigated thoroughly. They constitute a good source of vitamins (it was observed that the content of various vitamins increased markedly during germination, augmenting the nutrient values of the product). Mung bean sprouts have a total carotenoid content of 284 μg (100 g fresh wt)$^{-1}$ of which β-carotene represents 6% (Farhangi and Valadon, 1981). The carotenoid distribution is shown in Table 22.

The effect of light on carotenoids of etiolated mung bean seedlings was also investigated (Valadon and Mummery, 1969). The total carotenoid content of the seeds, 2.62 μg (g dry wt)$^{-1}$, increased to 71.4 μg g^{-1} after the seedlings had been grown in the dark for 8 days and to 926.5 μg g^{-1} in the light. The etiolated seedlings contained 10 different carotenoids, namely phytofluene, β-carotene and its mono- and diepoxides, lutein and its 5,6- and 5,8-epoxides, violaxanthin, and auroxanthin. As in the etiolated French bean seedlings, oxidized carotenoids are formed in the dark-grown seedlings. Upon exposure to light, chloroplast carotenoids, including violaxanthin, are intensively synthesized.

Soybeans
[*Glycine max* (L.) Merr.]
Leguminosae

The vitamin content as a function of germination time was followed in soybean sprouts (Wai et al., 1947). The initial carotene content of 1.2 μg (g dry wt)$^{-1}$ almost doubled in 48 h of germination. It showed a 2.8-fold increase after 54 h and a 3.4-fold increase after 72 h. The relative distribution in the hypocotyl and cotyledon (per soybean) after 54 h of germination was 0.13 and 0.44 μg carotene, respectively. Therefore the practice of discarding the bean involves a considerable loss in the provitamin content of the sprouted bean.

The changes of β-carotene content of soybeans during maturation and sprouting were followed (Bates and Matthews, 1975). The planting date did not influence β-carotene content [about 2.8 μg (g fresh wt)$^{-1}$] of the two studied varieties, Verde and Early Green. The β-carotene content decreased markedly as the beans matured, and increased upon sprouting: in green mature, 3.5 μg (g fresh wt)$^{-1}$ were found; in dry mature, 1.2 μg (g fresh wt)$^{-1}$, and in sprouted, 2.0 μg (g fresh wt)$^{-1}$.

Changes During Storage, Cooking, and Processing

Storage

The decrease of carotene content in green beans followed a linear course during storage at various temperatures (0–24°C), showing, however, a lower initial rate. The minimal rate of 1% loss per day occurs at 4°C, the maximum of 4.3% per day at 18°C, showing that the storage temperature has the preponderant effect on quality (Böttcher, 1986).

Cooking

The crude carotene content in fresh and cooked green beans was found to be 22.4 and 58.1 μg (g dry wt)$^{-1}$, respectively. The relative percentages of β-carotene isomers were 18.0% neo-β-carotene B, 12.2% neo-β-carotene U, and 68.9% all-*trans*-β-carotene in the raw beans, and 12.3%, 20.7%, and 66.9%, respectively, in cooked beans (Panalaks and Murray, 1970). The same carotene content in fresh and cooked green beans was also reported (Weits et al., 1970).

The effect of various cooking procedures on the qualitative and quantitative distribution of major carotenoid constituents of several green vegetables was studied. The carotenoids were analyzed by HPLC (Khachik et al., 1989). The results found for green beans are given in Table 21.

Table 21. Carotenoid Content of Green Beans

	Carotenoids in mg (100 g)$^{-1}$ Raw Food					
Cooking Method	*Neoxanthin*	*Violaxanthin*	*Lutein 5,6-epoxide*	*Lutein* cis + trans	*α-Carotene*	*β-Carotene*
Raw	0.13	0.23	0.20	0.60	0.08	0.47
Microwave	0.13	0.09	0.09	0.71	0.09	0.53
Boiling	0.13	0.08	0.07	0.61	0.08	0.54
			Long-term cooking study			
Raw	0.20	0.23	0.23	0.69	0.10	0.51
Boiling for 1 h	—	—	—	0.72	0.09	0.58

Source: Reproduced with permission, from Khachik et al. (1989); copyright © 1989 American Chemical Society.

Discussion. It appears that the xanthophyll epoxides are the least stable, whereas the carotenes and lutein remain unaffected, showing apparent increases.

Processing

Blanching. The carotene content in raw and blanched Tender green snap beans was unchanged (Lee et al., 1946). The carotene retention in blanched beans in relation to raw beans was 90%, 95%, and 93% in three processing plants (van Buren et al., 1982). The carotene levels in acidified or permeated green beans, Early Gallatin cv., were not affected by the four blanching methods applied (Kozup and Sistrunk, 1982).

Canning. The vitamin content of canned snap beans has been a matter of interest for 40 years. For β-carotene, a retention of 87% and 83% was reported (Cameron et al., 1949; Weits et al., 1970). Canned green beans contained four to six times more crude carotene than raw green beans and a higher percentage of all-*trans*-β-carotene (Panalaks and Murray, 1970). Carotene retention after canning and storage was 92% after 0 month and 76% after 12 months of storage (Elkins, 1979). In three plants, the β-carotene retention in canned beans in relation to raw beans was 77%, 84%, and 109% and 84%, 91%, and 118.6% in cooked beans in relation to blanched beans (van Buren et al., 1982). The intervarietal differences are considered the source of these variations.

Freezing. Carotene content showed no significant differences in snap beans, frozen at greatly different speeds, after freezing, after storage, or after cooking (Lee et al., 1946).

In green beans grown in Spain, β-carotene was 2 and 3 μg (g fresh wt)$^{-1}$ in fresh and frozen beans, respectively (Arqués et al., 1988). During a 22-month storage of frozen beans at −22°C, the total carotenoid content decreased, showing an irregular course difficult to explain. The overall loss was 54% (Aparicio-Cuesta et al., 1989).

Dehydration. Dehydrated green beans contained 37% of the crude carotene of raw beans and a higher percentage of all-*trans*-β-carotene (Panalakas and Murray, 1970). In air-dried and freeze-dried green beans, carotenoid retention was 102%, 55%, and 43% in air-dried green beans and 101%, 79%, and 74% in freeze-dried green beans, immediately after drying, after 6 months storage, and after 6 months storage and cooking (Weits et al., 1970).

Pickling. Carotene levels in green beans acidified with four organic acids (malic, lactic, citric, and acetic). Levels in fermented beans were slightly lower than the control, canned beans (Kozup and Sistrunk, 1982). There were no significant differences among the acid treatments except for beans acidified with malic acid. The highest retention occurred in fermented beans. The effect of acidified processing and storage on carotenoids in mung bean sprouts was investigated (Farhangi and Valadon, 1981). The results are shown in Tables 22 and 23.

Broccoli
(*Brassica oleracea* var. *italica* Plenck.)
Cruciferae

Early analytical data published on broccoli carotenoids relate only to its carotene content. At that time, the "crude carotene" of some vegetables was analyzed mainly to characterize the β-carotene isomers in order to make pos-

Table 22. Carotenoids of Fresh, Blanched, Bottled, and Canned Mung Bean Sprouts and Percentage Retention [$\mu g\ (100\ g)^{-1}$]

Carotenoids	*Fresh*	*Blanched*	*% Retention*	*Immediately after Bottling*	*% Retention*	*Immediately after Canning*	*% Retention*
β-Carotene	18	16	89	20	111	24	133
Cryptoxanthin	2	2	100	3	150	0	0
Violaxanthin	101	48	47	43	43	20	20
Lutein	109	172	158	185	170	190	174
Neoxanthin	8	5	62	6	75	4	50
Zeaxanthin	31	50	161	59	190	62	200
Flavoxanthin	15	2	13	3	20	0	0
Total carotenoids	284 ± 7	295 ± 9	104	319 ± 12	112	300 ± 8	106

Source: From Farhangi and Valadon (1981).

Table 23. Carotenoids of Bottled and Canned Mung Bean Sprouts and Percentage Retention after Storage for 6 Months at Different Temperatures

	*Bottled (10°C)**		*Bottled (RT)**		*Bottled (35°C)**	*Canned (RT)**	
Carotenoids	*μg × (100 g)$^{-1}$*	*% Retention*	*μg × (100 g)$^{-1}$*	*% Retention*	*μg × (100 g)$^{-1}$*	*μg × (100 g)$^{-1}$*	*% Retention*
β-Carotene	3	15	1	5	0	11	61
Cryptoxanthin	0	0	0	0	0	1	50
Lutein	28	15	12	6	0	104	95
Neoxanthin	0	0	0	0	0	2	25
Zeaxanthin	11	19	5	8	0	44	142
Total carotenoids	42 ± 2	13	18 ± 1	6	0	162 ± 10	57

Source: From Farhangi and Valadon (1981).
*Storage temperatures.

sible an exact assessment of the vitamin A potency of the food, since it was already known that isomers have different biological activity.

Italian broccoli was found to contain 12 $\mu g\ g^{-1}$ crude carotene, which was a mixture of 80% of all-*trans*-β-carotene, 1% neo-β-carotene U, and 6% neo-β-carotene B, and impurities (Kemmerer et al., 1945).

In studies concerned with the effect of freezing and cooking on the quality of fresh broccoli, carotene was determined in different parts of the raw vegetable (Sweeney et al., 1959; Martin et al., 1960). It was found that the distribution of carotene in broccoli is quite uneven. The content of the heads ranged from 7 to 25 $\mu g\ g^{-1}$, whereas the stems contained 1.2 to 4.0 $\mu g\ g^{-1}$. In stems the carotene was also unevenly distributed, the outer layers containing 2.3 to 4.6 $\mu g\ g^{-1}$ and the peeled stems 0.6 to 1.7 $\mu g\ g^{-1}$. Peeling removed the portion richest in carotene. The carotene content varied with the degree of maturity. Dark green and compact heads contained 7 $\mu g\ g^{-1}$; fully developed heads, 13.0 $\mu g\ g^{-1}$, and another lot contained 17 and 25 $\mu g\ g^{-1}$, respectively. Other lots ranged from 6.2 to 12.8 $\mu g\ g^{-1}$ with a mean value of about 9 $\mu g\ g^{-1}$ carotene.

The *cis–trans* isomers of β-carotene in vegetables known to be rich sources of provitamin A, including broccoli, were investigated with specially devised methods (Sweeney and Marsh, 1970; Sweeney and Marsh, 1971a). (See Appendix Table A19.)

In a review on the carotene content of vegetables, values based mostly on food composition tables, the carotene content of broccoli ranges between 8.3 and 24 $\mu g\ g^{-1}$ (Herrmann, 1975). The uneven repartition of carotene in the plant is exemplified: in leaves, florets, and stems it amounts to 96, 18, and 2.4 $\mu g\ g^{-1}$, respectively.

The pigments of three hybrid cultivars of broccoli were investigated in detail in order to find possible intercultivar differences in the total chlorophyll and carotenoid content—especially in the carotenoid pattern, which would be of taxonomic value (Gross, 1979). The pigments of two of these cultivars were reinvestigated (Yang and Henze, 1988b). The results of both reports are given in Table 24.

Discussion. As expected for a chlorophyll-containing vegetable, the carotenoid pattern was that of chloroplast carotenoids, β-carotene and lutein being the major carotenoids. Six other carotenoids were present, of which isolutein yielded a higher percentage (up to 18%) than the other two major chloroplast carotenoids, violaxanthin and neoxanthin.

The pigments, chlorophylls and carotenoids, of some green vegetables, including broccoli, were separated by HPLC (Khachik et al., 1986). The results are given in Table 25 (on page 160).

Discussion. The method permits the separation of several *cis* isomers that may be naturally present or artifacts. The β-carotene content is less than half of that found in broccoli cultivars grown in Bonn, Germany. Additional data are found in Appendix Tables A1 to A15 (passim).

Table 24. Qualitative and Quantitative Pigment Distribution in Three Broccoli Cultivars (*Brassica oleracea* var. *italica*)

Pigment	*Corvet*		*SG1*		*Gem*
Chlorophyll *a*	992	260.4	2016	425.0	1177
Chlorophyll *b*	366	74.5	775	137.8	536
Total carotenoids	171	45.7	371	68.4	302
Carotenoid Pattern	*Percent of Total Carotenoids*				
α-Carotene	tr	—	0.6	0.3	—
β-Carotene	25.8	23.2	21.5	26.7	22.5
Mutatochrome	tr	—	1.2	0.8	0.5
Cryptoxanthin	—	—	0.8	0.1	—
Mixture	1.5	0.7	—	0.9	1.2
Lutein	40.0	34.9	46.4	36.7	37.4
Isolutein	7.1	17.6	5.1	12.7	4.8
Chrysanthemaxanthin	1.8	—	2.7	2.3	1.2
Luteoxanthin	3.1	0.3	4.4	0.8	3.2
Violaxanthin	9.2	11.3	6.6	4.1	13.4
Neoxanthin	10.8	11.9	9.8	13.0	11.5
Neochrome	0.7	—	0.9	1.6	1.0

Notes: tr = trace. Data in the first column under each cultivar are in μg (g dry wt)$^{-1}$ and are from Gross (1979). Data in the second column are in μg (g fresh wt)$^{-1}$ and are from Yang and Henze (1988b).

Distribution in Other Parts of the Plant

The carotenoids of flowers of three broccoli hybrids were analyzed (Gross, 1979). The flower petals were of lemon yellow color. The results are given in Table 26 (on page 161).

Discussion. In the carotenoid pattern consisting of 10 to 12 carotenoids, two epoxides dominated, namely isolutein and violaxanthin, which together yielded 70% to 80% of the total carotenoids. This conforms to the carotenoid distribution in flowers, since many lemon yellow flowers contain large amounts of xanthophylls, particularly epoxides (Goodwin, 1980). The percentages of the two main carotenoids and of the additional carotenoids differed in each cultivar, so phenotypical differences are characteristically expressed also in the corolla carotenoids, thus proving additional taxonomic information on varietal differences. But, apparently, the petals had the same lemon yellow color.

Table 25. Qualitative and Quantitative Carotenoid Distribution in Green Vegetables [μg (g fresh wt)$^{-1}$]

	Broccoli	*Cabbage*	*Spinach*	*Brussels Sprouts*		*Kale*	
	Raw	*Raw*	*Raw*	*Raw*	*Cooked*	*Raw*	*Cooked*
			Xanthophylls				
all-*trans*-Neoxanthin	2.1	0.7	13.9	3.6	—	30.1	27.6
9′-*cis*-Neoxanthin	2.5	1.0	49.8	4.1	—	92.1	84.2
Violaxanthin	3.1	1.0	41.6	7.0	0.7	67.3	7.7
Neochrome	—	0.5	8.8	3.1	5.3	29.9	32.1
Lutein epoxide	8.2	1.4	12.7	6.6	1.5	15.0	—
cis-Lutein epoxide (13- or 13′-*cis*)	1.1	0.4	11.0	3.0	4.2	—	—
all-*trans*-Lutein	17.7	2.8	144.0	13.4	11.8	342.0	235.0
Neolutein B (9- or 9′-*cis*)	0.7	0.1	1.3	0.5	0.3	8.2	3.5
Neolutein B′ (9′- or 9′-*cis*)	1.1	0.1	5.1	0.7	0.2	21.9	6.9
Neolutein A (13- or 13′-*cis*)	1.1	0.1	9.0	1.3	0.6	23.4	10.5
Total	37.6	8.1	297.0	43.3	24.6	630.0	408.0
			Carotenes				
all-*trans* + 15,15′-*cis*-β-Carotene	4.8	0.8	67.1	5.3	4.5	146.0	126.0
Total carotenoids	42.4	8.9	364.0	48.6	29.1	776.0	534.0

Source: Reproduced with permission, from Khachik et al. (1986); copyright © 1986 by American Chemical Society.

Influence of Different Factors on Carotenoid Content

The variation of the total carotenoid content with planting date in the broccoli cultivars SG1 and Corvet, investigated by Yang and Henze (1988a), is shown in Table 27.

Discussion. As may be expected, the carotenoid content variations paralleled the variations of chlorophyll content given in Table 6. The cultivar SG1, with higher pigment content, appeared to be more affected by seasonal variations, whereas in Corvet the differences in carotenoid content were less noticeable.

Table 26. Qualitative and Quantitative Carotenoid Distribution in Flower Petals of Three Broccoli Cultivars

	Percent of Total Carotenoids		
Carotenoid Pattern	*Covert*	*SG1*	*Gem*
β-Carotene	2.6	1.0	1.3
Cryptoxanthin	1.2	0.6	—
Cryptoxanthin 5,6-epoxide	3.2	2.1	3.0
Lutein	2.5	2.1	2.5
Chrysanthemaxanthin *a*	2.0	4.1	1.9
Isolutein	43.6	55.9	46.0
Violaxanthin	33.7	24.6	36.8
Chrysanthemaxanthin *b*	—	2.0	—
Luteoxanthin	—	1.6	—
Auroxanthin-like mixture	5.8	1.8	2.8
Neoxanthin *a*	4.1	3.7	4.9
Neoxanthin *b*	1.3	0.5	0.8
Total carotenoids [μg (g fresh wt)$^{-1}$]	154.0	87.0	143.0

Source: From Gross (1979).

Table 27. Effect of Cultivar and Planting Date on Carotenoid Content in Two Broccoli Cultivars [μg (g fresh wt)$^{-1}$]

Cultivar	*Planting*	*Harvest*	*Total Carotenoids*
SG1	7.06	6.08	90
	15.07	6.09	45
	5.08	1.10	38
Corvet	7.06	2.09	36
	15.07	15.09	25
	5.08	6.10	30

Source: From Yang and Henze (1988a).

Changes During Storage, Cooking, and Freezing

Storage

In frozen broccoli stored for 61 weeks at −18°C no decrease in carotene content was found (Martin et al., 1960). The effects of various CA treatments on the total carotenoid content in broccoli was investigated by Yang and Henze (1988a). The total carotenoid content remained almost constant during 8 weeks of storage. An increase was observed in samples stored at 10% CO_2 and 3% O_2. The gradual yellowing of the broccoli was due to decrease of chlorophylls, which produced a partial unmasking of the carotenoids. The variation of the carotenoid pattern stored for 6 weeks under different controlled atmospheres was also studied (Yang and Henze, 1988b). The results are given in Table 28.

Table 28. Chlorophyll and Carotenoid Changes in Broccoli SG1 Stored in Air and CA for 6 Weeks

	Total Content [μg (g fresh wt)$^{-1}$]				
	Before Storage	*0.0% CO_2 21.0% O_2*	*0.0% CO_2 3.0% O_2*	*5.0% CO_2 3.0% O_2*	*10.0% CO_2 3.0% O_2*
Chlorophyll *a*	424.9	224.3	226.8	239.0	284.4
Chlorophyll *b*	137.8	82.1	78.6	84.3	101.2
Carotenoids	68.4	67.9	61.4	62.5	77.7
Carotenoid Pattern	*Percent of Total Carotenoids*				
α-Carotene	0.3	0.7	0.5	0.1	0.2
β-Carotene	26.7	22.9	22.0	21.1	23.5
Mutatochrome	0.8	0.1	0.2	0.1	0.1
Lutein	36.7	45.8	42.9	48.4	48.0
Isolutein	12.7	11.1	8.4	6.6	5.9
Chrysanthemaxanthin	2.3	—	2.3	0.4	0.2
Luteoxanthin	0.8	1.4	2.3	2.5	1.3
Violaxanthin	4.1	5.7	8.1	7.2	7.2
Neochrome	1.6	—	0.2	0.5	0.6
Neoxanthin	13.0	11.8	12.8	12.9	13.1

Source: From Yang and Henze (1988b).

Discussion. The total carotenoid content did not vary significantly. The β-carotene retention in both investigated broccoli hybrids, stored in four CA, ranged from 79% to 99%.

Cooking

The effect of various cooking methods on carotene retention in vegetables was followed in many reports. In the earliest, two cooking methods were applied: the "old-fashioned" procedure representative of methods in which no effort is made to retain the essential vitamins and minerals; a "new, improved" method designed to retain these nutrients to a maximal degree (Oser

Table 29. Effect of Cooking Methods on Carotenoid Distribution in Broccoli

Cooking Method	*Carotenoids [mg (100 g raw food)$^{-1}$]*				
	Neoxanthin	*Violaxanthin*	*Lutein 5,6-epoxide*	*Lutein* cis + trans	*β-Carotene*
Raw	0.65	1.20	0.65	2.60	2.10
Microwave	0.61	0.49	—	2.90	2.10
Steaming	0.58	0.45	—	2.90	2.40

Source: From Khachik et al. (1989).

et al., 1943). In broccoli, carotene was completely retained with both cooking methods.

In broccoli cooked by 13 different procedures, almost 100% of carotene was retained (Sweeney et al., 1959). The stereoisomers of β-carotene formed in broccoli during cooking increased with cooking time, especially neo-β-carotene U (Sweeney and Marsh, 1971). An additional *cis* isomer in small amounts was detected in stems (see Table A19).

The effect of microwaving and steaming on the carotenoids of broccoli were also studied (Khachik et al., 1989). The results are given in Table 29.

Freezing

Frozen broccoli stored for 61 weeks at −18°C had a carotene content ranging from 10.2 to 13.6 μg g^{-1} and that did not decrease during storage (Martin et al., 1960). Variations in the relative proportions of β-carotene isomers in frozen broccoli are shown in Table A19.

Cabbage
(*Brassica oleracea* var. *capitata* L.)
Cruciferae

Chlorophylls and carotenoids are unevenly distributed in the cabbage head leaves. The outer leaves are much richer in pigments than are the inner ones, which do not receive light. The outer leaves of cabbage contain 21 times more carotene than bleached inner leaves (Sheets et al., 1941). The total carotene content of the outer leaves was 146 μg (g dry wt)$^{-1}$, but only 5 μg (g dry wt)$^{-1}$ in the inner leaves (Bondi and Meyer, 1946).

The carotenoids of cabbage and their seasonal fluctuations during 12 months have been systematically studied in order to relate them to the development of the plant and of the insects feeding on them (Feltwell and

Valadon, 1972, 1974). Entire seedlings were used for carotenoid analysis for the first month; thereafter, representative samples of outer leaves were used. The results are given in Table 30.

Discussion. The carotenoid pattern was the chloroplast pattern found in leaves. The major carotenoids, β-carotene and lutein, were accompanied by small amounts of their epoxides. After the first 2 months the total carotenoids decreased gradually. The individual changes of carotenoids were most noticeable during the first 5 months; the greatest fluctuations were those of β-carotene and lutein, which started at high values initially and decreased in the first 8 weeks. According to the results, it appears that synthesis and degradation of carotenoids of the cabbage take place at different rates at different times of the year. The total carotenoids of cabbage may vary by a factor of 6 between plants of identical age and appearance. This variation may be due to the number of leaves present in the sample that had not received light.

The carotenoids of cabbage were analyzed by HPLC (Takagi, 1985). Cabbage grown in Japan (*B. oleracea* var. *capitata*) has a total carotenoid content of 173 μg (g fresh wt)$^{-1}$ comprising β-carotene (17.9%), lutein (36.3%), antheraxanthin (15.3%), violaxanthin (14.2%), and neoxanthin (13.5%).

The mean carotene content of cabbage grown in the United States was 0.8 μg (g fresh wt)$^{-1}$, varying significantly in different geographical areas (Klein and Perry, 1982; Khachik et al., 1986) (see Table 25). Additional data are found in Tables A1 to A15 passim.

Changes During Cooking

The retention of carotene in cooked cabbage was 82% (Klein and Perry, 1982).

Carrots
(*Daucus carota* L.)
Umbelliferae

Carrots are the most important vegetable source of provitamin A, contributing 14% of the total vitamin A consumption (Senti and Rizek, 1975). Carotenoids derive their name from the main representative of the group, β-carotene, which was isolated from carrots (*Daucus carota*) by Wackenroder in 1831.

Carotenoids of carrots have been exhaustively investigated. In the early 1930s it was found that the predominant "carotene" in carrots is a mixture of α- and β-carotene, which are the main pigments of the carrot (Karrer and Morf, 1931; Kuhn and Lederer, 1931; Strain, 1939). Additionally, minor

Table 30. Carotenoid Changes in *Brassica oleracea* var. *capitata* plants grown for 12 months [μg (g dry wt)$^{-1}$]

Carotenoid Pattern	*1971* Nov.	*1971* Dec.	*1972* Jan.	Feb.	Mar.	Apr.	May	June	July	Aug.	Sept.	O
α-Carotene	4.6	76.1	—	—	—	0.8	—	—	—	9.2	—	—
β-Carotene	865.2	320.5	350.6	681.7	345.3	409.8	335.7	245.1	77.9	162.1	125.9	19
β-Zeacarotene	5.4	3.8	—	—	—	2.8	—	—	—	—	—	—
β-Carotene 5,6-epoxide	47.6	64.9	84.7	55.0	30.6	14.4	6.2	31.2	4.0	16.9	8.9	
Diepoxy-β-carotene	—	—	—	—	—	—	—	—	—	5.2	—	—
Cryptoxanthin	6.2	7.9	—	—	—	—	—	—	—	—	—	—
Mutatochrome	7.4	2.5	9.9	—	—	—	—	—	—	—	—	—
Lutein epoxide	—	—	—	—	14.9	5.8	3.5	—	—	—	—	—
Lutein	1174.0	425.5	961.6	566.0	381.9	420.5	359.6	155.1	93.5	24.0	88.1	7
Neoxanthin	190.1	297.7	242.3	193.0	137.5	109.0	89.1	60.5	40.9	38.6	33.1	4
Violaxanthin	175.0	450.0	28.0	339.7	481.6	139.7	38.2	30.5	34.2	110.4	126.2	14
Chrysanthemaxanthin	78.6	271.5	31.6	19.4	—	—	—	68.0	—	50.8	69.4	—
Auroxanthin	34.2	568.8	89.1	95.6	—	—	—	47.5	—	16.9	—	1
Flavochrome	98.3	—	—	—	—	—	—	—	—	—	—	—
Unknowns	6.1	—	—	—	—	—	—	—	—	—	—	—
	3.3	—	—	—	—	—	—	—	—	—	—	—
Total carotenoids	2696.0	2489.1	1798.0	1950.4	1391.7	1102.7	832.2	637.9	250.5	434.1	450.8	48

Source: From Feltwell and Valadon (1974).

carotenes—γ-carotene, lycopene, and ζ-carotene, the last being widespread—were detected (Strain, 1939; Harper and Zscheile, 1945). Subsequently, phytofluene and phytoene were also isolated (Zechmeister and Sandoval, 1945; Rabourn et al., 1954). Complete analyses of carrot carotenoids by standard chromatographic methods or HPLC were carried out by several research groups (Baloch et al., 1977b; Ogunlesi and Lee, 1979; Edwards and Lee, 1986; Simon and Wolff, 1987). The results are compiled in Table 31.

Table 31. Carotenoids of Raw Carrots

Carotenoids	*Percent of Total Carotenoid*			
Phytoene	nd	7.9	nd	nd
Phytofluene	tr	4.2	2.3	nd
α-Carotene	17.8	25.7	27.4	38
β-Carotene	68.4	56.9	61.3	53
ζ-Carotene	5.6	3.1	6.5	4
β-Zeacarotene	—	0.8	1.3	2
γ-Carotene	2.3	0.9	1.2	2
cis-Neurosporene	0.7	—	—	—
Neurosporene	1.7	0.4	—	—
Lycopene	0.5	0.1	tr	1
Lutein	2.9	nd	—	—
Unidentified carotenoid	0.2	—	—	—

Sources: Column 1 from Baloch et al. (1977a); column 2 from Ogunlesi and Lee (1979); column 3 from Edwards and Lee (1986); column 4 from Simon and Wolff (1987).
Notes: nd = not determined; tr = trace.

Discussion. In these analyses the complexity of the hydrocarbon fraction is revealed as the colorless carotenes phytoene and phytofluene, and β-zeacarotene and neurosporene were separated and quantified. All the carotenes separated are the precursors of α- and β-carotene in the biosynthetic pathway. It was also shown that lutein is a constituent of the xanthophylls. By r-p HPLC, six carotenes (α-, β-, γ-, and ζ-carotenes, β-zeacarotene, and lycopene) can be routinely separated and quantified in typical and dark orange carrots (Simon and Wolff, 1987). The carotenoids of a deep orange selection are shown in the last column of Table 31. Generally, in these carrots β-carotene accounted for 45% to 80%, α-carotene for 15% to 40%, and ζ-carotene for 2% to 10%. The remaining 3% to 6% was distributed between β-zeacarotene, γ-carotene, and lycopene in a ratio of 1% to 3%:1% to 2%:1%.

Generally, β-carotene constitutes the major pigment in carrots, accounting for about 50% of total carotenoids. The second major pigment is α-carotene. The ratio β-carotene/α-carotene is very variable, depending on the phenotype and other factors, and ranges from 1 to 2 in orange and dark orange phenotypes (Laferriere and Gabelman, 1968; Simon and Wolff, 1987). The xanthophyll fraction usually amounts to 5% to 10% of the total carotenoids in orange carrots, but is much higher (75–94%) in yellow carrots (Dark and Booth, 1946; Laferierre and Gabelman, 1968).

Carotenoids are not uniformly distributed in the carrot root. The phloem has a 30% higher concentration than the xylem, which is evidenced by the different colors of the two parts. The concentration of the pigments also decreases longitudinally from the top to the tip of the root (Werner, 1941; Harper and Zscheile, 1945; Booth, 1951a).

Carrot carotenoids have been extensively investigated in various countries. (Carrots are probably the most analyzed vegetable.) In a study of 16 commercial carrot varieties and 18 strains grown in the United States, the total carotene average was 54 $\mu g\ g^{-1}$ over a range of about twice the minimum: Nantes 52 and Chantenay 108 μg (g fresh wt)$^{-1}$ (Harper and Zscheile, 1945).

Over 70 carrot varieties grown in England had an average 'carotene' content of 124 μg (g fresh wt)$^{-1}$ (Dark and Booth, 1946). The varieties of German origin had a very high total carotenoid content [210 to 350 μg (g fresh wt)$^{-1}$]. The values have been further improved by selection, and strains have been obtained with a mean total carotenoid content of 430 μg (g fresh wt)$^{-1}$.

The α- and β-carotene content of common carrot cultivars grown in Romania, investigated by Bodea et al. (1954), are shown in Table 32.

Table 32. Carotene Content of Some Carrot Cultivars Grown in Romania [μg (g fresh wt)$^{-1}$]

Cultivar	*α-Carotene*	*β-Carotene*
Chantenay	12.0	50.0
Bulgarian	13.5	45.2
Nantes	5.0	20.0
Amsterdam	6.4	18.8
Flacke	3.0	9.8

Source: From Bodea et al. (1954).

Total carotenoids in carrots grown in Switzerland ranged between 80 and 112 $\mu g\ g^{-1}$ (Gnaegi et al., 1977).

The carotenoid contents of two carrot varieties grown in Japan, the red-rooted 'Kintoki' and the orange-rooted 'Kokubu', were compared (Yamaguchi and Sugiyama, 1960). Although Kintoki is widely cultivated in Japan, its carotenoids were never analyzed. The results are shown in Table 33.

Table 33. Carotenoid Distribution in Kokubu and Kintoki Cultivars of Japanese Carrots [μg (g fresh wt)$^{-1}$]

Carotenes	*Kokubu*	*Kintoki*
α-Carotene	23.7	0.5
β-Carotene	153.0	32.8
ζ-Carotene	3.7	7.7
Lycopene	2.8	124.3

Source: Reproduced with permission, from Yamaguchi and Sugiyama (1960); copyright © 1960 by American Chemical Society.

The color of the red roots is imparted by lycopene, the predominant carotene. In all red-rooted Japanese carrot varieties, lycopene is the major pigment, as confirmed in other reports (Sugano and Hayashi, 1967). β-Carotene is present in smaller amounts; ζ-carotene, γ-carotene, and phytofluene were also detected, but α-carotene was absent (Umiel and Gabelman, 1971; 1972).

A comparison between several European and Asiatic carrots showed that European carrots were richer in carotene (Gill and Kataria, 1974). This fact was attributed to the higher core size and high moisture percentage of the Asiatic carrots.

Recently, the data on the provitamin A content of food, including carrots, have been updated by HPLC methods. The methods focus on assessing the predominant provitamin A carotenoids (80–90%), the minor pigments being considered negligible. Some recent results are shown in Table 34.

Table 34. α- and β-Carotene Content in Carrots Determined by HPLC Method [μg (g fresh wt)$^{-1}$]

α-Carotene	*β-Carotene*	*Reference*
24–55	34–110	Sweeney and Marsh (1971a)
20–59	46–125	Bushway and Wilson (1982)
26–60	48–60	Dietz et al. (1988)
38	76	Bureau and Bushway (1986)
38–40	80–112	Bushway (1986)
11	60	Pepping et al. (1988)
35	105	Pepping et al. (1988)

An average carrot in the United States contains 65 to 90 μg (g fresh wt)$^{-1}$ total carotenoids according to Adams (1975) and 70 to 120 μg (g fresh wt)$^{-1}$ according to Simon et al. (1985). In early reports on carrot carotenoids, the *cis* isomers of α- and β-carotene were investigated, both for the exact assessment of the vitamin A value and to follow the alterations induced by processing. It was found that the carotenes of fresh carrots from the market were a mixture of all-*trans*-α-carotene (29–32.5%), neo-βcarotene U (0.7–1%), all-*trans*-β-carotene (59.6–62%), and neo-β-carotene B (1.7–3%) (Kemmerer and Fraps, 1945; Kemmerer et al., 1945). In Canadian carrots the percentages were all-*trans*-β-carotene (84.2%), neo-βcarotene B (14.0%), and neo-β-carotene U (1.8%) (Panalaks and Murray, 1970).

By a special method of separation (Sweeney and Marsh, 1970), the *cis* and *trans* isomers of α- and β-carotene were separated (Sweeney and Marsh, 1971a). The results are given in Table A20.

A similar mixture of α- and β-carotene and their *cis* isomers was again reported in Chantenay carrots (Ogunlesi and Lee, 1979) (see Table 43). Raw carrots commercialized in São Paulo contained 20.7 to 21.6 μg (g fresh wt)$^{-1}$ α-carotene and 34 to 46 μg (g fresh wt)$^{-1}$ β-carotene (Almeida and Penteado, 1987). These main carotenes were accompanied by their *cis* isomers neo-α-carotene B and neo-β-carotene B at the level of 3.2% and 3.5% of the respective all-*trans* isomers; the neo-*cis* isomers U were absent. By r-p HPLC, the total β-carotenes of fresh carrots were found to contain 13-*cis*-β-carotene (neo B) and 9-*cis*-β-carotene (neo U) at the levels of 1.2% and 0.4% of the total carotenes (Quackenbush, 1987).

A special HPLC method that allows the separation of 11 different *cis* isomers of β-carotene was applied in the analyses of plant extracts (Vecchi et al., 1981). The results for a carrot extract are shown in Table 35.

Table 35. Qualitative and Quantitative Isomer Composition of the Carotene Fraction of a Carrot Extract

	Percent of Total Carotenes
β-Carotene	
all-*trans*	58.4
9-*cis*	2.0
13-*cis*	5.7
15-*cis*	0.8
9,9′-di-*cis*	1.4
9,13-di-*cis*	1.2
9,13′-di-*cis*	0.3
13,13′-di-*cis*	0.5
9,13,13′-tri-*cis*	0.2
Other *cis*	0.8
Unknown comp.	7.0
α-Carotene	21.7

Source: From Vecchi et al. (1981).

As in previous reports, the isomers 9- and 13-*cis* are present at higher levels, the 13-*cis* isomer being found at higher amounts.

Distribution in Other Parts of the Plant

Carotenoids of carrot leaves were also investigated. In earlier reports, only the carotenes were analyzed. Total carotene content was 60 to 170 μg (g fresh wt)$^{-1}$, the concentration remaining nearly constant through the season. The concentration of leaf total carotenes was independent of the variety (Booth, 1957). α-Carotene constituted 10% of the total (MacKinney, 1935). Later, a higher value of 13% was determined, namely in the leaves of carrots having appreciable total carotenes in the roots; it was already detectable in their cotyledons (Booth, 1956, 1957).

The presence of α-carotene in leaves along with β-carotene probably indicates that α-carotene is also synthesized in other parts of the plant, such as fruit, seeds, or roots. So far, this hypothesis is true for carrots, maize, banana (Gross et al., 1976), and avocado (Gross et al., 1973).

The carotenoids of carrot leaves grown in Japan were analyzed by HPLC (Takagi, 1985). The results are shown in Table A5. The carotenes were not separated with the method used.

Influence of Different Factors on Carotenoid Content

In addition to intervarietal variations that are genetically determined, environmental factors and maturity influence carotenoid distribution in carrots.

Environmental Factors

Carrots grown at low temperature have subnormal low total carotenoid content because the roots remain larger than normal in an unripe stage (Banga et al., 1955).

Color differences of various carrot varieties between planting dates were attributed mainly to temperatures in the 3–6 weeks preceding harvest (Bradley and Smittle, 1965). Temperatures from 10°C to 16°C were considered optimal for producing the best color and the highest β-carotene content (Bradley et al., 1967). These findings were true for a number of varieties and lines, whereas others had fair or poor color in both seasons (Bradley and Rhodes, 1969). Yet the influence of environmental factors was illustrated by the variation of the carotenoid content in the same carrot variety grown in different locations. For example, Chantenay carrots grown in different regions in the United States were reported to contain 67 to 99, 80, or 306 μg (g fresh wt)$^{-1}$ (Bills and McDonald, 1932; Barnes, 1936; Hansen, 1945). Seeds of several varieties of carrots sown in different localities in England yielded carrots with different total carotene contents (Booth and Dark, 1949).

Variations in total and individual carotene quantities in seven carrot lines due to genotype and environment investigated by Simon and Wolff (1982) are given in Table 36.

Table 36. Carotenes in Seven Carrot Lines Grown in Several Years and Locations

	Location and Year				
Line	*CA 1984*	*CA 1985*	*WI 1985*	*FL 1986*	*CA 1986*
HCM	533 (53:38:4)*	548 (65:25:5)	452 (44:40:10)	251 (68:20:7)	517 (60:28:7)
B6274	101 (59:34:3)	132 (74:21:2)	90 (57:35:4)	63 (67:27:2)	104 (73:21:2)
B2566	98 (69:25:2)	144 (74:16:4)	111 (64:29:3)		108 (71:23:2)
F524		113 (76:18:2)	142 (53:40:3)		176 (61:32:2)
B3080	104 (65:27:4)	156 (79:13:3)	137 (60:33:3)		145 (72:20:3)
B6139		173 (74:21:2)	137 (57:36:4)		186 (61:30:3)
B9692	209 (66:29:2)	213 (78:15:2)			

Source: From Simon and Wolff (1987).
*Sum of carotenes from HPLC μg (g fresh wt)$^{-1}$ (% β-carotene: % α-carotene: % ζ-carotene), average of five replicates.

Discussion. Some lines tended to vary greatly between location and year, but no general trends were observed. The low carotene content is probably attributable to cold weather, which yielded smaller and less physiologically mature roots.

Maturity

The carotenoid content of carrots increases with maturity up to approximately 90 to 100 days of growth in the field, attaining a maximum and then decreasing or remaining constant for some time, and decreasing afterwards. Results of an early investigation (Pepkowitz et al., 1944) are shown in Table 37.

Discussion. The table illustrates both intervarietal variations and variation with maturity. Color decreases from the deeply pigmented Danvers Half-Long to the light yellow Hutchinson. Maximal color was attained 90 days from seeding, and then it decreased.

In two varieties, Red Cored Chantenay and Nantes, the variation of the carotene content during growing was followed separately in phloem and in xylem (core) (Werner, 1941). It reached a maximum in both root parts after about 100 days and remained relatively constant until harvest. Similar results, a highest total carotene level after 105 days and then a relatively constant level up to a 135-day period, were reported for other orange-fleshed carrot varieties grown in Wyoming (Brown, 1947).

Table 37. Total Carotene Content of Ten Carrot Varieties at Four Different Periods of Growth [μg (g fresh wt)$^{-1}$]

	Days from Seeding to Harvest			
Variety	*76*	*83*	*90*	*97*
Danvers Half-Long	44.5	50.8	71.2	57.3
Tendersweet	33.8	42.9	63.5	51.2
Chantenay Oregon Long	37.6	37.5	63.2	49.7
Nantes	43.2	39.7	62.6	51.4
Coreless Chantenay	38.4	41.4	60.6	57.7
Imperator	39.0	33.4	59.7	51.3
Chantenay	38.8	39.6	58.9	48.9
Touchon	40.1	47.6	56.1	49.1
Improved Nantes Coreless	37.0	43.8	56.0	40.6
Hutchinson	26.6	28.3	37.0	32.8

Source: From Pepkowitz et al. (1944), by permission of the copyright owner, ASPP.

In accord with these observed variations during maturity, some recommendations have to be followed (Booth and Dark, 1949). To develop their full total carotenoid concentration by autumn, carrot varieties with normal total carotenoid values must be sown before the end of May. High-carotene varieties must be sown at least a month earlier. However, the increase of total carotenoid content in growing carrots until a maximum is reached may vary with season and locality (Banga and de Bruyn, 1954).

Two carrot cultivars grown in Pakistan developed the highest carotene concentrations between 130 to 140 days after sowing (Khan et al., 1975).

In a more recent report it was found that during the growing season of carrots, the level of most carotenes, present in the root, increased up to harvest time (140 days after planting) (Lee, 1986). However, the maximum α-carotene and β-carotene levels were attained after 110 days, in two of the three investigated cultivars. Since α- and β-carotene are the major provitamin A carotenes in carrots, their changes during the growing season have a significant effect on their vitamin A value. The variations of the vitamin A value of carrots during growth expressed in RE, including all the provitamins A, α- and β-carotene, γ-carotene, and β-zeacarotene present in carrots, are shown in Figure 33.

The average vitamin A value for three cultivars reached a maximum of 1070 RE (100 g fresh wt)$^{-1}$ at 116 days and decreased to 990 RE (100 g fresh wt)$^{-1}$ at harvest time. Consequently, the variation of the total carotenoids in carrots may be used as an index of maturity to determine optimal harvest time, as was proposed for fruits (Gross, 1987). For carrots, the optimum harvest date seems to be reached when the carotene and sucrose content is high-

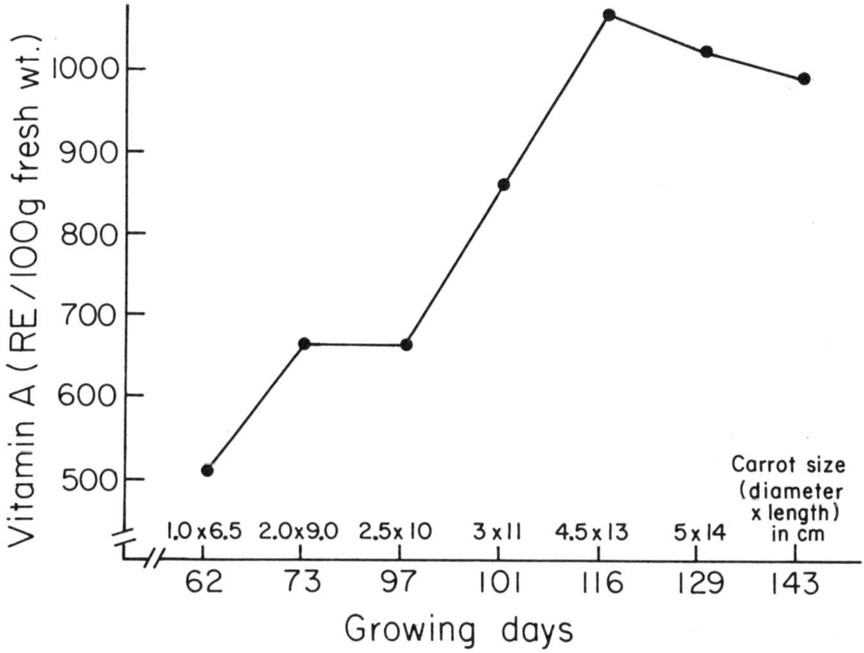

Figure 33. Changes in vitamin A value of carrots during growth. (*From Lee, 1986.*)

est and the monosaccharide content and the respiration are decreasing (Fritz and Habben, 1975).

Selection

Extensive studies over about 60 years revealed the wide varietal differences in color intensity of carrots produced by a highly variable carotenoid content that includes highly variable carotene proportions within and between different cultivars. The primary factors determining the varietal differences are genetic. Extensive studies on the inheritance of root color in carrots were carried out by Gabelman and his group (Laferriere and Gabelman, 1968; Umiel and Gabelman, 1971, 1972; Buishhand and Gabelman, 1979). These genetic studies helped to bring about improved carrot strains with a higher carotenoid content. Selection for dark orange roots was initiated in 1970 (Simon et al., 1985). Whereas typical U.S. carrot cultivars contain 60 to 120 μg (g fresh wt)$^{-1}$, the selections have a higher content (Simon, 1987). Some examples are shown in Table 38.

Table 38. Carotene Content of Several Carrot Cultivars and Germ Plasma Releases

Name	*Total Carotene Content [μg (g fresh wt)$^{-1}$]*
Danvers 126	71
Chantenay	41
Nantes	59
Imperator 58	78
Hicolor 9	65
A Plus	168
Beta III	270
HCM	475

Source: From Simon (1987).

Eight generations of mass selection have resulted in a gradual increase in total carotenoids, with an average of 452 μg (g fresh wt)$^{-1}$ and individual roots ranging from 340 to 640 μg (g fresh wt)$^{-1}$. These carotenoid concentrations represent the maximal levels reported in higher plants. The carotene content of the deep orange carrot HCM is detailed in the last column of Table 31. Carrot HCM was developed for use in breeding, genetic, and biochemical studies (Simon et al., 1989). Carrot Beta III was released recently primarily for its value as a garden cultivar and as an improved source of provitamin A carotenes (Peterson et al., 1988).

Another hybrid, A plus, developed in California and analyzed by HPLC, had a total carotene content of 305 μg (g fresh wt)$^{-1}$, a mixture of *trans*- and *cis*-β-carotene (59.6%), α-carotene (34.9%), and *trans*- and *cis*-ζ-carotene (5.5%) (Khachik and Beecher, 1987).

Changes During Storage, Cooking, and Processing

Storage

Carrots can be stored easily for use as a fresh vegetable during winter and spring. In stored carrots the carotene content is maintained or usually increased until sprouts start (Werner, 1941; Brown, 1947; Rygg, 1949; Booth, 1951b). In some cases the increase course was not continuous, but showed an initial decrease at the beginning of storage (Brown, 1949; Strmiska and Holčiková, 1970; Khan et al., 1975; Gnaegi et al., 1977).

In all these reports the total carotene content in the stored carrots was determined. In a more recent report the levels of individual provitamin A carotenoids were measured during storage to assess accurate values of vitamin A (Lee, 1986). The carrots were stored at 2°C and 90% relative humidity. The results are shown in Table 39.

Table 39. Changes in Provitamin A Carotenes of Carrots During Storage [μg (g fresh wt)$^{-1}$]

	Storage Days						
Carotene	*5*	*27*	*50*	*70*	*100*	*125*	*155*
α-Carotene	21.1	21.4	22.1	22.5	25.5	21.5	18.5
β-Carotene	54.9	58.9	61.3	62.3	62.8	63.1	58.2
β-Zeacarotene	0.7	0.8	1.3	1.1	1.2	1.6	1.6
γ-Carotene	0.8	0.9	1.9	1.9	1.3	1.5	1.3

Source: From Lee (1986).

Discussion. The content of α- and β-carotene increased slowly up to 100 to 125 days and then decreased. The other two provitamins showed an increase that was not as continuous. The colorless precursors phytoene and phytofluene were detected only after 70 days, accumulating as in ripe fruits when the turnover slows down (Gross, 1987). The vitamin A value rose from 11 RE (g fresh wt)$^{-1}$ to 12.5 after 100 days of storage, decreasing to the initial value after 160 days.

Cooking

During cooking carotene is lost and *trans-cis* isomerization increases, both of which produce a color change from red-orange to yellow. Retention was 91% in carrots cooked the old-fashioned way and 96% for the new, improved method (Oser et al., 1943). In cooked carrots the *cis* isomers neo-β-carotene U and neo-β-carotene B increased from their low level in raw carrots to 2% and 11%, respectively (Kemmerer et al., 1945).

The cooking procedure also affects carotene loss. Loss was lower in steamed carrots than in fried (Simeonova, 1965). In fresh and dried carrots the length of cooking time affected the carotene loss, which was 17% after half an hour and 25% after keeping the cooked carrots at 90°C for 3 h (Schillinger and Zimmermann, 1965).

The color changes in differently cooked carrots, related to their carotene content, were followed by Borchgrevink and Charley (1966), and the results are shown in Table 40.

Table 40. Carotenes in Differently Cooked Carrots [μg (g raw carrot)$^{-1}$]

Treatment	*Crude Carotene*	*α-Carotene*	*All*-trans-*β-carotene*	*Neo-β-carotene B*
Saucepan, 19 min	118.5	31.6	53.8	10.8
Pressure saucepan				
50 s	107.8	28.6	48.1	9.4
2 min	102.0	26.3	40.3	13.6

Source: From Borchgrevink and Charley (1966).

Discussion. Carrots cooked under normal pressure had the best color (measured by the Hunter color difference meter) and the highest carotene content. Overcooked carrots had a dull yellow color and the lowest carotene content.

During cooking considerable α-carotene was lost. The relative percentages of the β-carotene isomers neo-β-carotene B, neo-β-carotene U, and all-*trans*-β-carotene were 14%, 1.8%, and 84.2% in fresh carrots, 33.3%, 53%, and 61% in canned carrots, 13.3%, 6.7%, and 79.9% in dehydrated carrots, and 9%, 3.9%, and 87% in cooked carrots (Panalaks and Murray, 1970).

Formation of stereoisomers of α- and β-carotene as affected by cooking time was followed in both fresh and frozen carrots (Table A20). The percentage of stereoisomers increased with cooking time: four and five times neo-α-carotene B increased in fresh and frozen carrots after 25 and 15 min, respectively; neo-β-carotene B increased three times in both fresh and frozen carrots, whereas the variation of neo-β-carotene U was negligible. The effect of cooking on individual carotenoids was investigated (Baloch et al., 1977a), and the results are shown in Table 41.

Discussion. During cooking a loss of 7% of total carotenoid occurred. β-Carotene appeared to be the most labile carotenoid.

In four carrot hybrids the effects of cooking varied with the cultivar (Simon and Lindsay, 1983). Carotenoid levels diminished 9% to 12% in fresh-cooked carrots and 6% to 19% in frozen cooked carrots. In another carrot hybrid, the fluctuation of individual carotenes *cis* and *trans* (ζ-carotene, α-carotene, and β-carotene) after cooking was minimal (Khachik and Beecher, 1987).

In carrots grown in São Paulo, the carotene loss during cooking was followed (Almeida and Penteado, 1987), and the results are given in Table 42.

Processing

Carrots can be canned, frozen, dehydrated, and pickled.

Table 41. Effect of Autoclaving for 30 min at 121°C on Carotenoid Pigments of Carrots

Carotenoid Pigments	*Carotenoid Content (mg/g, wet weight basis)* Raw	Autoclaved	*Loss of Pigment in Relation to Total Pigment Content (%)*	*Loss of Pigment in Relation to Total Pigment Loss (%)*
α-Carotene	22.4	17.7	3.7	35.6
β-Carotene	86.0	77.1	7.0	66.6
ζ-Carotene	7.0	7.3	−0.2	−2.3
γ-Carotene	2.9	2.8	0.1	0.8
cis-Neurosporene	0.9	0.9	0.0	0.0
Neurosporene	2.1	2.0	0.1	0.0
Lycopene	0.6	0.6	0.0	−1.5
Lutein	3.6	3.8	−0.2	
Unidentified carotenoid	0.3	0.3	0.0	0.0
Total	125.8	112.6	10.5	—

Source: From Baloch et al. (1977a).

Canning. Due to leaching of soluble solids during processing and storing, an apparent increase of the carotene content was observed (Weckel et al., 1962). Generally, the carotenoid concentration was 160% higher in canned carrots than in frozen carrots. Due to canning, isomerization of α- and β-*trans* isomers to *cis* isomers occurred. The relative percentages of carotenes over 4 years ranged as follows: neo-α-carotene B, 2.1% to 6.4%; all-*trans*-α-carotene, 24.2% to 30.9%; neo-α-carotene U, 1.6% to 2.7%; neo-β-carotene B, 6.0% to 13.1%; all-*trans*-β-carotene, 41.4% to 46.4%; and neo-β-carotene U, 3.8% to 4.7%, the *cis* isomer B predominating. The estimated loss in total provitamin A activity of processed canned carrots due to isomer-

Table 42. Carotene Content and Vitamin A Value of Raw and Cooked Carrots from São Paulo

	Raw		*Cooked*	
	μg (100 g)$^{-1}$	*Vitamin A Value [RE (100 g)$^{-1}$]*	*μg (100 g)$^{-1}$*	*Vitamin A Value [RE (100 g)$^{-1}$]*
α-Carotene	2215	179.6	1894	157.8
β-Carotene	3401	566.8	2911	485.2
Total	5556	746.4	4805	643.0

Source: From Almeida and Penteado (1987).

ization of α- and β-carotene was 7.3% to 11.7%. Generally, processing of green vegetables lowers vitamin A values by 15% to 20%, and lowers those of yellow vegetables by 30% to 35% (Table A21).

In canned carrots grown in Pakistan, carotene losses were 19% and 17% for the Long Red and Black varieties (Khan et al., 1975). During storage at room temperature for 100 days, 96% of the carotene content was retained in both varieties, and color and taste were unchanged.

Canning increased the carotenoid content of carrots (Ogunlesi and Lee, 1979; Edwards and Lee, 1986) due to leaching loss of soluble solids (35% of the total solids). However, when leaching of soluble solids into the canning liquid medium was taken into account, a decrease was observed.

The composition of the major provitamin A carotenes in carrots, α- and β-carotene isomers, is affected by thermal processing. The results are given in Table 43.

Table 43. Stereoisomers of α- and β-Carotene Isolated from Fresh and Canned Chantenay Carrots

Isomer	*Fresh* $[\mu g\ (g\ dry\ wt)^{-1}]$	*Canned* $[\mu g\ (g\ dry\ wt)^{-1}]$
Neo-α-carotene B	1.8	65.0
all-*trans*-α-Carotene	145.8	108.0
Neo-α-carotene U	0.7	3.8
Neo-β-carotene B	9.2	98.2
all-*trans*-β-Carotene	291.8	191.0
Neo-β-carotene U	2.5	18.4

Source: From Ogunlesi and Lee (1979).

Discussion. The *cis* isomer concentration substantially increased, whereas the all-*trans* isomers decreased. Consequently, a 15% decrease in the vitamin A value of the processed carrots was noted.

In four carrot hybrids the carotenoid level diminished 7% to 12% (Simon and Lindsay, 1983) for canned–cooked carrots. The thermal destruction of β-carotene in carrot juice heated from 104°C to 132°C follows first-order reaction kinetics (Hojilla et al., 1985).

Freezing. Fresh and frozen carrots grown in Spain contained 129.5 and 116.0 $\mu g\ (g\ fresh\ wt)^{-1}$ β-carotene, respectively (Arqués et al., 1988).

Dehydration. It was observed almost 50 years ago that a true assessment of the nutritional value of dehydrated carrots depends not only on the total carotene content but also on the degree of isomerization of the carotenes (Beadle and Zscheile, 1942).

Degradation of carotenoids has been reported to play an important role in determining the overall storage life of dehydrated carrots. Off-flavors became noticeable when 20% to 45% of the total carotenoids had been destroyed (Tomkins et al., 1946). In dehydrated carrots stored in the presence of oxygen, β-carotene is destroyed and an off-flavor develops. This investigation clearly showed for the first time that in freeze-dried carrots the loss of carotene is closely related to development of an off-flavor (Falconer et al., 1964). Nine components with a violetlike odor were isolated from dehydrated carrots, three of which are α-ionone, β-ionone, and β-ionone 5,6-epoxide (Ayers et al., 1964).

Comparison of β-carotene content in dried carrots prepared by three dehydration processes (puffing, air drying, and vacuum freeze-drying) was carried out by Della Monica and McDowell (1965). Few differences were found in the total β-carotene content of the three products after processing, with the highest level found in the freeze-dried product. The latter retained 80% of the all-*trans*-β-carotene. After cooking, the retention of all-*trans*-β-carotene was 76% for vacuum-freeze-dried carrots, 64% for puffed, and 52% for air-dried carrots.

The effects of water activity, blanching, and antioxidants on the stability of carotenoids in stored dehydrated carrots was investigated by Arya et al. (1979). Carotenoid pigments were found to be most stable at $0.43a_w$ water activity. The rate of carotenoid destruction was higher in blanched freeze-dried carrots than in unblanched. This suggested that some substances that stabilize carotenoids are either degraded or leached out during blanching. Adding sodium chloride and sodium metabisulfite significantly reduced the rate of carotenoid destruction.

The effect of different drying techniques and storage of dried carrots on the sensory changes, volatile constituents, and carotenoid stability was reinvestigated by Kaminski et al. (1986). Carotenoid losses increased with the length of storage. The rate of destruction was higher in freeze-dried carrots stored in air; after 2 months the loss was 89%. In samples dried by hot air or microwave hot air, the loss amounted to 36%. Storage under nitrogen was about the same for all samples, amounting to 20% to 27% after 2 months storage and 52% to 59% after 4 months storage.

The effect of blanching and sulfur dioxide on the stability of carotenoids of dehydrated carrots was also reevaluated (Baloch et al., 1987). Sulfiting was found to have a marked effect on the stability of unblanched and blanched carrots during dehydration and storage at 37°C. The combination of sulfiting and blanching to a level sufficient to inactivate peroxidase activity is the most effective method for enhancing the storage life of dehydrated carrots.

Cassava
(*Manihot esculenta* Crantz)
Euphorbiaceae

Cassava, also known as manioc, mandioc, and yuca, has its origin in the American tropics. It is cultivated throughout the tropical world for its tuberous roots, from which cassava flour, breads, tapioca, laundry starch, and alcoholic beverages are derived. In some countries, cassava is used instead of cereals to supply carbohydrates.

Cassava is cultivated especially in Africa and Latin America. Its native Brazil produces about a third of the world's total. The pulp of the tubers is light yellow due to the presence of carotenoids, which were recently analyzed for the first time (Penteado and Almeida, 1988). The carotenoids of five cassava cultivars from São Paulo state consisted mainly of β-carotene accompanied by small amounts of xanthophylls. The β-carotene fraction further analyzed was a mixture of β-carotene and two of its isomers. The results are given in Table 44.

Table 44. β-Carotene Isomers in Roots of Five Cassava Cultivars [μg (100 g fresh wt)$^{-1}$]

		Cultivars				
β-Carotene Isomers		*IAC 576-70*	*Ouro do Vale*	*Pioneira*	*IAC 289-70*	*Branca de Sta. Catarina*
Neo-*cis* B	r	21 ± 3	31 ± 17	17 ± 6	25 ± 7	—
	c	10 ± 2	22 ± 5	11 ± 4	23 ± 7	—
all-*trans*	r	44 ± 2	60 ± 15	49 ± 19	56 ± 18	14 ± 1
	c	19 ± 1	38 ± 2	17 ± 1	45 ± 15	—
Neo-*cis* U	r	30 ± 14	28 ± 10	18 ± 9	28 ± 6	7 ± 3
	c	11 ± 4	15 ± 8	9 ± 3	19 ± 3	—
RE	r	11.0	13.9	10.8	13.4	2.8
	c	9.0	9.3	5.1	10.7	—

Source: From Penteado and Almeida (1988).

Discussion. Major carotenoids in cassava roots were all-*trans*-β-carotene accompanied by high amounts of the *cis* isomers neo-β-carotene B and neo-β-carotene U. The sum of their concentration was approximately as high as that of β-carotene, an uncommon pattern. The vitamin A value ranged between 2.8 and 13.9 RE (100 g fresh wt)$^{-1}$.

The leaves of cassava are also a source of provitamin A (see Tables A9 and A12).

Changes During Cooking

The decrease of β-carotene and its isomers is shown in Table 44. The vitamin A decrease varied from 22% to 55%.

Cauliflower
(*Brassica oleracea* var. *botrytis* L.)
Cruciferae

The carotenoids of cauliflower bud tissue, which is composed of undifferentiated cells, were analyzed because it is an interesting example of undifferentiated tissue whose carotenoids were not yet investigated (Prebble and West, 1975). The top 1 mm to 3 mm of the white bud tissue of commercial cauliflower contains about 10 μg (g dry wt)$^{-1}$ total carotenoids; the stalk, less than 1 μg (g dry wt)$^{-1}$. The qualitative and quantitative distributions are given in Table 45.

Table 45. Carotenoids of Cauliflower Bud and Stalk Tissues

Carotenoid Pattern	*Bud % of Total*	*Stalk Carotenoids*
β-Carotene	12	21
Lutein	33	43.5
Violaxanthin	44.5	14.5
Neoxanthin	10.5	21

Source: From Prebble and West (1975).

Discussion. The carotenoids were chloroplast carotenoids, their relative proportions differing in the analyzed parts. In the stalks the relative proportions were typical for chloroplasts; for buds they differed from proportions found in chloroplasts because they contained a high percentage of violaxanthin, the predominant pigment. Similar relative proportions of β-carotene (14.6%), lutein (30.5%), violaxanthin (50%), and neoxanthin (5%) were found in the chloroplast envelope of spinach leaves (Douce et al., 1973). This impressive similarity suggested that the proplastid envelope may be the site of the bud carotenoids, an assumption confirmed in a study on preparation and characterization of envelope membranes from nongreen plastids (Alban et al., 1988). Envelope membranes from cauliflower bud plastids and other nongreen plastids containing typical plastid components, such as galactolipids and carotenoids, are highly homologous with chloroplast envelope membranes. These results demonstrate that the biosynthetic capabilities of envelope membranes are not restricted to chloroplasts with a well-developed thylakoid network, but are also present in plastids whose membrane structures are almost limited to their envelope membrane.

Cucumber
(*Cucumis sativus* L.)
Cucurbitaceae

Cucumber carotenoids were first analyzed in a study that investigated pigment changes during brining (Jones et al., 1962). As expected, they are chloroplast carotenoids that were individually characterized qualitatively. The total carotenoid content was 17.2 μg (g fresh wt)$^{-1}$, of which β-carotene accounted for 12.8%.

Applying a new HPLC method that resolves *cis-trans*-β-carotene isomers, investigators found that the fresh cucumber β-carotene fraction is a mixture of all-*trans*-β-carotene (74.9%), 13-*cis*-β-carotene (10.5%), and 9-*cis*-β-carotene (14.5%) (Schwartz and Patroni-Killam, 1985; Chandler and Schwartz, 1987). The average provitamin A content per gram of several cucumber cultivars grown in Australia is 0.2 μg α-carotene, 2.5 μg β-carotene, and 0.2 μg cryptoxanthin (Wills et al., 1987).

Changes During Pickling

During brining and brine storage the xanthophylls and carotene underwent appreciable loss (Jones et al., 1962). After 207 days 72% of the total xanthophylls and 35% of the carotene were left. The relative proportions of β-carotene isomers present in raw cucumbers are different in pickled cucumbers: all-*trans*-β-carotene (72.9%), 13-*cis*-β-carotene (7.3%), and 9-*cis*-β-carotene (19.8%) (Chandler and Schwartz, 1987).

Lettuce
(*Lactuca sativa* L.)
Compositae

Of the carotenoids found in lettuce leaves, the β-carotene content was determined in order to assess its nutritional value. Reported values reveal pronounced quantitative variations due to intervarietal differences, stage of maturity, and applied analytical methods.

In early reports β-carotene ranges between 1.4 and 42 μg (g fresh wt)$^{-1}$ and between 10.8 and 24.5 μg (g fresh wt)$^{-1}$, the latter values (recalculated) in May and December, respectively (Fitzgerald and Fellers, 1938; Bondi and Meyer, 1946). The β-carotene of Iceberg lettuce, determined by HPLC, was

1 to 35 μg (g fresh wt)$^{-1}$ in November, 1 to 4.8 μg (g fresh wt)$^{-1}$ in March, and 0 to 3 μg (g fresh wt)$^{-1}$ in July (Bureau and Bushway, 1986).

The β-carotene content of various lettuce cultivars grown in Japan was also determined by HPLC (Izaki et al., 1986). The results are given in Table 46.

Table 46. β-Carotene Content in Various Lettuce Types (*Lactuca sativa* L.)

Type	*β-Carotene [μg (g fresh wt)$^{-1}$]*
Butterhead (mini)	12.8
Butterhead (salad-na)	39.0
Leaf (red)	5.1
Leaf (sunny)	17.1
Leaf (green)	32.2
Leaf	10.9
Stem	6.3
Cos (romaine)	5.7
Crisp head	2.7

Source: From Izaki et al. (1986).

The carotenoids of Reskia lettuce grown in Belgium were analyzed by standard methods (Rouchaud et al., 1984a, 1985). Total content was about 68 μg (g fresh wt)$^{-1}$, the chloroplast carotenoids being β-carotene (25%), lutein (26%), violaxanthin (11%), and neoxanthin (4%). The monohydroxy carotenoid fraction (2.5%) contained both α- and β-cryptoxanthin, and lutein was accompanied by small amounts of zeaxanthin.

The β-carotene stereoisomers were all-*trans*-β-carotene (77%), neo-β-carotene B (5%), and neo-β-carotene U (18%). The maximum total carotenoid content of fresh lettuce decreased when the weight of the lettuce increased during development.

A complete HPLC analysis of the leaf carotenoids of the prickly lettuce (*Lactuca sativa* L. var. *sativa* Bisch) grown in Japan was carried out in winter (Takagi, 1985). In addition to the usual four chloroplast carotenoids, a fifth pigment was identified as antheraxanthin. The results are given in Table A5.

Carotene content of Butterhead lettuce produced by various cultivation methods ranged from 13.6 to 31.9 μg (g fresh wt)$^{-1}$ in soil-grown samples and from 18.1 to 27.6 μg (g fresh wt)$^{-1}$ in hydroponically grown samples (Kobayashi et al., 1989). The β-carotene content of Brazilian leafy vegetables, including lettuce, was analyzed by conventional methods (Ramos and Rodriguez-Amaya, 1987). The variation with maturity in young and mature lettuce and endive leaves is illustrated in Table 47. In order to eliminate vari-

Table 47. β-Carotene Content and Vitamin A Value as a Function of Maturity of Leafy Vegetables

Vegetable	*β-Carotene [μg (g dry wt)$^{-1}$]*	*Vitamin A Value (RE kg^{-1})*
Lettuce		
Sample 1		
young leaves	3.8	633
mature leaves	12.3	2050
Sample 2		
young leaves	3.2	533
mature leaves	11.6	1933
Endive		
Sample 1		
young leaves	5.8	967
mature leaves	16.2	2700
Sample 2		
young leaves	2.7	450
mature leaves	11.2	1867

Source: From Ramos and Rodriguez-Amaya (1987).

ations due to sunlight, temperature, and soil composition, the young (inner) and mature (outer) leaves were taken from the same bunch.

Discussion. The results indicate a pronounced increase of β-carotene with the maturation process.

An unusual xanthophyll was also isolated from *L. sativa.* It was called, according to its source, lactucaxanthin (ε,ε-carotene-3,3′-diol) (Siefermann-Harms et al., 1981). In a chemosystematic screening, lactucaxanthin was restricted to the *Lactuca* genus: *L. sativa capitata* var. Manoa and Corelli, *L. sativa crispa, L. serriola,* and *L. perennis,* and a few closely related genera within the tribe of *Cichoriae* of the *Compositae.*

Changes During Storage

Carotenoid changes during storage of head lettuce cultivars Dickkopf, Ramses, and Gravitas grown in the Halle district, Germany, were followed (Böttcher, 1988). The mean carotene content was 13.5 μg (g fresh wt)$^{-1}$. Carotene loss increased linearly with storage time, whereas the decreasing rate depended on storage temperature. At 0.3°C, a loss of 29.4% occurred after 20 days of storage; at 4°C, a loss of 19% after 12 days; at 12°C to 16°C, a loss of 17% after 6 days; and at 18°C to 20°C, a loss of 10.1% after 3 days; the optimum keeping quality being a function of temperature.

Lupin
(*Lupinus* spp.)
Leguminosae

Lupin seeds are a source of high-quality protein and edible oil. The carotenoids of the seeds of four *Lupinus* species (*L. albus, L. angustifolius, L. luteus,* and *L. mutabilis*) were investigated (El-Difrawi and Hudson, 1979). All contained significant quantities of β-carotene [0.2 to 1.4 μg (g dry wt)$^{-1}$] and zeaxanthin [4.5 to 14.0 μg (g dry wt)$^{-1}$] as the major pigment. *L. luteus* contained a further pigment, which was tentatively identified as β-carotene 5,6,5′,6′-diepoxide.

Maize
(*Zea mays* L.)
Gramineae

Maize or corn is a cereal plant bearing large ears with kernels that are used for food or fodder. Unripe maize is eaten as a vegetable. The structure of the mature corn kernel was described in Wolf et al. (1952). The corn kernel is a fruit composed of a thin pericarp enclosing a single seed. The pericarp is the mature ovary wall. Along the inner surface it adheres closely to the seed coat. The latter encloses the germ and the endosperm, with the three forming the seed. This type of single-seeded fruit is called *caryopsis.*

The color of corn kernels is variable; the pattern may be of one color or variegated. Pigments responsible for the coloration are located in one or more tissues; the pericarp can be colorless, orange, cherry red, red, dark red, or variegated; the aleuron layer can be colorless, red, red–purple, purple, or brown; the endosperm is either colorless or of various shades of yellow; and the germ can be colorless, yellow, orange–red, or purple.

The pigments of maize are mainly carotenoids. Some of the color may be due to anthocyanins. Maize carotenoids were extensively investigated, and individual carotenoids were identified about 60 years ago. Most reports are concerned with the carotenoids of dried field maize. Only a few studies were published on the carotenoids of fresh sweet corn; most were analyzed in frozen or canned sweet corn. Maize was the first source of many carotenoids, and some of them, such as zeaxanthin, zeinoxanthin, and α- and β-zeacarotene, received names derived from this source.

Zeaxanthin and cryptoxanthin were first isolated from maize (Karrer et al., 1929; Kuhn and Grundmann, 1934). From Italian and Hungarian varieties β-carotene (0.5–0.7 μg g^{-1}), cryptoxanthin (5–7 μg g^{-1}), and zeaxanthin (13–15 μg g^{-1}) were isolated and determined. Subsequently, α-carotene, γ-carotene, lutein (Fraps and Kemmerer, 1941; White et al., 1942), phytoene, and phytofluene (Zechmeister and Sandoval, 1945; Rabourn and Quackenbush, 1953) were identified. Then α-zeacarotene (7′,8′-dihydro-ε, ψ-carotene) and β-zeacarotene (7′,8′-dihydro-β, ψ-carotene) were isolated from maize (Petzold et al., 1959), followed by zeinoxanthin (Petzold and Quackenbush, 1960).

As usual, in early investigations only β-carotene was determined. Sweet corn varieties were found to have a carotene content from 1 to 5 $\mu g\ g^{-1}$, and varied considerably between the varieties (Zimmerman et al., 1940; Scott and Belkengren, 1944). In 1961 a procedure was worked out for routine analyses of the complex mixture of maize carotenoids (Quackenbush et al., 1961). The pigments were separated by column chromatography on magnesia from which seven fractions were eluted. The first carotene fraction comprised the colorless phytoene and phytofluene, α- and β-carotene, α- and β-zeacarotene, and ζ-carotene. The next fractions contained either a single pigment or a mixture: (2) zeinoxanthin, (3) cryptoxanthin, (4) esters, (5) lutein, (6) zeaxanthin, and (7) polyoxy carotenoids. Although the method permitted the separation of a- from b-carotene, a-carotene was rarely reported to be present.

In five analyzed corn samples from different parentage, the total carotenoid content varied between 30.2 and 90.8 $\mu g\ g^{-1}$. Lutein was the dominant carotenoid (16–20 $\mu g\ g^{-1}$), followed by zeaxanthin (4–11 $\mu g\ g^{-1}$). The three provitamin A carotenoids (β-carotene, β-zeacarotene, and cryptoxanthin) were found at levels from 0.1 to 4.4 $\mu g\ g^{-1}$. A remarkable variation with the genotype existed in maize carotenoids.

In a comprehensive study, the carotenoids of 125 corn inbreds were analyzed with the foregoing method (Quackenbush et al., 1963). The carotenoid pattern was the same in all strains, but the total carotenoid content and the proportions of the individual carotenoids varied considerably in each strain. The diol levels especially show considerable variations among the inbreds, a fact confirmed in all subsequent investigations.

Total carotenoids ranged from 0.5 to 57 $\mu g\ g^{-1}$. The following pigments were separated: phytoene (5.6–45.8 μg), phytofluene (0.8–10.2 μg), β-carotene (0.1–5.0 μg), β-zeacarotene (0.1–4.7 μg), ζ-carotene (0.1–4.2 μg), α-cryptoxanthin (0.1–7.8 μg), β-cryptoxanthin (0.3–5.1 μg), esters of zeaxanthin (0.3–4.1 μg), lutein (2.0–33.1 μg), zeaxanthin (0.6–27.4 μg), and polyoxy carotenoids (0.4–3.9 μg).

The complex maize carotenoid pattern may be characterized as a mixture of carotenes and mono- and dihydroxy carotenoids of both the β,ε- and β,β-series with lutein and zeaxanthin dominating. This carotenoid pattern is unusual. It is not encountered in other fruits, and is perhaps specific for a caryopsis. Also rare is the accumulation in some maize kernels of five provitamin A carotenoids: α-, β-, and γ-carotene, β-zeacarotene, and β-cryptoxanthin.

In hybrids and in inbred and exotic corn lines, two fractions (carotene and xanthophylls) were determined (Blessin et al., 1963a). In commercial lines, carotene varied from 1 to 4 $\mu g\ g^{-1}$, xanthophylls from 10 to 30 $\mu g\ g^{-1}$; in inbred lines of yellow corn, carotene varied from 0 to 7 $\mu g\ g^{-1}$, xanthophylls from 5 to 50 $\mu g\ g^{-1}$. Some exotic strains from South America, particularly Argentina, contained approximately 60 $\mu g\ g^{-1}$ xanthophylls. It was concluded that carotene and xanthophyll contents are influenced by the genetics and environment of the plant. Consequently, it is possible to breed special corn varieties with high xanthophyll content for feeding or with high provitamin A content for food, as demonstrated by Brunson and Quackenbush (1962) and Grogan and Blessin (1968). In the latter work, the carotenoids of 15 lines, 5 high-carotenoid exotic strains, and 5 crosses of maize were analyzed and showed carotenes (6–25%), α-cryptoxanthin (1–11%), β-cryptoxanthin (7–16%), lutein (6–38%), zeaxanthin (13–67%), and polyoxy carotenoids (4–12%). Total carotenoid content was 6 to 50 $\mu g\ g^{-1}$.

The results confirmed the high variability of the xanthophyll profile in each line. For pigmentation of broilers and egg yolk, various strains of yellow corn containing different levels of the principal xanthophylls lutein and zeaxanthin were developed (Dua et al., 1967). They were designated as high-zeaxanthin–low-lutein (3:1), equal-zeaxanthin–equal-lutein, and low-zeaxanthin–high-lutein (1:2).

The carotenoids of three different Hungarian maize varieties of white, yellow, and red grains were investigated at the unripe and ripe stages. (Zsolt et al., 1963). Total carotenoid content in the white, yellow, and red ripe grains was 2.8, 12.6, and 23 $\mu g\ g^{-1}$, respectively. During ripening a tenfold increase of the total carotenoids occurred, and a more complex carotenoid pattern appeared.

In hybrid corn grown in different Spanish regions, the total carotenoid content was 82.2 to 279.5 μg (g fresh wt)$^{-1}$ (Carballido et al., 1971).

The carotenoids of maize grown in Romania, a country with a high maize production, were investigated in much detail. To breed new lines with good color and high provitamin A value, researchers analyzed the carotenoids of a series of inbred lines with known genetic background by a combination of column chromatography and TLC (Neamţu et al., 1982). The results are given in Table 48.

Discussion. The highest carotenoid content occurred in both varieties of *Z. mays indurata* (var. *rubra* and *dulcis*) with a total carotenoid content of 35 and 23 $\mu g\ g^{-1}$. In all varieties the major pigments were lutein and zeaxanthin. In *Z. mays indurata* the percentage of zeaxanthin was higher than that of lutein. Generally, it was observed that a carotenoid pattern dominated by carotenoids of the β,β-series exists in grains of orange color, whereas if carotenoids of the β,ϵ-series dominate, the strains are yellow.

Table 48. Carotenoid Distribution in Seven Maize Varieties Grown in Romania

Carotenoid Pattern	*Percent of Total Carotenoids*						
	1	*2*	*3*	*4*	*5*	*6*	*7*
α-Carotene	4.1	3.0	0.3	6.2	2.1	6.6	1.1
β-Carotene	3.0	2.6	0.5	6.2	2.5	1.1	1.8
α-Zeacarotene	—	0.7	—	—	0.5	—	—
β-Zeacarotene	—	0.3	—	—	1.8	—	—
α-Cryptoxanthin	3.2	2.8	1.4	4.6	6.6	3.0	4.7
β-Cryptoxanthin	2.7	2.3	0.9	—	4.4	1.2	2.2
Lutein	46.5	36.0	49.4	39.3	25.1	37.8	45.4
Zeaxanthin	21.8	44.0	40.0	36.5	45.2	51.0	39.1
Diol epoxides (β, ϵ)	5.8	4.3	3.5	4.0	6.0	3.7	3.1
Diol epoxides (β, β)	12.9	4.0	4.0	3.2	5.8	2.6	2.6
Total carotenoids $\mu g\ g^{-1}$	2.6	23.0	13.3	1.3	35.3	4.8	1.4

Varieties

1. *Zea mays dentiformis* var. *pyrodon*
2. *Zea mays indurata* var. *rubra*
3. *Zea mays dentiformis* var. *xantodon*
4. *Zea mays evertra* var. *crysoides*
5. *Zea mays indurata* var. *vulgata*
6. *Zea mays rugoza* var. *dulcis*
7. *Zea mays dentiformis* var. *alborubra*

Source: From Neamţu et al. (1982).

A HPLC procedure that permits the determination of both carotenoids and tocols (tocopherols and tocotrienols) in the same extract of maize grains has been developed (Weber, 1987). Fifteen yellow corn inbreds were analyzed; carotenes were determined as a fraction and β-cryptoxanthin, lutein, and zeaxanthin separated individually. The results are shown in Table 49.

Discussion. Total carotenoids were 30 to 77 $\mu g\ g^{-1}$, except that one inbred had a content of 16 $\mu g\ g^{-1}$. Although the overall pattern was the same, the relative proportions of the carotenoids varied markedly among the inbreds. So lutein, the major pigment in almost all the inbreds, varied from 8.0% to 64.6%. Zeaxanthin, the second major pigment, varied between 4.8% and 71.9%. In some inbreds the carotene fraction constituted half of

Table 49. Carotenoids of Yellow Corn Inbreds

Inbred	*Kernel dry wt (mg)*	*Carotenes (%)*	*β-Crypto-xanthin (%)*	*Lutein (%)*	*Zeaxanthin (%)*	*Total ($\mu g\ g^{-1}$)*
A619	230	50.5	8.3	24.7	16.4	69.3
A632	234	10.7	9.4	8.0	71.9	15.8
B37	253	25.5	7.0	52.7	14.7	59.0
B73	222	11.6	14.7	60.3	13.4	28.5
B84	312	10.6	6.6	64.6	18.2	27.4
C103D	247	6.5	7.0	54.5	31.9	32.3
H51	214	18.4	12.5	51.9	17.2	48.2
Mo17	370	9.0	7.0	44.4	39.5	34.4
NY16	158	17.5	8.3	60.1	14.1	31.2
Oh43	236	28.8	7.8	55.2	8.1	49.2
Oh45	288	32.8	5.0	54.9	7.3	76.9
R802A	219	50.0	10.5	34.7	4.8	73.6
R806	305	28.1	12.0	43.4	16.5	49.9
T220	184	39.4	8.0	38.7	13.9	51.2
W64A	177	12.2	13.2	46.0	28.6	36.1

Source: From Weber (1987), by permission of AOCS.

the total carotenoids or exceeded the zeaxanthin levels. Inbreds could be selected for high percentages of carotenoids or zeaxanthin.

The distribution of carotenoids in the hand-dissected corn kernel fractions (horny endosperm, floury endosperm, and germ + tip cap) was followed in four inbreds. The results are shown in Table 50.

Discussion. The weight percentages of the hand-dissected kernel parts exhibited little variation. The horny endosperm account for the largest proportion of the kernel weight (66–69%), the floury endosperm for 15% to 20%, and the germ + tip cap 13% to 16%.

The carotenoids were localized mainly in the horny endosperm with 78% to 87%, floury endosperm contained 11% to 20%, and the germ only 1–4%. These results corroborate earlier data on xanthophyll and carotene distribution in hand-dissected fractions of yellow dent corn (Blessin et al., 1963b).

Lutein was the predominant carotenoid in all of the kernel parts. In the floury endosperm the ratios of the carotenoids were different from those in the whole kernel; the proportion of β-cryptoxanthin was greater and that of lutein was less. In the germ + tip cap, the only consistent change was an increase in the proportion of zeaxanthin. In three of the four inbreds the carotene proportion decreased.

Table 50. Distribution of Carotenoids in Hand-Dissected Corn Kernel Fractions

Inbred	Kernel Fraction	Dry Weight (%)	Carotenes (%)	β-Cryptoxanthin (%)	Lutein (%)	Zeaxanthin (%)	Total Carotenoids $\mu g\ g^{-1}$	Total Carotenoids %
B37	Whole kernel	100.0	28.2	10.7	50.8	10.3	59.0	100.0
B84		100.0	15.3	7.5	64.0	13.2	27.4	100.0
Oh45		100.0	37.9	8.5	48.2	5.4	76.9	100.0
Canadian Flint		100.0	23.1	14.2	53.7	9.0	19.8	100.0
B37	Horny endosperm	69.5	25.0	6.8	45.3	10.2	47.7	87.3
B84		66.3	10.1	4.0	53.6	11.9	19.1	79.4
Oh45		69.3	31.3	5.4	37.6	3.8	54.0	78.1
Canadian Flint		65.7	21.2	12.0	43.5	9.1	15.2	85.8
B37	Floury endosperm	14.9	3.2	2.2	4.6	1.0	6.0	11.0
B84		18.6	2.6	1.9	9.9	2.5	4.1	16.9
Oh45		17.4	7.4	3.3	8.5	0.9	13.9	20.1
Canadian Flint		19.8	3.6	3.1	5.3	0.9	2.3	12.9
B37	Germ + tip cap	15.6	0.4	0.2	0.8	0.3	0.9	1.7
B84		15.0	0.6	0.3	2.0	0.7	0.9	3.6
Oh45		13.3	0.5	0.1	1.0	0.2	1.2	1.8
Canadian Flint		14.5	0.2	0.1	0.8	0.2	0.2	1.3

Source: From Weber (1987), by permission of AOCS.

Distribution in Other Parts of the Plant

In an early report only the carotene content of the whole corn plant was analyzed (Porter et al., 1946). The carotene concentration was 20 to 50 times higher in the leaf blades than in the stalks, ears, and husks. Carotene concentration in the leaves reached a peak at the pollen shedding stage and then declined rapidly. Subsequently, the corn seedling carotenoids were analyzed by a specially developed method (Moster et al., 1952), involving column chromatography first on $Ca(OH)_2$, followed by rechromatography of the various fractions on other suitable absorbents [MgO, sucrose, or $Ca(OH)_2$]. The major carotenoids were the four chloroplast carotenoids along with α-carotene and zeaxanthin. Five other carotenoids were tentatively identified to be cryptoxanthin, monohydroxy-α-carotene, violeoxanthin, and two flavoxanthins. In addition, small quantities of four unknown pigments were present, probably artifacts. Additional data are given in Table A5 (Takagi, 1985).

Maize pollen was first investigated by Karrer et al. (1950). It contained small amounts of carotenoids consisting probably of xanthophyll esters and very little carotene. Epoxides were not detected. Pollen of maize grown in Romania contains 57 $\mu g\ g^{-1}$ total carotenoids consisting of lutein (63.2%), zeaxanthin (28.4%), β-carotene (7.0%), and cryptoxanthin (1.4%) (Tămaş et al., 1970). In the maize root tip during seed germination, violaxanthin was shown to be the major carotenoid, accompanied by small amounts of β-carotene (Maudinas and Lematre, 1979).

Changes During Storage, Cooking, and Processing

Storage

It was generally known that yellow maize gradually loses its carotenoid pigments during storage. The effects of temperature and moisture on maize carotenoid losses during storage were studied (Quackenbush, 1963). Maize was stored in the dark for 3 years, dried to 3% and 11% moisture at 7°C and 25°C. Total carotenoid losses were most rapid during the early part of the storage. Temperature exerted a greater influence than did moisture content, the losses being higher at 25°C. Carotenes decreased at higher rates than did xanthophylls.

The stability of carotenoids in four corn inbreds during 6 months storage at room temperature was recently investigated (Weber, 1987). The results are given in Table 51.

Table 51. Recovery of Carotenoids in Four Corn Inbreds after Storage at Room Temperature for 6 Months

			Recovery of Initial Value of Individual Carotenoids			
Inbred	*Initial Total Carotenoid Content ($\mu g\ g^{-1}$)*	*Recovery of Total Carotenoids (%)*	*Carotenes (%)*	*β-Cryptoxanthin (%)*	*Lutein (%)*	*Zeaxanthin (%)*
B84	27.4	66.7	64.6	76.1	64.7	68.7
H51	48.3	53.7	56.5	63.5	51.0	52.0
Oh45	76.9	62.7	74.2	82.0	54.0	53.8
R802A	73.6	49.8	48.2	61.1	49.7	58.6
Mean		58.2	60.9	70.7	54.8	58.3

Source: From Weber (1987), by permission of AOCS.

Discussion. The recovery of total carotenoids after storage was 49.8% to 66.7% of the harvest values. The inbred with a lower total carotenoid content showed a lower loss. Among individual carotenoids, lutein appeared to be least stable, which contradicts other reports on carotenoid stability.

Cooking

Provitamin A carotenoids in raw corn from different geographical areas ranged from 0.6 to 2.1 μg (g fresh wt)$^{-1}$ and were higher in fall varieties than in spring varieties. Cryptoxanthin represented 8.0% to 8.5% of the total carotenoids contributing substantially to the vitamin A value. The retention of provitamin A carotenoids in cooked, husked corn on the cob was 98% (Klein and Perry, 1982).

Processing

The individual carotenoids and the vitamin A value of four cultivars of commercially processed sweet corn were studied by Lee et al. (1981). The distribution of the major carotenoids is shown in Table 52.

Table 52. Major Carotenoids in Four Cultivars of Canned Sweet Corn [μg (100 g)$^{-1}$]

Carotenoids	*Jubilee*	*Iosweet*	*70-2499*	*Stylpack*
Hydrocarbon	89.0	41.4	45.0	45.6
Monohydroxy	98.5	92.7	101.6	89.5
Polyhydroxy	58.5	19.6	35.0	51.5
α-Carotene	24.6	5.0	8.4	8.1
β-Carotene	9.8	5.3	14.0	8.4
β-Zeacarotene	15.7	16.8	14.5	17.0
γ-Carotene	15.1	8.9	7.5	10.6
Zeinoxanthin	79.2	66.0	77.2	68.9
β-Cryptoxanthin	18.9	18.3	15.1	18.4

Source: Reprinted with permission, from Lee et al. (1981); copyright © 1981 American Chemical Society.

Discussion. Unlike other maize cultivars, the monohydroxy carotenoid fraction was the major fraction and was dominated by zeinoxanthin (α-cryptoxanthin). Not fitting in the general maize carotenoid pattern was the level of polyhydroxy carotenoids, which was lower than that of the hydrocarbons. The variations of the various carotenoid levels are due to varietal differences. The cultivar Jubilee, which has the deepest yellow color, contained the highest levels of total carotenoids, α-carotene, and α-cryptoxanthin.

Provitamin A carotenoids and vitamin A values of some samples of commercial sweet corn (frozen and canned) were also analyzed (Table 53).

Table 53. Provitamin A Carotenoids and Vitamin A Values of Commercial Sweet Corn [μg (100 g)$^{-1}$]

	Frozen Samples		*Canned Samples*			
Carotenoids	*A*	*B*	*A*	*B*	*C*	*D*
α-Carotene	34.2	23.2	3.3	19.7	5.6	4.1
β-Carotene	14.7	13.0	8.1	12.1	6.9	4.7
β-Zeacarotene	57.7	19.1	40.6	12.6	19.9	24.4
γ-Carotene	8.6	10.7	0.4	5.9	1.9	4.2
β-Cryptoxanthin	8.9	11.7	3.2	8.7	17.1	11.2
Vitamin A value						
RE	11.5	7.5	5.3	5.9	4.8	4.4

Source: Reprinted with permission, from Lee et al. (1981); copyright © 1981 American Chemical Society.

Discussion. β-Zeacarotene was the predominant provitamin A in all samples. The vitamin A value in all sweet corn samples ranged from 4.4 to 7.7 RE (100 g)$^{-1}$.

Freezing

No significant differences in carotene content were found between fresh and frozen sweet corn (Zimmerman et al., 1940). The carotene level remained unchanged in frozen, blanched sweet corn stored for 3 to 4 months at -18°C and -40°C.

A more detailed analysis of the carotenoids of two frozen and refrigerated sweet corn varieties, Evertender C and Royal Gold, was carried out in 1965 (Tichenor et al., 1965). The β-carotene content of both varieties ranged from 0.9 to 1.3 μg g^{-1} and remained stable during the length of storage (4–280 days). The cryptoxanthin content was 2.4 to 3.0 μg g^{-1} and was higher in frozen samples. Diols and polyols made up 44% to 51% of the total carotenoids of frozen Evertender and 51% to 55% of the total carotenoids of frozen Royal Golden sweet corn.

Pea
(*Pisum sativum* L.)
Leguminosae

Complete analysis of the carotenoids of pea seeds, which are supposed to be chloroplast carotenoids, has never been reported. Generally, β-carotene was determined. Then total xanthophylls and, more recently, β-cryptoxanthin (a provitamin A) have been determined. As usual, the β-carotene content among cultivars varied with maturity stage and seed size. Carotene content was 3 to 5 μg (g fresh wt)$^{-1}$. Some examples follow: 3.6 μg (g fresh wt)$^{-1}$ (Stimson et al., 1939); 3.3 and 4.4 μg (g fresh wt)$^{-1}$ in Nott's Excelsior and Little Marvel peas (Pepkowitz et al., 1944); 3.4, 5.2, and 7.2 μg (g fresh wt)$^{-1}$ in Walah, Wando, and Miracle peas (Heinze et al., 1947); 3.7, 4.4, and 3.0 μg (g fresh wt)$^{-1}$ in Pride, Shasta, and Thomas Laxton peas (Kelley et al., 1950). Whole green peas contained 13.9 μg (g dry wt)$^{-1}$, yellow peas 14.8 μg (g dry wt)$^{-1}$ (Murray, 1948). A detailed analysis of the "crude carotene" fraction of fresh peas revealed that it contains α-carotene and all-*trans*-β-carotene accompanied by two *cis*-β-isomers (Panalaks and Murray, 1970) (see Table 55).

Besides chlorophyll, the carotenoids of six pea cultivars, wrinkled and smooth, grown in Hungary were analyzed (Aczél, 1973). The pigments were separated chromatographically; the carotenoids were separated into two fractions, β-carotene and total xanthophylls, which were determined. The results are given in Table 54.

Table 54. Carotenoid Content in Various Pea Cultivars (raw and processed) [μg (g dry wt)$^{-1}$]

	β-carotene			*Total Xanthophylls*		
Cultivars	*a*	*b*	*c*	*a*	*b*	*c*
Wrinkled						
Express	5.1	4.6	5.3	9.2	8.8	10.1
Rhein	4.4	4.2	4.2	11.2	12.0	12.1
Debrecener	1.9	2.0	2.6	8.9	9.9	9.4
Gloria	3.6	3.3	3.9	10.0	9.4	12.8
Smooth						
Malenkoye Csudo	2.9	2.2	4.1	13.8	12.9	13.0
Kobri	4.9	4.8	5.4	9.9	10.2	9.9
Perfection	3.3	4.1	3.3	7.0	10.8	11.2

Source: From Aczél (1973).
Notes: a = raw; b = blanched; c = sterilized.

The carotene content of green peas grown in the Halle district, Germany, was 4.4 μg (g fresh wt)$^{-1}$ (Böttcher, 1987).

The carotene content of dry peas seeds and sprouts was determined in a study on the nutritive value of the so-called natural foods before and after sprouting (Fordham et al., 1975). The dry seeds had a very low carotene content, 10 to 100 times less than the green peas. The intervarietal differences were considerable: 3.16, 13.0, and 37.4 μg (100 g)$^{-1}$ in Mammoth Meltin, Thomas Laxton, and Early Alaska peas. The sprouts contained 3 to 10 times less carotene. Lee has measured the vitamin contents of green peas at various stages of raw product handling on different cultivars, on different sizes of peas, and at various stages of processing (Lee et al., 1982). The carotene content from four different harvester loads from a single field showed minor variations [19.3 to 20.9 μg (g dry wt)$^{-1}$]. Intervarietal variations were observed in the cultivars Early Sweet, Trojan, Reva, and Freezer, which contain 14.0, 15.0, 18.0, and 12.6 μg (g dry wt)$^{-1}$, respectively. The carotene content also varied with size. Size grades 3 and 5 of Early Sweet peas contained 15.3 and 11.4 μg (g dry wt)$^{-1}$, proving definitely that smaller peas have a higher carotenoid content. In a more detailed study, the carotenoids of the green pea were separated chromatographically into three main fractions—hydrocarbons, monohydroxy carotenoids, and polyoxy carotenoids—which were then determined. The total carotenoid content was 38.5 μg (g dry wt)$^{-1}$, with hydrocarbons accounting for 15.8, monohydroxy carotenoids 1.2, and polyoxy carotenoids 15.3 μg (g dry wt)$^{-1}$ (Edwards and Lee, 1986). The hydrocarbon carotenoid fraction consisted of β-carotene only. Spanish green peas contain only 2 μg (g fresh wt)$^{-1}$ β-carotene (Arqués et al., 1988).

For additional new data see Tables A8 and A11 (Izaki et al., 1986; Wills, 1987; Heinonen et al., 1989).

Chick-peas
(*Cicer arietinum* L.)
Leguminosae

Chick-peas are grown and consumed as a seed legume in the Middle East, Asia, and Africa. Asia contributes 92% to the world production. In the Middle East, chick-peas are eaten as a green vegetable at various stages of maturity when the seeds are still green and tender. These immature seeds contain carotene, which is almost absent in the dry seeds. The β-carotene of chick-pea seeds at five stages of seed development was analyzed by Abu-Shakra et al. (1970). The level of β-carotene decreased continuously with maturity. At 14 days after flowering the seed β-carotene was 1.23 μg (g fresh wt)$^{-1}$ or 52.4 μg (g dry wt)$^{-1}$, at 28 days after flowering it was 6.8 μg (g fresh wt)$^{-1}$ or 29 μg (g dry wt)$^{-1}$, and at 42 days after flowering it decreased to 3.3 μg (g fresh wt)$^{-1}$ or 13.8 μg (g dry wt)$^{-1}$.

Changes During Storage, Cooking, and Processing

Storage

Carotene content changes in green peas stored under various conditions, after harvest and before canning, were investigated (Kelley et al., 1950). A variation in carotene losses existed between the three studied cultivars, the retention being highest in Pride peas and decreasing in Shasta and Thomas Laxton cultivars. A minimum retention of 73% occurred in Pride peas when washed and held 5 h in the shade, 82% when washed and held in cold storage for 48 h, and 85% when held for 26 h in snow ice. In Shasta peas the retention was about 80% for all storage conditions. In Thomas Laxton peas the retention was considerably less: a minimum of 57% occurred in cold storage, and other retention values ranged between 73% and 88%.

The effects of postharvest handling on vitamin contents of peas were also investigated (Lee et al., 1982). During transport from the field to the plant and the off-loading in the plant (about 1 h), the carotene retention was 95%.

Quality changes of green peas during storage were investigated by Böttcher (1987). Peas with and without pods were stored at 0°C, 4°C, and 12°C to 18°C for up to 32 days. Carotene losses were especially influenced by storage temperature. The rate of carotene content decrease was highest at higher temperatures (up to 40%) and lowest at temperatures $\leq$4°C ($\leq$26%). The influence of other factors such as varietal differences, form of crop, or growing conditions was negligible compared with the effect of temperature.

Cooking

Two cooking methods were studied (Oser et al., 1943). The retention of carotene in peas following the new, improved method of cooking was 94%, whereas after old-fashioned cooking the retention was 82%. A carotene retention of 78%, 95%, and 104% in fresh cooked peas was also reported (Guerrant and O'Hara, 1953; Panalaks and Murray, 1970; Weits et al., 1970).

Processing

Blanching. There was little or no loss of carotene from peas as a result of blanching (Lee et al., 1946; Guerrant and O'Hara, 1953; Lynch et al., 1959; Aczél, 1973).

Canning. Generally, carotene retention in canned green peas was very good—100% or even more (Kelley et al., 1950; Guerrant and O'Hara, 1953; Aczél, 1973) (see Table 54). On the other hand, carotene losses in canned peas after 2 years of storage were reported (Guerrant et al., 1946).

The effects of processing on the carotene isomers content was studied in some vegetables (Panalaks and Murray, 1970). The results for green peas are given in Table 55.

Table 55. Contents of Carotene Fraction of Peas before and after Processing and Cooking

*Sample**	*Crude Carotene*	*α-Carotene*	*Neo-β-carotene B*	*Neo-β-carotene U*	*all*-trans-*β*-*Carotene*
Fresh peas	15.7	1.1	1.7	1.5	10.7
β-isomers			12.2	10.8	77.0
Canned peas	14.1	0.6	2.4	1.9	8.5
β-isomers			18.8	14.8	66.4
Dehydrated peas	7.8	0.4	0.9	1.1	5.6
β-isomers			11.8	14.5	73.7
Cooked peas	16.4	1.3	1.6	2.0	11.3
β-isomers			10.7	13.4	75.8

Source: From Panalaks and Murray (1970).
*Contents of peas are in μg (g dry wt)$^{-1}$. Values of β-isomers are in percents.

Discussion. In fresh peas the *cis*-β-isomers constituted about 11% and 12% of the β-carotene fraction. Their relative proportions changed randomly.

Recently, the effect of processing on vitamin contents of peas was investigated (Lee et al., 1982; Edwards and Lee, 1986).

The carotene content of peas was 14.9, 17.5 and 13.3 μg (g dry wt)$^{-1}$ in raw, blanched and canned peas. It was different in cans of different size, #300 cans and #10 containing 19.2 and 14.2 μg (g dry wt)$^{-1}$. The loss may be due to longer heat exposure of the #10 cans.

The different results of vitamin contents in peas processed by three plants are attributed to differences in processing practices, which therefore are susceptible to improvements.

In the second report, the pea carotenoids were separated in three fractions and determined, and the contents of all fractions were higher in the canned peas than in fresh green peas. The increase was attributed to the loss of carotenoids due to lipoxygenease activity in fresh peas during extraction. Fresh green peas blended and allowed to stand for 2 h before extraction showed a 68% decrease in total carotenoids. The loss was hindered by adding an antioxidant (pyrogallol).

Freezing. Generally, losses of carotene in peas during freezing and frozen storage are small (Fitzgerald and Fellers, 1938). Some losses occur during frozen storage of unblanched peas (Zscheile et al., 1943). During frozen storage for 11 months at −40°C, carotene losses were negligible but ranged from 7% to 26% at −18°C (Stimson et al., 1939).

Few differences in carotene content were observed in Thomas Laxton peas frozen at different rates, after freezing, after storage, or after cooking. Their carotene content ranged from 4.4 to 5.8 μg (g fresh wt)$^{-1}$ (Lee et al., 1946).

No differences in carotene retention were found between blanched, frozen peas and blanched, frozen, and cooked green peas, nor were differences noted during a storage of 12 months at −23°C (Guerrant and O'Hara, 1953).

Recently, frozen green peas, grown in Spain, were found to contain twice as much β-carotene as fresh peas, namely 1.5 and 3.2 μg (g fresh wt)$^{-1}$ (Arqués et al., 1988). See also Table A26.

Dehydration. Different results are reported about the carotene levels in dehydrated peas. In one report, dehydrated peas contained only half the amount of fresh peas, whereas another reported the same amounts of β-carotene in fresh, air-dried, or freeze-dried peas (Panalaks and Murray, 1970; Weits et al., 1970).

Pepper
(*Capsicum annuum* L.)
Solanaceae

The pepper is cultivated for its edible fruits. The fruits of the various *Capsicum* cultivars vary greatly in shape (conical, cylindrical, spherical, and bell-shaped) and in length and thickness. The mild-flavored cultivars, bell- or tomato-shaped, are usually red when ripe (also eaten at the unripe stage) and are used in salads, pickles, or cooking. The European species provide the paprika of commerce, a powdered red seasoning and food colorant that may be mild or pungent.

The beautiful red color of *Capsicum* fruits, a puzzle to scientists at the end of the nineteenth century, is due to ketocarotenoids of uncommon structure: capsanthin (3,3′-dihydroxy-β,κ-caroten-6′-one), capsorubin (3,3′-dihydroxy-κ,κ-carotene-6 ,6′-dione), and cryptocapsin (3′-hydroxy-β,κ-caroten-6′-one). For structure see under "Unique Carotenoids."

In 1927, the deep red pigment capsanthin was first isolated from the red pepper fruit in crystalline form (Zechmeister and Cholnoky, 1927), and one year later carotene was isolated and crystallized (Zechmeister and Cholnoky,

1928). One kilogram of dried pericarp meal yielded 4.0 g capsanthin and 0.55 g carotene. Subsequently, by applying Tswett's method of adsorption chromatography, Zechmeister and Cholnoky reported the separation of another red ketocarotenoid, *capsorubin* (Zechmeister and Cholnoky, 1934). Xanthophylls such as zeaxanthin, lutein, cryptoxanthin, β-carotene, and unknown polyenes were also separated. The esterifying fatty acids of the various xanthophylls were identified as myristic, palmitic, carnaubic, oleic, and unknown.

Investigation of the carotenoids of Hungarian ripe red paprika, with a specially adapted method, revealed that capsanthin is the major red pigment, followed by capsorubin (Cholnoky, 1937, 1939). Pigment separation was achieved by a combined Tswett chromatographic analysis based on selective adsorption on $CaCo_3$ and $Ca(OH)_2$. Ripe red paprika contains ketocarotenoids, zeaxanthin, cryptoxanthin, and β-carotene, with traces of lutein and α-carotene. Total carotenoids in 1 kg were 4–5.5 g and consisted mainly of capsanthin (2.2–3.5 g) and capsorubin (0.4–1.0 g). Paprika is also a valuable source of provitamin A (400–1300 IU g^{-1}). In capsaicin-free paprika (that is, sweet paprika) total carotenoids were 1.8–5.2 g, with capsanthin 0.8–3.2 g kg^{-1} and capsorubin 0.2–0.6 g kg^{-1}.

The pigments of another Hungarian red, tomato-shaped pepper (*C. annuum* var. *lycopersiciforme rubrum*) were investigated at the unripe green and the fully ripe red development stages (Cholnoky et al., 1955). The green fruits contained the chloroplast carotenoids; neoxanthin was then named *foliaxanthin,* an epoxide of unknown structure. In ripe fruits the following carotenoids were found: β-carotene accompanied by the *cis* isomers B and U, its epoxides mutatochrome and aurochrome, cryptoxanthin, cryptocapsin, capsanthin, violaxanthin, antheraxanthin, lutein epoxide, zeaxanthin, and *cis* isomers of these pigments. The carotenoids of the leaves of paprika were also analyzed, and their pattern was similar to that of the unripe fruits but with a content 65 times higher. In the ripe fruit the carotenoid content was 35 times higher than in the green fruit, increasing from 6.7 to 235 μg (g fresh wt)$^{-1}$.

Cryptocapsin, the third *Capsicum* ketone, was separated and mentioned for the first time in this report. Also the biosynthesis of the ketocarotenoids via xanthophyll epoxides by "ring cleavage" was suggested for the first time.

The carotenoids of the Hungarian yellow pepper (*C. annuum* var. *lycopersiciforme flavum*) were also investigated in unripe green and in ripe yellow fruits (Cholnoky et al., 1958b). In green fruits as in green leaves, which were also analyzed, the same chloroplast carotenoid pattern was detected. In contrast to ripe red peppers, yellow peppers are devoid of ketocarotenoids, containing instead carotenoids of the β,ε- and β,β-series, such as α- and β-carotene, α- and β-cryptoxanthin, and epoxides of the β,β-series, such as β-carotene 5,6-epoxide, antheraxanthin, and violaxanthin. Ripening involved a 12-fold increase from 4.8 to 68.0 μg (g fresh wt)$^{-1}$.

The carotenoids of red bell peppers grown in the United States were investigated in green unripe and red ripe fruits (Curl 1962, 1964). In green peppers the total carotenoid content was about 10 μg (g fresh wt)$^{-1}$. The carotenoid pattern was that of chloroplast carotenoids; minor pigments included phytoene, phytofluene, α-carotene, ζ-carotene, hydroxy-α-carotene, and cryptoxanthin. In red peppers the total carotenoid content was 127 to 284 μg (g fresh wt)$^{-1}$. The major pigment was capsanthin (35% of the total). Other identified carotenoids were β-carotene and violaxanthin (about 10% each), cryptoxanthin and capsorubin (about 6% each), and cryptocapsin (4%). Of the numerous other carotenoids found in amounts of 2%, at least eight were apparently undescribed previously, such as capsanthin 5,6-epoxide.

The carotenoids of *Capsicum* pericarp were investigated without saponification so that the different xanthophyll esters could be separated and identified (Egger, 1968). A specially adapted method of partition chromatography on oil-impregnated cellulose layers were used (Egger, 1964). All xanthophylls were combined with palmitic, myristic, and lauric acids in every possible way (i.e., simple and mixed diesters), and a complex chromatogram of more than 60 different components was obtained from at least 14 xanthophylls and 3 fatty acids.

Another report deals only with the capsanthin esters in paprika (Philip et al., 1971). Capsanthin, which amounted to 35% of the total carotenoids, occurred as the dilaurate ester.

The carotenoids of different-color cultivars (white, yellow, orange, and red fruiting ornamental pepper) were investigated and compared (Davies et al., 1970). The total carotenoid content was 855, 25, 22, and 0.7 μg (g fresh wt)$^{-1}$ in the red, orange, yellow, and white fruits, respectively, proving that enormous quantitative differences were responsible for the color differences. In addition, the three ketocarotenoids were found in the red and orange fruiting cultivars, but were absent in the yellow one. Their respective percentages in the red and orange cultivars were as follows: capsanthin, 31.7% and 35.4%; capsorubin, 7.5% and 14.3%; and cryptocapsin, 5% and 8%, respectively. Capsanthin 5,6-epoxide was also detected (4% and 6%). In the yellow fruiting cultivar four yellow carotenoids dominated: violaxanthin (31.3%), lutein (28.1%), neoxanthin (20.0%), and antheraxanthin (15.9%). Hydroxy-α-carotene (5.7%) was also found.

Based on the carotenoid distribution in the three colored *Capsicum* cultivars, a scheme of the biosynthetic steps leading to the formation of the red ketocarotenoids was outlined.

Spanish *C. annuum* fruits contain the following major carotenoids: β-carotene, cryptoxanthin, lutein, capsanthin, and capsorubin (de la Torre Boronat and Farré Rovira, 1975).

Fully green and mature red fruits of the Yolo Wonder A cultivar of *C. annuum* were analyzed for their carotenoid content (Camara and Monéger, 1978). The results are given in Table 56.

Table 56. Carotenoid Distribution in *Capsicum annuum* Fruit at Two Stages of Maturity

		Fully Green		*Mature Red*	
Carotenoid		*μg (g dry wt)$^{-1}$*	*%*	*μg (g dry wt)$^{-1}$*	*%*
β-Carotene		54.9	24.0	271.8	15.4
β-Cryptoxanthin	F,M*	20.8	9.1	217.0	12.3
Cryptocapsin	F	—	—	90	5.1
Lutein	F	79.9	34.9	—	—
Zeaxanthin	M	—	—	54.7	3.1
Antheraxanthin	D	—	—	162.4	9.2
Violaxanthin	F	33.8	14.8	125.3	7.1
Capsanthin	F,M,D	—	—	587.7	33.3
Capsanthin 5,6-epoxide	D	—	—	30.0	1.7
Capsorubin	D	—	—	181.7	10.3
Neoxanthin	F	39.1	17.1	35.3	2.0
Total		228.5		1755.9	

Source: From Camara and Monéger (1978).
*F = free; M = monoester; D = diester.

Discussion. The results are similar to those reported earlier in other red *Capsicum* fruits (Cholnoky, 1955; Curl, 1962; Davies, 1971). Carotenoid esters were found only in the ripe fruit. Capsanthin was the main carotenoid found as diester (28.8%), monoester (7.6%), and free (2.9%), whereas cryptocapsin was found only in the free form.

Total red and total yellow carotenoids were determined separately in five ripe Spanish red peppers (the *Capsicum* genus) (Mínguez Mosquera and Fernández Díez, 1981). Relationships between both values and percentage of global red pigments were determined in the fresh fruit and peel. These parameters may be used to detect possible adulteration of the commercial derived products. The results are shown in Table 57.

Discussion. In the ripe peeled fruit the percentage of red pigments is about 76%, the ratio of red carotenoids to yellow carotenoids is about 3. In the peel and in the whole fresh fruit this ratio is higher because the percentage of the global red pigments is 80% to 83%.

The carotenoids of six cultivars of Spanish paprika fruits were subsequently analyzed in detail (Mínguez Mosquera et al., 1984). Total carotenoid content as well as total red carotenoids and total yellow carotenoids were determined. Total carotenoid content in μg (g fresh wt)$^{-1}$ was 379 to 688 in the whole fruit, 482 to 964 in the peeled fruit, and 1346 to 3600 in the peel, values several times higher than those in "traditional" paprika used as sea-

Table 57. Red and Yellow Carotenoids in Various Ripe Spanish Red Pepper Cultivars [μg (g fresh wt)$^{-1}$]

	Whole Fruit		*Peeled Fruit*		*Peel*	
Cultivar	*Red*	*Yellow*	*Red*	*Yellow*	*Red*	*Yellow*
Campana	112.8	22.3	102.9	33.0	859.9	197.1
Custro cascas	92.5	19.0	138.9	53.2	968.1	224.4
Chato	75.5	15.7	101.8	31.1	1035.6	233.2
Morron	96.6	29.3	189.7	58.6	849.1	198.8
Bola	245.3	52.2	286.8	78.7	984.8	188.3

Source: From Mínguez Mosquera and Fernández Díez (1981).

soning. The following carotenoids were separated: phytoene, phytofluene, β-carotene, ζ-carotene, cryptoxanthin, zeaxanthin, a mixture of capsanthin and its isomers, a mixture of capsorubin and its isomers, and violaxanthin.

This carotenoid pattern was similar to that of Hungarian ground paprika (Baranyai et al., 1982). The carotenoids of peppers and their products were analyzed by a specially developed HPLC method. Ground paprika has a total carotenoid content of 6040 μg (g dry wt)$^{-1}$. The predominant pigment was capsanthin (55%), followed by zeaxanthin (24%), violaxanthin (6.6%), capsorubin (5.5%), β-carotene (4.6%), β-cryptoxanthin (3.9%), and cryptocapsin (0.4%). Other pigments, including antheraxanthin, mutatoxanthin, and capsanthin epoxide, were also separated along with four unidentified additional pigments. The ripe pepper *C. annuum* var. *lycopersiciforme rubrum* has a total carotenoid content of 392 μg (g fresh wt)$^{-1}$. Its carotenoid distribution is shown in Table 58 and compared with data from other reports previously discussed.

Discussion. The typical red *Capsicum* carotenoid pattern is a mixture of pattern 6, characterized by the predominance of unique carotenoids, namely the red ketocarotenoids, and pattern 4, relatively large amounts of β-carotene and its hydroxy derivatives cryptoxanthin and zeaxanthin.

In Mexico, where consumption of peppers is a tradition, peppers are eaten at the immature green stage. The provitamins α-carotene, β-carotene, and β-cryptoxanthin were determined by HPLC, and the vitamin A value was calculated in five commonly eaten cultivars (Mejía et al., 1988). In four different cultivars the provitamin ranges were as follows: α-carotene, 20 to 109 μg (100 g fresh wt)$^{-1}$; β-carotene, 170 to 599 μg (100 g fresh wt)$^{-1}$; and cryptoxanthin, 1.6 to 7.0 μg (100 g fresh wt)$^{-1}$. The vitamin A value was 26–156 RE (100 g fresh wt)$^{-1}$.

Table 58. Carotenoid Distribution in Red *Capsicum annuum* Fruits

	Percent of Total Carotenoids			
Carotenoid Pattern	*(1)*	*(2)*	*(3)*	*(4)*
Capsanthin	34.7	31.7	33.3	38.1
β-Carotene	11.6	12.3	15.4	18.6
Violaxanthin	9.9	9.8	7.1	7.9
Cryptoxanthin	6.7	7.8	12.3	4.2
Capsorubin	6.4	7.5	10.3	9.5
Cryptocapsin	4.3	5.0	5.1	1.8
Zeaxanthin	2.3	6.5	3.1	4.0
Antheraxanthin	1.6	9.2	9.2	5.0
Capsanthin epoxide	0.9	4.2	1.7	2.6

Sources: Column 1 from Curl (1962); column 2 from Davies et al. (1970); column 3 from Camara and Monéger (1978); column 4 from Baranyai et al. (1982).

Changes in Ripening Peppers

Systematic investigations of pigment changes in ripening fruits have been carried out only recently, since they provide a basis for detailed studies on carotenogenesis as a function of ripening. An exception was the pepper fruit whose pigment changes during ripening were followed very early even when the importance of this investigation was not yet established. It was done mainly because the red pigments isolated from the ripe red pepper were recognized as uncommon and their structure was unknown. After structure elucidation the investigations were intensified to elucidate the biosynthetic pathway leading to the formation of the unique pepper carotenoids. Carotenoids in both green and ripe pepper were investigated (Cholnoky et al., 1955, 1958; Curl, 1962, 1964; Camara and Monéger, 1978).

The first systematic investigation of carotenoids of peppers of increasing maturity was done to clarify certain biosynthetic steps leading to capsanthin and related compounds (Valadon and Mummery, 1977). Five stages of fruit maturity (green, yellow, pale orange, deep orange, and red) of *C. annuum* var. *grossum* were investigated. The total carotenoid content underwent a tenfold gradual increase from 227 to 2278 μg (g dry wt)$^{-1}$. The qualitative change was the transformation of the chloroplast carotenoid pattern into a typical *Capsicum* carotenoid pattern dominated by the unique ketocarotenoids. At the yellow stage the predominant carotenoid was violaxanthin (41%); the other chloroplast carotenoids decreased. The first red ketocarotenoid, capsanthin (8.8%), was present in the yellow fruit and increased gradually to 31% to become the major pigment in the red fruit. Capsorubin was found at low levels and stayed constant at 4%. Capsanthin 5,6-epoxide was detected first at the pale orange stage. Cryptocapsin was absent, and antheraxanthin, the immediate precursor of capsanthin, was not found either, probably due to a rapid turnover. Instead, *trans*- and *cis*-zeaxanthin accumulated in the red fruit. Biosynthetic pathways of capsanthin and related compounds modified from Davies et al. (1970) were proposed in correlation with the obtained results.

A similar investigation was carried out in which the carotenoid and chlorophyll changes during ripening were followed in five Australian *Capsicum* cultivars at four ripening stages (Rahman and Buckle, 1980). The results of three cultivars, two with red fruits (Long Red Cayenne and Pacific Bell) and one with ripe yellow fruit (Golden Californian Wonder), are shown in Table 59.

Discussion. In all three cultivars a substantial increase of the total carotenoid content occurred during ripening, 30-fold in the red fruits and 16-fold in the yellow ones. At the green stage the carotenoid pattern was that of chloroplast carotenoids. During development all four carotenoids decreased, but β-carotene was still found at relatively high levels in the ripe fruits. The transformation of the chloroplast carotenoid pattern to a typical *Capsicum* carotenoid pattern began at the pink stage, with enough capsanthin and capsorubin present to influence the color. In the red fruit red pigments dominated, their total levels being 55% and 62%.

In the yellow cultivar the chloroplast pigments are still the major carotenoid pattern, but in the greenish yellow stage the pattern becomes complex as more than 15 other carotenoids accumulate at relatively high levels, especially the β,β-series, including their epoxides.

With a specially adapted method of overpressured layer chromatography, the formation of the main ketocarotenoids during ripening was followed in the Hungarian *Capsicum* cultivar SZ-20 (Aczél, 1986, 1988). Capsanthin and capsorubin were detected in the half-ripe fruit and increased spectacularly to 62% in the ripe fruit.

A systematic study following the carotenoid changes at five ripening stages in cultivar SZ-20 was also carried out (Biacs et al., 1989). The pigments were separated by r-p HPLC without saponification. The presence of phytofluene at high levels at the immature stage and its decreasing during ripening, the presence of neurosporene (never reported in pepper fruits), and the increase of the β-carotene level during ripening are all unusual. The major carotenoids were capsanthin and capsorubin, which increased during the process and were already esterified when they appeared. Capsanthin dominated, its monoester containing mostly unsaturated fatty acids $C_{18:2}$. Diesters of both ketocarotenoids contained fatty acids such as C_{12}, C_{14}, and C_{16}.

Table 59. Carotenoid Changes in *Capsicum* Cultivars During Ripening*

	Long Red Cayenne				*Pacific Bell*				*Golden Californian Wonder*			
	Green *1*	*Green* *2*	*Pink* *3*	*Red* *4*	*Green* *1*	*Green* *2*	*Pink* *3*	*Red* *4*	*Green* *1*	*Green* *2*	*Greenish-yellow* *3*	*Yellow* *4*
Carotenoid Pattern	*Percent of Total Carotenoids*											
Phytoene	—	—	—	—	—	—	—	—	—	-	1.0	1.0
Phytofluene	—	—	0.3	0.4	—	—	tr	—	—	1.0	2.4	7.4
α-Carotene	4.2	7.1	2.9	1.2	—	—	0.5	0.5	—	—	1.0	2.1
β-Carotene	33.3	28.3	23.1	14.8	28.6	24.7	4.0	7.5	32.3	30.6	22.0	16.6
ζ-Carotene	—	—	—	—	—	—	—	—	—	—	2.0	3.1
β-Carotene 5,6-epoxide	—	—	—	—	—	—	—	—	tr	1.0	0.5	0.2
β-Carotene 5,6,5′6′-diepoxide	—	—	—	—	—	—	—	—	tr	4.1	2.4	3.1
Mutatochrome	—	—	1.2	1.2	—	—	1.3	1.5	—	—	—	—
Hydroxy-α-carotene	—	—	—	—	—	—	—	—	—	—	0.7	2.1
Cryptoxanthin	—	—	2.3	5.1	—	—	1.3	—	—	—	1.0	7.4
Cryptoxanthin 5,6-epoxide	—	—	—	—	—	—	—	2.5	—	—	2.4	3.1
Cryptoxanthin 5,6,5′6′- diepoxide	—	—	—	—	—	—	—	—	—	—	0.5	2.1
Lutein	33.3	28.3	0.3	—	28.6	24.7	0.3	—	48.4	40.8	21.9	9.4
Cryptocapsin	—	—	—	—	—	—	2.6	1.5	—	—	—	—

(continued)

Table 59. Continued

	Long Red Cayenne				*Pacific Bell*				*Golden Californian Wonder*			
	Green 1	*Green 2*	*Pink 3*	*Red 4*	*Green 1*	*Green 2*	*Pink 3*	*Red 4*	*Green 1*	*Green 2*	*Greenish-yellow 3*	*Yellow 4*
Carotenoid Pattern	*Percent of Total Carotenoids*											
Zeaxanthin	tr	0.9	13.9	11.3	tr	1.2	2.6	9.0	3.2	4.1	7.3	5.2
Antheraxanthin	—	—	1.7	2.9	—	—	0.8	3.5	—	—	1.0	5.2
Mutatoxanthin	—	—	0.5	0.7	—	—	0.5	1.5	—	—	0.5	4.1
Violaxanthin	20.8	25.9	9.2	5.8	28.6	24.7	4.0	7.5	9.7	10.2	17.1	10.4
Luteoxanthin	—	—	—	—	—	—	—	—	—	—	2.4	4.1
Auroxanthin	—	—	—	—	—	—	—	—	—	—	2.4	6.2
Neoxanthin	8.3	9.4	2.3	0.5	14.2	24.7	1.3	1.0	6.4	8.2	1.7	1.0
Capsanthin	—	—	24.9	31.3	—	—	38.4	32.5	—	—	—	—
Capsorubin	—	—	5.8	9.3	—	—	15.9	13.5	—	—	—	—
Capsanthin isomer	—	—	7.5	13.3	—	—	18.5	14.0	—	—	—	—
Capsorubin isomer	—	—	1.2	1.0	—	—	4.0	2.0	—	—	—	—
Unidentified	—	—	2.9	1.2	—	—	4.0	2.0	—	—	9.8	6.2
Total carotenoid content [μg (g fresh wt)$^{-1}$]	24.0	42.4	173.0	728.0	7.0	8.1	75.6	200.0	6.2	9.8	41.0	96.2

Source: Adapted from Rahman and Buckle (1980).

Note: tr = trace.

*Ripening stage: 1 = immature; 2 = mature; 3 = half-ripe; 4 = fully ripe.

Changes During Storage and Dehydration

Storage

The retention of carotene in Yolo Wonder green bell peppers was 94% and 78% after 7 and 14 days of storage at 7.2°C and 83% after 7 days at 21°C (Matthews et al., 1975). These changes are explained by the influence of temperature on the biosynthetic rate and that during ripening the carotene levels gradually decrease.

Dehydration

The most important product of peppers is paprika, the dry powder obtained by grinding ripe dry pods. Ground paprika loses color, with subsequent browning, on storage at a rate influenced by cultivar, harvesting conditions, drying and storage conditions, and light and temperature (Lease and Lease, 1956a, 1962).

The quantitative changes of the main carotenoids capsanthin, capsorubin, β-carotene, and zeaxanthin were followed both in red pepper powder and dried red pepper during a 6-month storage (Malchev et al., 1982). Drying temperatures of 60°C, 70°C, 80°C, and 90°C were used. The total carotenoid content varied only slightly with drying temperature. Nonetheless, its effect was revealed during the storage, with carotenoid degradation occurring at a higher rate in samples dried at higher temperatures. At all drying temperatures the carotenoids in the dried red peppers were more stable than in the powder. Capsanthin and capsorubin were the most stable.

To understand the mechanism of the photocatalyzed oxidation of the paprika carotenoids, various investigations were carried out. According to earlier reports, retardation of color loss by antioxidants indicated an autoxidation mechanism, or the color loss was attributed to an oxidative attack catalyzed by light (van Blaricom and Martin, 1951; Lease and Lease, 1956b, 1962). Commercial paprika was subjected to severe bleaching (3 months exposure to sunlight), and the carotenoids were compared in fresh and bleached paprika (de la Mar and Francis, 1969). They were separated by countercurrent and chromatographic methods.

The bleaching effect accounted for a loss of almost 96% of the carotenoid content. There were 54 and 37 carotenoids isolated from bleached and unbleached paprika, respectively, of which 33 and 21 were completely or tentatively identified. Sixteen of the known pigments of unbleached paprika were still present in the bleached sample. The bleached sample contained an increased number of isomers and oxygenated derivatives of capsanthin and capsorubin, which appeared to contain additional keto groups.

The oxidation of capsanthin by molecular oxygen at 40°C was studied by Philip and Francis (1971). The process involved the oxidation of the hydroxy group at the cyclopentanol ring into carbonyl, leading to the transformation of capsanthin into capsanthone (3-hydroxy-β,κ-carotene-3',6'-dione). A further oxidation at the carbon 7',8' double bond occurred, and citraurin in small amounts was tentatively detected. The oxidation products did not accumulate but were rapidly oxidized to low-molecular compounds.

The mechanism of the photocatalyzed oxidation of paprika pigments was reported by Carnevale et al., (1980). The rate curves for color loss indicate an autocatalytic reaction with an induction period dependent on the intensity of light and susceptible to protection by antioxidants. A free-radical-type reaction was suggested.

On the other hand, bleaching of paprika carotenoids may be catalyzed by enzymes, a hypothesis not yet proved.

Two water-soluble factors with carotene bleaching activity were isolated from pepper: ascorbic acid and a peroxidase-like protein (Kanner et al., 1976, 1977). When paprika peels are ground together with seeds that form about 15% of the product, a remarkable decrease in color intensity was observed during storage of these products (Biacs et al., 1989). Followed in parallel in paprika with and without seeds, the HPLC profiles were different; seeds reduced the total pigments and degraded the most important carotenoids. This effect is attributed to lipoxygenase, which was found to be present in paprika seeds (Daood and Biacs, 1986). The degradation of individual paprika carotenoids incubated with paprika seed lipoxygenase and linoleic acid was followed also by HPLC. The main carotenoids were tested in free and esterified forms. The carotenoid esters were more stable than free pigments. Furthermore, capsanthin esters containing saturated fatty acids better resisted the enzymatic oxidation.

Evidence for the enzymatic degradation of the paprika carotenoid has to be brought. In connection with this topic it should be pointed out that peppers are a good source of ascorbic acid, and the glycerides of whole ground paprika contain 66% linoleic acid (Philip et al., 1971; Matthews et al., 1975).

Potato
(*Solanum tuberosum* L.)
Solanaceae

The potato (common potato, white potato, or Irish potato) is one of the main food crops of the world. The edible part of the plant is a tuber (i.e., the swollen end of an underground stem). The potato is a native of South America, where it was cultivated about 2000 years ago. It was introduced in Europe in the sixteenth century. In the seventeenth century it was a major crop in Ireland, and in the following century it was cultivated particularly in Germany and western England.

The skin color of the tuber varies from brownish to deep purple, and the flesh is normally white to yellow, but occasionally purple. The color of potatoes, white to yellow, is considered an important quality index both for the potato and its products. The potatoes, even white ones, contain carotenoids, although at very low levels. This fact was not recognized until about 1940 when some contradictory data was published (Lampitt and Goldenberg, 1940).

Investigation of potato carotenoids progressed slowly probably due to the numerous technical problems posed by material with a high-starch, low-carotenoid content. Therefore the early published data differ so much that they are hardly comparable. To allow the reader to readily comprehend the quantitative results, they are given as μg (100 g wt)$^{-1}$.

The first proof that the yellow color of the potato flesh was imparted by carotenoids was given when the Kipfler potato was analyzed in connection with "yellow fleck" disease (Schmid and Lang, 1939). It was found that three carotenoids are present: one with the absorption spectrum of β-carotene was identified as α-carotene, another with the absorption spectrum of lutein was identified as β-carotene, and a third was "taraxanthin or violaxanthin."

The total carotenoid content of yellow-fleshed varieties grown in northern Sweden was 80 to 260 μg (100 g fresh wt)$^{-1}$ with a carotene content between 7 μg and 10 μg (von Euler et al., 1943). The potatoes were steamed because the carotenoids could not be extracted completely from raw material.

The relation between the color of the potato flesh and the total carotenoid content was first given by Caldwell et al. (1945). Total carotenoids in 19 white-fleshed potato varieties ranged from 14 to 54 μg (100 g fresh wt)$^{-1}$, and in three yellow-fleshed varieties they ranged from 110 to 187 μg (100 g fresh wt)$^{-1}$.

The carotenoids of the Katahdin potato grown in Maine and Louisiana in the United States, harvested in a relatively immature condition, were analyzed qualitatively and quantitatively (Brunstetter and Wiseman, 1947). The total carotenoid content was 60 μg (100 g fresh wt)$^{-1}$ (recalculated values) and contained β-carotene (6 μg), lutein (10–16 μg), and ζ-carotene (2.2 μg). Flavoxanthin, auroxanthin, and isolutein were also tentatively identified.

The carotenoids of nine German potato varieties were analyzed in detail (Kasim, 1967a,b). They were separated by Craig's countercurrent distribution method followed by paper chromatography. The following pigments were identified: β-carotene 5,6,5,6'-diepoxide [33–108 μg (100 g fresh wt)$^{-1}$], lutein [30–119 μg (100 g fresh wt)$^{-1}$], lutein 5,6-epoxide [81–257 μg (100 g fresh wt)$^{-1}$], violaxanthin [8–29 μg (100 g fresh wt)$^{-1}$], and tentatively identified neoxanthin. The total carotenoid content ranged from 199 to 560 μg (100 g fresh wt)$^{-1}$.

In Canadian white potatoes the carotenoid pattern was less complex (Lepage, 1968). Nearly half of the carotenoids was lutein (48.5%); the rest were α-carotene (6.4%), β-carotene (16.3%), an unidentified pigment (14.2%), and lutein 5,6-epoxide (14.8%).

Reliable results were obtained only in the last decade due to the use of a specially adapted HPLC (Iwanzik et al., 1983). The carotenoid distribution of 13 German potato cultivars grown in Saxony under identical conditions was analyzed. The results are given in Table 60.

Discussion. The total carotenoid content was 27 to 343 μg (100 g fresh wt)$^{-1}$. Cultivars with about 300 μg (100 g fresh wt)$^{-1}$ were intense yellow, whereas the cultivars with 30 to 70 μg (100 g fresh wt)$^{-1}$ were white-fleshed. The carotenoid pattern is a modified chloroplast carotenoid pattern, characteristic for amyloplasts and resembling that found in etioplasts—that is, a low carotene content and xanthophylls dominated by epoxides. With the method used, lutein could not be separated from zeaxanthin, nor lutein epoxide from antheraxanthin. All-*trans*-neoxanthin was accompanied by an isomer that was not further characterized. In yellow potatoes the carotenoid pattern was dominated, up to 80%, by a mixture of epoxides, with violaxanthin (40–70%) as the major carotenoid, followed by lutein epoxide (up to 30% in some cultivars). Lutein was the major pigment in white potatoes, even exceeding the levels found in chloroplasts, and the second major pigment in yellow potatoes. Both β-carotene and neoxanthin were found at much lower levels than in chloroplasts.

In a yellow potato tuber a typical carotenoid distribution in μg (100 g fresh wt)$^{-1}$ is the following: β-carotene, 3–5; lutein (+ zeaxanthin), 40–70; lutein epoxide (+ antheraxanthin), 15–18; violaxanthin, 80–110; neoxanthin, 4–6; and neoxanthin A, 8–10 (Tevini et al., 1984b).

Xanthophylls in the potato tuber are partially esterified (Costes et al., 1976; Fishwick and Wright, 1980). A specially adapted HPLC method permitted the quantitative determination of the mono- and diester carotenoid fractions in various potato cultivars (Tevini et al., 1984a). The results are given in Table 61.

Discussion. In all yellow potatoes, the carotenoid diesters dominated, ranging from 20% to 30%, whereas in white potatoes (e.g., Sommerstaerke) esters were almost absent. The following acylating fatty acids were identified and quantified: palmitic (38%), linoleic (21%), stearic (20.4%), linolenic (9.4%), myristic (4.9%), and oleic (0.9%). Individual esters were not identified.

Potato carotenoids are not uniformly distributed within the tuber (Tevini et al., 1984b). This variation is illustrated in Table 62.

Table 60. Carotenoid Distribution in 13 German Potato Cultivars (*Solanum tuberosum*)

	Monza	*Granola*	*Assia*	*Roxi*	*Irmgard*	*Culpa*	*Terrine*	*Tempora*	*Saturna*	*Kri-stalla*	*Sommer-stärke*	*Illse*	*Kero*
Carotenoid Pattern	*Percent of total carotenoids*												
β-Carotene	—	—	—	—	—	—	—	0.9	—	—	—	—	—
Lutein	25.7	15.5	23.0	22.0	25.9	17.4	32.9	37.9	28.8	54.7	51.1	35.1	57.3
Lutein epoxide	14.5	5.7	26.5	6.5	19.8	20.9	25.6	18.0	29.9	10.9	7.5	15.3	9.9
Violaxanthin	55.2	74.4	46.0	67.8	44.8	56.2	24.9	39.5	36.2	29.6	20.6	43.7	28.8
Neoxanthin	4.6	3.5	4.5	3.7	9.5	5.5	16.6	3.8	5.1	4.8	20.8	5.9	4.0
Total carotenoids [μg (100 g fresh wt)$^{-1}$]	342.7	328.9	192.6	171.2	157.0	150.2	113.5	112.8	85.9	73.5	65.5	60.9	27.4

Source: From Iwanzik et al. (1983).

Table 61. Distribution of Free and Esterified Carotenoids in Various Potato Cultivars

	Percent of Total Carotenoids		
Cultivars	*Free*	*Monoesters*	*Diesters*
Irmgard (Lüben)	71.1	1.8	27.1
Irmgard (Ohrdorf)	68.6	3.0	28.4
Saturna (Lüben)	75.1	2.0	22.9
Saturna (Ohrdorf)	76.5	1.7	21.8
Ilse	83.4	2.6	13.9
Hertha	71.6	2.0	26.4
Sommerstärke	~98.0	—	~1.0

Source: From Tevini et al. (1984a).

Table 62. Carotenoid Distribution Within the Tuber of Potato cv. Saturna [μg (100 g fresh wt)$^{-1}$]

Carotenoid Pattern	*Skin (1–2 mm)*	*Cortex (5 mm)*	*Pith*
Neoxanthin	21.8	30.9	21.0
Violaxanthin	75.0	96.8	74.8
Lutein epoxide	11.8	14.4	7.3
Lutein	80.5	68.3	48.0
Carotenoid diester	24.6	30.1	28.6
Total carotenoids	213.7	240.5	179.7

Source: From Tevini et al. (1984b).

Localization

The carotenoid distribution in subcellular fractions of potato tuber cells was investigated (Costes et al., 1976). All fractions contained carotenoids, the highest concentrations being in the microsomal fraction, light membranes, and mitochondria. Phytofluene, ζ-carotene, and α-carotene were identified in the microsomal fractions. Lutein, violaxanthin, and neoxanthin were found free or esterified in the light membranes and mitochondria.

Later, the envelope membranes of amyloplasts from potatoes were isolated and characterized (Fishwick and Wright, 1980). The envelope lipid fraction contained 0.7% to 1% w/w carotenoids. The following carotenoids were identified and quantified: β-carotene 5,6-epoxide (7%), cryptoxanthin (8%), lutein and zeaxanthin (24%), antheraxanthin (11%), violaxanthin (30%), and neoxanthin (20%); 17% to 22% of the xanthophylls were esterified.

The total carotenoid content of whole tubers was 1.6 μg (g fresh wt)$^{-1}$, with a similar carotenoid distribution and the same proportion of carotenoid esters.

The potato amyloplast envelope represents an additional example that demonstrates that there is a high degree of uniformity in the structure and chemical composition of the envelope membranes among the different plastid types. In all these envelopes a modified chloroplast carotenoid pattern is found, with epoxides such as violaxanthin and lutein 5,6- or 5,8-epoxide predominant. This pattern exists in etiolated French bean seedlings (Goodwin and Phagpolngarm, 1960), in etiolated mung bean seedlings (Valadon and Mummery, 1969), in cauliflower bud tissue (Prebble and West, 1975), and in the etioplasts of avocado fruit pulp (Gross, 1987). This special oxidative metabolism may be caused by the absence of light. Chloroplast envelopes prepared from dark-adapted spinach leaves have a similar epoxide/nonepoxide ratio (2:1) (Jeffrey et al., 1974). The oxidative metabolism in the envelope membranes of nongreen plastids, such as etioplasts and amyloplasts, still needs to be investigated.

Influence of Different Factors on Carotenoid Distribution

Besides intra- and intervarietal differences, the changes during development and environmental factors were also followed. A single early investigation followed the carotenoid distribution changes in six British varieties over three maturity stages (Pendlington et al., 1960). The eight most abundant pigments were common to all varieties and were independent of maturity. They were tentatively identified as β-carotene, β-carotene 5,6-monoepoxide, cryptoxanthin 5,6,5′,6′-diepoxide, lutein, *cis*-violaxanthin, *cis*-antheraxanthin, *cis*-neoxanthin, and an unidentified carotenoid. The carotenoids of King Edward potatoes were analyzed during ten maturity stages. Total carotenoid content paralleled climatic conditions favoring rapid growth until the seventh harvest (127 days after planting).

Examination of the levels of individual carotenoids showed that during rapid growth the epoxides were especially abundant, particularly *cis*-violaxanthin. The epoxide/nonepoxide ratio for these periods varied from 2.5 to 3.5. As the potatoes matured, the value approached unity.

Variations in individual and total carotenoids in two German potato cultivars grown in two locations were investigated by Tevini et al. (1984b). The results are shown in Table 63.

Discussion. The location had no significant effect on carotenoid distribution.

Table 63. Carotenoid Distribution in Two Potato Cultivars Grown in Different Locations

Carotenoid Pattern [μg (100 g fresh wt)$^{-1}$]	Cultivars			
	Irmgard (Lüben)	*Irmgard (Ohrdorf)*	*Saturna (Lüben)*	*Saturna (Ohrdorf)*
Neoxanthin	23.0	23.4	15.5	16.1
Violaxanthin	141.0	151.7	83.2	88.6
Lutein epoxide	24.7	26.6	15.4	15.5
Lutein	41.2	47.2	43.7	46.9
β-Carotene	3.4	3.7	3.4	3.8
Carotenoid diester	89.6	105.6	48.8	48.4
Total carotenoids	322.9	358.2	210.0	219.3

Source: From Tevini et al. (1984b).

Changes During Storage, Cooking, and Processing

Carotenoid changes during storage, cooking, and processing in potatoes were investigated systematically in German potatoes by HPLC. The results will be discussed further.

Storage

In an early report it was found that the ratio of epoxide to nonepoxide carotenoids, which in mature potatoes is 1, was 0.8 in stored potatoes. A similar change occurred in stored immature potatoes (Pendlington, 1960).

The carotenoid changes during a 6-month storage were followed in eight German potato cultivars (Tevini et al., 1984a,b). The levels of total carotenoids, individual carotenoids, and carotenoid diesters were determined. In some yellow potato cultivars (Tempora, Irmgard, and Saturna) the total carotenoid content increased by 15%, whereas in Clivia and Ilse it decreased by about 5% to 10% but remained unchanged in white Sommerstärke.

In Clivia, with 500 μg (100 g fresh wt)$^{-1}$ total carotenoids, the level of all free xanthophylls (lutein, lutein 5,6-epoxide, violaxanthin, and neoxanthin) decreased, especially violaxanthin. On the other hand, the level of the carotenoid diesters increased by 80%. This increase of xanthophyll diesters was also observed in seven other potato cultivars. In Hertha and Saturna it almost doubled, whereas in Irmgard and Tempora it increased by about 30% and remained unchanged in Ilse and Sommerstärke.

The irregular fluctuations of the levels of free and esterified carotenoids during storage may be attributed to intercultivar differences or to different initial maturity stages of the harvested potatoes. This is also valid for the esterification degree, which is known to increase in ripening fruits and in senescent tissue.

Cooking

Carotene content was determined in potatoes after two heat treatments (96–98°C and 190–200°C) (Simeonova, 1965). An apparent increase in carotene content was observed. The carotene content was 14.6, 18.3, and 16.7 μg (100 g dry wt)$^{-1}$ in raw, steamed, and baked spring potatoes, and 10.2, 37.6, and 21.1 μg (100 g dry wt)$^{-1}$ in raw, steamed, and baked autumn potatoes.

For testing the thermostability of potato carotenoids during cooking, Saturna potatoes were cooked at 90°C and 100°C and for different times (Tevini et al., 1984a,b). At those temperatures all free xanthophylls were rapidly degraded, especially violaxanthin, the main pigment. After 10 min cooking at 90°C or 100°C more than half the violaxanthin was destroyed, and after 30 min it totally disappeared. In contrast, the carotenoid diesters were much more stable. After 10 min and 30 min the retention was 88% and 60%, respectively.

Changes of individual carotenoids during cooking may be followed in Table 64, which shows the preparation of potato granules, the initial two processing stages involving cooking. Violaxanthin is the least stable. It is reduced to half its level at the precooking stage and disappears totally at the cooking stage, either being destroyed or partially converted into luteoxanthin and auroxanthin. Lutein, in contrast, is stable and decreases only slightly during cooking. Xanthophyll diesters were relatively stable; 15% of their initial level was destroyed after precooking and 65% during cooking. After the first "wet phase," 60% of the initial total carotenoid amount still exists in the processed potatoes.

Processing

Freezing and dehydration are used for potatoes.

Dehydration. Although minor constituents, the lipids and carotenoids, being highly susceptible to autoxidation, are responsible for the autoxidative degradation of the dehydrated potato products, which results in bleaching and off-flavor (Stute, 1981).

The effect of drying on the total carotenoid content was followed in the German cultivars Grandifolia diced and Culpa sliced (Tevini et al., 1984b). High carotenoid losses (up to 50%) occurred in both cultivars, independent of the product's form. During a 7-week storage of the dehydrated products, the carotenoid losses continued, being greater at higher storage temperatures.

Later, the carotenoid changes were followed during all processing stages involved in the preparation of dehydrated potatoes (Tevini and Bergthaller, 1985). The cultivar Grandifolia with an initial total carotenoid content of 600 μg (100 g fresh wt)$^{-1}$ was studied.

Peeling of the potatoes caused a loss of 20% to 30% of the total carotenoids, which is explained by the higher carotenoid concentration existing in peels. During cutting, a further loss of about 10% occurs, due to the enzymatic activity of peroxidase and lipoxygenase. Blanching in water or steaming caused a further loss of 10% to 20% with no differences between the two methods, the total carotenoid retention being 40%. The losses were produced exclusively by heat; the enzyme was denatured in the blanching process. At this stage unidentified carotenoids, probably *cis* isomers, appeared. Their concentration increased in succeeding steps. During drying, which involved prolonged heating in air, a large amount of carotenoids is oxidized, and retention decreases to 8%. Under optimal drying conditions, the retention may be increased to 25%.

During storage of dehydrated potatoes at 35°C, the carotenoid content decreased gradually, the retention being 30% and 10% after a 9-week and 36-week storage with a higher degradation rate at the beginning.

Qualitative and quantitative carotenoid changes during the preparation of potato granules were investigated by Tevini et al. (1986). The results are given in Table 64.

Discussion. After the "wet" steps of cooking, carotenoid retention was about 60%. During drying, various degradation products, such as 5,8-epoxides (luteochrome, mutatoxanthin), were formed. In the end product the total carotenoid content fell to 20% to 40% of that found in fresh potatoes. The most stable carotenoid was lutein during processing and storage of the granules. Storage in inert atmosphere or vacuum cannot hinder carotenoid degradation.

Sweetpotato
(*Ipomoea batatas* L.)
Convolvulaceae

The sweetpotato is a food plant native to tropical America and widely cultivated in tropical America, in the southern United States, the warmer islands of the Pacific, in Japan, and the USSR. The edible part is the much enlarged tuberous root, which varies in shape from fusiform to oblong or pointed. A valuable source of carbohydrates in the human diet, the sweetpotato is eaten cooked.

Table 64. Qualitative and Quantitative Carotenoid Changes During Preparation of Potato Granules [μg (100 g dry wt)$^{-1}$]

	Processing Stages				
Carotenoid Pattern	*Raw*	*Precooking*	*Cooking*	*Drying*	*End Product*
Neoxanthin	96.2	60.2	15.5	17.5	9.2
Neoxanthin isomer	39.4	19.0	13.6	26.9	8.8
Violaxanthin	582.2	260.2	—	22.5	19.9
Luteoxanthin	—	—	29.8	33.1	—
Antheraxanthin	52.9	44.6	—	—	—
Auroxanthin	—	—	31.4	68.5	30.6
Mutatoxanthin	—	16.0	39.3	51.5	16.6
Mutatoxanthin isomer	—	18.8	59.5	74.5	20.7
Lutein	267.7	233.9	246.8	260.1	129.6
β-Carotene	20.2	—	—	—	—
Total diesters	146.6	124.2	143.6	103.5	19.9
Total carotenoid (HPLC)	1213.3	777.0	496.0	629.5	245.9
Total carotenoids (spectrophotom)	1149.0	1052.7	745.1	651.6	264.6

Source: From Tevini et al. (1986).

The color of the root flesh ranges from white through light yellow, deep yellow to orange and occasionally purple, and the color of the peel is brown or rose to purple. All yellow hues are due to carotenoids at various levels, even in the white roots. The investigation of the carotenoids began in the early 1930s and is still going on.

Since β-carotene is the major carotenoid, its content only was determined, and reported to be 12, 51, 37 to 51, and 32 to 70 μg (g fresh wt)$^{-1}$ (Fraps, 1933; Miller and Covington, 1942; Kimbrough et al., 1946; O'Connor et al., 1946). The Porto Rico sweetpotato, the most studied cultivar, was used as a source for commercial preparation of pure β-carotene (Sherman and Koehn, 1948). The carotenoids of other yellow sweetpotato varieties were investigated, and, for the first time, both total carotenoids and carotene were determined (Ezell and Wilcox, 1946, 1948). The results are given in Table 65.

Discussion. They showed that sweetpotatoes contain appreciable amounts of other carotenoids. Tuber flesh of a more intensive yellow color has a higher total carotenoid content, the decreasing depth of color corresponding to the decreasing levels of total carotenoids and β-carotene per-

Table 65. Carotenoid Distribution in Different Sweetpotato Varieties

Variety	*Total Carotenoids [μg (g fresh wt)$^{-1}$]*	*Carotene [μg (g fresh wt)$^{-1}$]*	*Carotene (% tot. carot.)*
Maryland Golden	58.4	51.8	88.7
Porto Rico	43.7	36.0	82.4
Nancy Hall	18.1	10.0	55.2
Vineland Bush	7.3	2.4	32.9
Southern Queen	4.7	1.3	27.7

Source: From Ezell and Wilcox (1946); copyright ©1946 by the AAAS.

centage. Hence the color intensity of the flesh may be considered a reliable indication of the vitamin A value. Thus it is appropriate to apply color difference meter measurements, especially of the *a* value, for rapid estimation of carotenoid content (Lauber et al., 1967).

Besides intervarietal differences, there are intravarietal differences between sweetpotatoes. In 27 individual roots of Porto Rico sweetpotatoes, total carotenoid content was 21 to 69.5 (g fresh wt)$^{-1}$ (Ezell and Wilcox, 1946). Individual roots from the same plant varied in carotene as much as 47% in Orange Little Stem and 82% in Yellow Jersey sweetpotatoes (Ezell and Wilcox, 1958).

Moreover, carotenoids are not uniformly distributed within the roots: total carotenoids [μg (g fresh wt)$^{-1}$] were 31.5 to 67.5 in the stem end, 22.6 to 44.6 in the center, and 14.3 to 34.4 in the root end; β-carotene varied accordingly.

Genetic studies initiated 30 years ago resulted in improved cultivars with higher carotenoid levels (Cordner et al., 1959). The carotenoid contents of such sweetpotatoes, as affected by cultivar and size, were investigated by several authors (Lanier and Sistrunk, 1979; Reddy and Sistrunk, 1980). The results are given in Table 66.

Table 66. Effect of Cultivar and Size on the Total Carotenoid Content of Sweetpotatoes [μg (g fresh wt)$^{-1}$]

Cultivar		
Georgia Set	80	98.9
Goldrush	118	141.0
Jasper	155	193.1
Centennial	185	202.5
Size		
small	126	159.2
large	142	158.5

Sources: Column 2 from Lanier and Sistrunk (1979); column 3 from Reddy and Sistrunk (1980).

Total carotenoid content of orange-fleshed sweetpotatoes, the four major cultivars grown in the United States, was 50 to 95 μg (g fresh wt)$^{-1}$ (Picha, 1985) (see Table 68).

The first detailed analysis of the carotenoids of yellow Goldrush sweet potatoes was done by Purcell in 1962. The total carotenoid content was 462 μg (g dry wt)$^{-1}$. The carotenoids were fractionated chromatographically into carotene (98%), monohydroxy carotenoids (1.6%), and polyhydroxy carotenoids (0.84%). Seven pigments were identified and determined: phytoene (2.6%), phytofluene (0.8%), β-carotene (89.9%), ζ-carotene (1.2%), β-carotene 5,8-epoxide (2.5%), γ-carotene (0.7%), and hydroxy-ζ-carotene (0.5%). About 20 additional separated pigments were characterized by their absorption spectra and epoxide tests, but were not identified.

The carotenoids of the orange Centennial sweetpotato, the cultivar most commonly grown in southeastern United States, were analyzed by the same method (Purcell and Walter, 1968). The total carotenoid content was 130 μg (g fresh wt)$^{-1}$, which corresponds to 450 μg (g dry matter)$^{-1}$, a content very close to that of Goldrush sweetpotatoes. The carotenoid pattern was more complex than the former, 41 pigments being separated of which 19 were identified and determined. The remainder, mainly in amounts of less than 0.05% and characterized by absorption spectra and epoxide test, were determined but not identified. The carotenes were phytoene (2.5%), phytofluene (2%), α-carotene (0.9%), β-carotene (86.4%), and ζ-carotene (1.8%). Some epoxides, such as β-carotene 5,6,5′,6′-diepoxide, luteochrome, and aurochrome, were identified. Epoxides of cryptoxanthin, lutein 5,6-epoxide, and violaxanthin were found in trace amounts.

The carotene fraction of raw sweetpotatoes was a mixture of all-*trans*-β-carotene (96.7%), neo-β-carotene B (0.1%), and neo-β-carotene U (3.2%) (Sweeney and Marsh, 1970).

Complete analyses of the various yellow sweetpotato cultivars need to be carried out with modern methods to fill the gap that exists. Limiting the analysis to β-carotene is not satisfactory.

The carotenoids of white-fleshed sweetpotato cultivars were studied by Martin (1983) and Almeida and Penteado (1988). All sweetpotato cultivars that are not orange are called white-fleshed, although their colors vary from white to white yellow. In the first study, the carotenoids of 15 white-fleshed potato cultivars grown in Puerto Rico were identified as phytofluene, ζ-carotene, neurosporene, β-zeacarotene, and β-carotene. β-Carotene was the main carotene in six cultivars, β-zeacarotene in seven cultivars, and neurosporene in two cultivars. The relative amounts were visually estimated.

The green color of cooked sweetpotatoes was associated with an epoxide of these carotenoids.

White-fleshed sweetpotatoes are also eaten in São Paolo, Brazil. The carotenoid distributions in the Brazilian white potatoes investigated by Almeida and Penteado (1988) are given in Table 67.

Table 67. Carotenoid Distribution in Brazilian White-Fleshed Sweetpotatoes

Carotenoid Pattern	*μg (100 g fresh wt)$^{-1}$*	*Percent of Total*
Luteochrome	191	43.0
β-Carotene 5,6,5′,6,-diepoxide	114	25.6
β-Carotene	61	13.5
β-Carotene 5,6-epoxide	25	5.6
ζ-Carotene	22	4.9
α-Zeacarotene	19	4.3
Neurosporene	13	2.9

Source: From Almeida and Penteado (1988).

Discussion. The total carotenoid content of cultivars Monalisa and IAC 271 was very low, about 5 μg (g fresh wt)$^{-1}$. On the other hand, the carotenoid pattern was uncommon, containing β-carotene and its two 5,6-epoxides (the mono and diepoxide), and luteochrome, a 5,6,5′,8′-diepoxide. (For structure see under "Unique Carotenoids.") The acyclic precursors of β-carotene, ζ-carotene, neurosporene, and the uncommon α-zeacarotene were also identified. The main carotenoid was luteochrome, which yielded about half the total pigments the first time it was found at higher concentrations in plants.

The presence of α-zeacarotene, which is seldom found in nature, is also uncommon. It was reported for the first time in maize (Petzold et al., 1959). Its presence in Brazilian white sweetpotatoes, which did not contain any carotenoid of the β,ε-series, is difficult to explain. In other various sweetpotato cultivars β-zeacarotene was reported (Martin, 1983).

Influence of Different Factors on Carotenoid Content

Besides the intervarietal variations reported in sweetpotatoes, some studies on the effect of the harvest time were carried out. It had been observed that Porto Rico sweetpotatoes, produced by late planting, contained less carotene than those produced by early and midseason planting. In general, the color of the flesh increased as time of harvest was deferred. A systematic study of the carotenoid changes during an entire season involved four sweetpotato cultivars with different colored flesh; Yellow Jersey (creamy white), Nancy Hall (light yellow), Porto Rico (deep yellow), and Orange Little Stem (deep orange) (Ezell et al., 1952). The roots were harvested at nine dates between mid-August and late November. Total carotenoids and carotene were determined.

Both carotene and total carotenoid content increased in all cultivars as the season advanced, reached a maximum, and then declined late in the season as climatic conditions became less favorable. In the more deeply colored cultivars the carotenoid increased much more rapidly than in the paler ones. Wide differences were found between locations and seasons. Sweetpotatoes from the same field averaged more than twice as much carotene one season as the next (Ezell and Wilcox, 1958). The cultivar Orange Little Stem contained 68 μg (g fresh wt)$^{-1}$ total carotenoid in 1951 and 34 μg (g fresh wt)$^{-1}$ in 1952; Yellow Jersey, 7.2 and 4.7 μg (g fresh wt)$^{-1}$ in the respective seasons. Potatoes grown with limited soil moisture were higher in carotene than those grown with abundant soil moisture.

Changes During Storage, Cooking, and Processing

Storage

Carotenoid changes in stored sweetpotatoes showed all possible differences (increase, decrease, or no change), depending on the influencing factors, but in general the increasing trend predominated. Porto Rico and Yellow Jersey sweetpotatoes had a vitamin A value three or four times higher after 2 months of storage than that at harvest (MacLeod et al., 1935). In Porto Rico sweetpotatoes stored at 37°C the carotene content increased rapidly during the first month of storage (47%), gradually during the second month (13%), and then remained constant (Miller and Covington, 1942).

Subsequently, the various factors that affect the carotenoid changes during storage were investigated systematically, especially by Ezell and Wilcox. First the variation with cultivar was studied (Ezell and Wilcox, 1948). In all cultivars investigated the carotene and total carotenoid content increased during curing at 29.4°C and storage at 15.6°C. The rate of increase varied widely with cultivar. In Nancy Hall sweetpotatoes the carotene content increase was short and rapid, in Porto Rico it was slow and gradual, whereas in Yellow Jersey it increased rapidly throughout the storage period. In some of the 14 sweetpotato cultivars, an increase of the β-carotene content was observed during the first month of storage at 24°C, and a decrease after 4 months' storage (Miller et al., 1949). Storage temperature affected the carotenoid content changes differently in different cultivars (Ezell and Wilcox, 1952). In Nancy Hall sweetpotatoes the carotenoid content decreased at all the applied temperatures. In Yellow Jersey, Porto Rico, and Orange Little Stem cultivars, the total carotenoid content decreased at 10°C, changed very little at 12.8°C, and it increased appreciably at 15.6°C and 21°C.

Storage humidity has no effect on the carotenoid changes in sweetpotatoes (Ezell et al., 1956). In Orange Little Stem and Yellow Jersey, cured and stored at 15.6°C with relative humidity from 70% to 75%, 80% to 85%, and 95% to 100%, the carotenoid content increased.

The carotene content at harvest was not a determining factor in amount of increase during storage, but in a season when lots were relatively high in carotene the increase during storage was also high (Ezell et al., 1958). In Orange Little Stem sweetpotatoes containing 68 μg (g fresh wt)$^{-1}$ total carotenoids, the increase after 159 days of storage was 21%. The next year the total carotenoids were 31.7 μg (g fresh wt)$^{-1}$, and the increase after 151 days was 9%. In sweetpotatoes stored for 7 months at 16°C, the total carotenoid content increased by 36% (Reddy and Sistrunk, 1980).

The effect of different storage temperatures on the total carotenoid content of orange-fleshed sweetpotatoes was studied by Picha (1985). The results are given in Table 68.

Table 68. Total Carotenoid Content of Four Orange-Fleshed Sweetpotato Cultivars at Harvest and after Different Storage Treatments

	*Total Carotenoids [μg (g fresh wt)$^{-1}$]**			
Storage Treatment	*Centennial*	*Travis*	*Jasper*	*Jewel*
At harvest	95	63	54	50
Cured† + 4 wk 15.6°C	105	74	63	60
Cured + 16 wk 15.6°C	115	80	65	68
Cured + 5 wk 26.6°C	106	88	70	55
Cured + 4 wk 7°C + 1 wk 15.6°C	112	86	68	65

Source: From Picha (1985).

*All data were corrected for weight loss during storage and expressed on an at-harvest weight basis.

†Cured = 10 days at 32°C, 90% RH.

Discussion. Carotenoids were higher, although not significantly, after curing and short-term storage at different temperatures, than at harvest, in all four cultivars.

Cooking

Carotenoid changes in orange-fleshed sweetpotatoes during cooking, as affected by cultivar, root size, storage period, and cooking methods, were studied. β-Carotene retention was especially followed, it being the major pigment (about 90%) and the most effective provitamin A.

A single study followed the changes of individual carotenoids of Goldrush sweetpotatoes in cooked puree (Purcell, 1962). The relative amounts of the carotenoids did not change appreciably after preheating, lye peeling, slicing, and steam cooking.

Retention of carotene in large roots during cooking was better than that of small roots (MacNair, 1956; Lanier and Sistrunk, 1979) or else was not affected by root size (Reddy and Sistrunk, 1980).

The effects of cooking methods (baking, boiling, steaming, and microwaving) were different. Generally, the total carotenoid content was lowest in boiled potatoes and highest in baked potatoes, but cultivars responded differently (Lanier and Sistrunk, 1979; Sistrunk, 1979; Reddy and Sistrunk, 1980).

In some studies the isomerization of β-carotene induced by heat in sweetpotatoes was followed. In raw sweetpotatoes the relative percentages of β-carotene isomers were neo-β-carotene B (0.1%), all-*trans*-β-carotene (96.7%), and neo-β-carotene U (1.3%). After 20 min cooking the percentages were 24.9%, 69.7%, and 5.4%, respectively. The same percentages were found after 40 min cooking also. The formation of large amounts of the neo-β-carotene B isomer is considered characteristic for cooked red and yellow vegetables (Sweeney and Marsh, 1970). Cooking sweetpotatoes produced a 25% loss of crude carotene, but the relative proportions of the three β-carotene isomers remained almost unchanged (Panalaks and Murray, 1970).

Processing

Recently, the isomerization and losses of all-*trans*-β-carotene in sweetpotatoes as affected by processing treatments were studied in detail by applying a specially adapted HPLC method (Chandler and Schwartz, 1988). The results are shown in Table 69.

Table 69. *Cis* and *trans* β-Carotene Isomers in Raw and Processed Sweetpotatoes [μg (g dry wt)$^{-1}$]

Sweetpotato Treatment	*% Solids*	*Total β-carotene*	*13*-cis	*all*-trans	*9*-cis	*% Isomers* (cis)
Raw product	20.8	440.0	22.5	417.5	—	5.1
Strips (2-min blanch)	21.7	499.2	39.1	460.1	—	7.8
Strips (10-min blanch)	23.7	457.8	69.9	387.9	—	15.3
Puree	17.8	485.9	25.2	460.7	—	5.2
Steam inj (starch gelat)	16.7	495.3	34.5	460.8	—	7.0
Steam inj (enz inact)	15.6	455.9	36.9	419.0	—	8.1
Canned	15.8	390.3	56.7	323.0	10.6	17.2
Dehydration (drum drying)	94.9	349.8	101.0	248.8	tr	28.9
Microwaved	31.0	340.2	56.3	283.9	tr	16.5
Baked	31.3	301.8	69.5	232.3	tr	23.0

Source: Reprinted with permission, from Chandler and Schwartz (1988); copyright © 1988 American Chemical Society.
Note: tr = trace.

Discussion. Carotene content was altered during processing, the changes depending on the treatment applied. After 2 min blanching, an increase of 13% occurred, which decreased to 4% after 10 min. This was attributed to an enhanced extraction efficiency following heat treatment, which causes ultrastructure changes, namely disruption of the plastids and dissolution of carotene in surrounding cellular lipids (Purcell et al. 1969; Walter and Schadel, 1982). Lye peeling and pureeing produced a 10.4% increase in carotene; steam injection produced an 8% loss, and, in contrast to other data reported, microwaving and baking produced losses of 22.7% and 31.4%, respectively. Heat processing induced the formation primarily of the 13-*cis*-β-carotene isomer, thus confirming earlier published results (Sweeney and March, 1970). The quantity formed was related to the severity and length of the heat treatment.

In white-fleshed sweetpotatoes, 10 min cooking produced a 26% loss of β-carotene and a 56% loss of β-carotene 5,6-epoxide, which corresponds to a decrease of vitamin A value from 11.1 to 7.9 RE $(100\ g)^{-1}$ (Almeida and Penteado, 1988).

Canning. In canned sweetpotatoes a loss of 40% of the carotene content was reported (Elkins, 1979), whereas other studies found that the total carotenoid content did not change significantly after canning (Lanier and Sistrunk, 1979; Reddy and Sistrunk, 1980). The notable effect of canning was not the destruction of the crude carotene content of sweetpotatoes (10%) but the induced isomerization (Panalaks and Murray, 1970). All-*trans*-β-carotene underwent a loss of 40%, whereas neo-β-carotene B increased 1600%, and neo-β-carotene U 300%.

The carotene stereoisomerization in sweetpotatoes as affected by rotating and still retort canning processes was also studied (Lee and Ammerman, 1974). Maximum retention of all-*trans*-β-carotene was in canned roots receiving the 116°C still and the 127°C and 132°C agitating processes. Neo-β-carotene B was found in significantly larger amounts in canned sweetpotatoes receiving the 121°C still or agitating processing. Neo-β-carotene U content was not significantly affected by the heat processes studied.

The same retention of total β-carotene (89%) as reported earlier was found recently in sweetpotatoes (Chandler and Schwartz, 1988) (see Table 69). The *cis* isomers found in canned samples were preponderantly 13-*cis*-β-carotene (14.5%) along with 9-*cis*-β-carotene (2.7%), the latter being detected uniquely in canned samples at appreciable levels.

Freezing. No change in total or isomeric carotene content was observed in freezing sweetpotato samples (Chandler and Schwartz, 1988).

Dehydration. Carotenoids were not destroyed during drum drying of sweetpotato puree to a moisture content of 5% to 7% (Purcell, 1962). Antioxidant added to the flakes, storage under nitrogen, and storage at low temperatures decreased losses of carotenoids in storage. The lowest loss was 7% after a storage of 120 days.

The lipid autoxidation of dehydrated sweetpotato flakes stored in air occurs via a dual-reaction mechanism, with surface and bound lipids oxidizing concurrently at very different rates. Surface carotene was lost 100 times more rapidly than bound carotene (Walter and Purcell, 1974).

Dehydrated sweetpotatoes contained all-*trans*-β-carotene (81.7%), neo-β-carotene B (12.7%), and neo-β-carotene U (5.6%) (Panalaks and Murray, 1970).

During drum drying, 20.5% of the carotene content was lost through oxidation, and the remaining carotene contained the highest amount of 13-*cis*-carotene (28.9%) (Chandler and Schwartz, 1988) (see Table 69).

Pumpkins and Squashes (*Cucurbita pepo* L., *Cucurbita moschata* Lam., *Cucurbita maxima* Duch.) Cucurbitaceae

The edible pumpkins and squashes are fruits of the species *C. pepo, C. moschata,* and *C. maxima.* The names of pumpkin and squash are applied inconsistently to certain species, so here the names given in the reports will be used.

C. pepo includes summer squash, some varieties of winter or baking squash, and field pumpkins, *C. moschata* includes winter or baking squash and field pumpkins, and *C. maxima* includes winter squash (Hubbard and Banana groups), large pumpkins, and turban squash. The fruits are remarkably diverse in size, shape, and color. They can be round, oval, elongated, flattened, or crooked, and range from white through cream, yellow to deep orange, or light green to dark green and variegated.

Pumpkins and squashes are consumed as a vegetable around the world. The differences between summer and winter squashes is the stage at which the fruits are harvested and their ability to last well in storage.

Summer squash (e.g., zucchini) is eaten wholly, whereas from other fruits, such as winter squash or pumpkin, which have a hard thick skin, the pulp is consumed. Being rich in carotenoids, these fruits are a good source of provitamin A, and their carotenoids were widely investigated. In one of the earliest investigations, two carotenoids of the flesh of ripe squash were obtained in crystalline form (Suginome and Ueno, 1931). They were considered different from any other known carotenoids and were named "cucurbitene" ($C_{40}H_{56}$) and "cucurbitaxanthin" ($C_{40}H_{56}O_2$). A few years later the pigments of the same fruits were reanalyzed (Zechmeister and Tuzson, 1934). It was proved that cucurbitene is a mixture of lutein and zeaxanthin.

About 1940 it was recognized that pumpkins and squashes are a good source of vitamin A, and their carotene fraction was determined assuming that the main pigment is β-carotene. The carotene of fresh winter squash ranged from 16.8 to 120 μg (g fresh wt)$^{-1}$ (Fitzgerald and Fellers, 1938). Seven varieties of summer squash ranged from 0.6 to 62.1 μg (g fresh wt)$^{-1}$ (Holmes et al., 1945). Between-sample variations were observed even in squashes harvested from plants cultivated under the same conditions (Holmes et al., 1948). The carotenoid contents in μg (g fresh wt)$^{-1}$ in the cultivars were 17 to 41 for Blue Hubbard, 30 to 46 for Buttercup, 69 to 118 for Butternut, 35 to 83 for Golden Cushaw, and 6 to 13 for Des Moines.

Pumpkins of unspecified varieties grown in Romania had a β-carotene content from 10.1 to 19.6 μg (g fresh wt)$^{-1}$, the pigment being separated chromatographically (Bodea et al., 1954). The carotene content in pumpkins (*C. pepo* L.) grown in Israel was 59 μg (g fresh wt)$^{-1}$, whereas in squashes (*C. moschata* Duch.) only trace amounts were found (Halevy et al., 1957). In six varieties of winter squashes the β-carotene content at harvest was 4 to 29 μg (g fresh wt)$^{-1}$ (Hopp et al., 1960). Other carotenoids then were determined (Lewis and Merrow, 1962). The carotenoids of Butternut squash were separated into three fractions: hydrocarbons, monohydroxy carotenoids, and polyhydroxy carotenoids. From the hydrocarbon fraction α- and β-carotene were determined to be 3.7 and 4.4 μg (g fresh wt)$^{-1}$, and a third pigment was not identified. Fractions II and III were determined together to be 8.8 μg g^{-1}.

As part of a comprehensive investigation in stereoisomers of α- and β-carotene in fresh and frozen vegetables, two squash cultivars and one pumpkin cultivar were analyzed (Sweeney and Marsh, 1971a). The results are given in Table 20.

The carotenoids of the Japanese edible snake gourd (*Trichosanthes cucumeroides*), which is widely consumed in Asia, were investigated by thin-layer chromatography (Matsuno et al., 1970). The flesh contained only β-carotene, whereas the epiphasic pigments of peels were β-carotene, γ-carotene, and lycopene.

A complete analysis of all the individual carotenes in several cultivars of winter squash was carried out by Lee et al. (1984). The investigated fruits were two cultivars of *Cucurbita maxima* (Golden Delicious and NK 530), a cultivar Aiguri, the F_1 hybrid of *C. maxima* × *C. moschata,* and three breeding lines (81-586, 81-587, and 81-621), representing advanced generation of the cross *C. maxima* cv. Buttercup × *C. moschata* cv. Butternut. Standard column chromatography was applied. The results are given in Table 70.

Discussion. The main hydrocarbon fraction in all six cultivars consisted exclusively of carotenes, their epoxides being totally absent. The major pig-

Table 70. Carotenoid Distribution in the Pulp of Six Cultivars of Winter Squashes [μg (100 g fresh wt)$^{-1}$]

Carotenoids	*81-568*	*81-587*	*81-621*	*Golden Delicious*	*NK 530*	*Aiguri*
Total carotenoids	2628	1213	1757	2618	2558	1340
Hydrocarbon	2062	1008	1420	1873	1819	1145
Monohydroxy	246	43	62	196	93	82
Polyhydroxy	188	43	89	283	173	84
Phytoene	38	47	37	64	54	55
Phytofluene	56	10	tr	20	65	tr
α-Carotene	9	214	6	tr	10	16
β-Carotene	1745	702	1243	1646	1742	995
ζ-Carotene	146	21	41	36	87	28
β-Zeacarotene	32	17	10	13	23	10
γ-Carotene	33	75	90	74	32	68
Neurosporene	46	25	34	38	42	22
Lycopene	tr	tr	tr	tr	tr	tr
RE (100 g)$^{-1}$	297	142	216	281	296	174

Source: From Lee et al. (1984).
Note: tr = trace.

ment was β-carotene, which yielded about 80% of the fraction. α-Carotene was found at low levels (0.4% in the line 81–568), and was highest in the cultivar 81-587, which had the lowest total carotenoid content. Lines 81-568, NK 530, and Golden Delicious had the most intense color; they containing higher amounts of total carotenoids and β-carotene. Breeding line 81-568 contained an unusually high amount of ζ-carotene. Four of the nine isolated carotenes are provitamins A: α-, β-, γ-carotene, and β-zeacarotene. Their average distributions in the total carotenoids were, approximately, β-carotene (67%), γ-carotene (3%), α-carotene (2%), and β-zeacarotene (1%).

In the calculation of the fruits' vitamin A value, cryptoxanthin, which may be present in the monohydroxy carotenoid fraction, was not taken into account.

Only recent studies involve complete carotenoid analyses. The polyhydroxy carotenoid fraction, which was ignored all this time, proved to be the source of two xanthophylls, which is extremely interesting from a scientific point of view. From the flesh of pumpkin (*C. maxima*) grown in Japan, two new carotenoids, to which the old name cucurbitaxanthin (Suginome and Ueno, 1931) A and B was assigned, were recently isolated and their structure elucidated (Matsuno et al., 1986). For structure, see under "Unique Carotenoids." The carotenoid distribution of the Japanese pumpkin investigated by Matsuno (1988) is shown in Table 71.

Table 71. Carotenoid Distribution in the Flesh of Japanese Pumpkin (*Cucurbita maxima*)

Carotenoid Pattern	*Percent of Total Carotenoids*
β, β-Carotene	16.8
(3*R*)-β-Cryptoxanthin	1.5
Lutein A	26.1
(3*R*,3′*R*)-Zeaxanthin	4.1
(3*S*,5*R*,6*S*,3′*R*,6′*R*)-Lutein-5,6-epoxide	1.2
(3*S*,5*R*,6*S*,3′*R*)-Antheraxanthin	2.1
(3*S*,5*R*,6*S*,3′*S*,5′*R*,6′*S*)-Violaxanthin	11.1
(3*S*,5*R*,6*R*,3′*S*,5′*R*,6′*S*)-Neoxanthin	3.4
Cucurbitaxanthin A	25.0
Cucurbitaxanthin B	0.8
Total carotenoids [μg (g fresh wt)$^{-1}$]	108.0

Source: From Matsuno, personal communication (1988).

The carotenoids of yellow- and orange-fleshed Japanese pumpkins *C. moschata* and *C. maxima* were investigated by Hidaka et al. (1987). The carotenoids were separated by liquid–liquid partition followed by standard column chromatography. The purpose of the study was to find out what differences in the carotenoid composition produce the color diversity and to assess the exact vitamin A value of the parts, taking into account all the provitamins A found in the fruit. The results are given in Table 72.

Discussion. According to the classification of carotenoid patterns found in fruits (Gross, 1987), the pumpkin carotenoid pattern may be characterized as a modified chloroplast carotenoid pattern. The major carotenoid is lutein, followed by β-carotene. Instead of violaxanthin, its isomers luteoxanthin and auroxanthin are present, whereas neoxanthin is absent. The carotenoid pattern is nearly similar in both species. The orange color of *C. maxima* is due to its higher total carotenoid content. The higher vitamin A value of the yellow-fleshed *C. moschata* is an example that depth of color does not always reflect nutritive value.

Similar results were found on the carotenoids of the fully ripe Japanese squash (*C. maxima*) (Kon and Schimba, 1988). The total carotenoid content was 62 μg (g fresh wt)$^{-1}$. The carotenoids were lutein (63.9%), α-carotene (10.9%), β-carotene (7.8%), antheraxanthin (6.0%), violaxanthin (4.3%),

Table 72. Carotenoid Distribution in the Flesh of Two Different Colored Pumpkins

Carotenoid Pattern	*Percent of Total Carotenoids*	
	C. moschata *yellow*	C. maxima *orange*
α-Carotene	0.5	0.3
β-Carotene	21.6	16.3
ζ-Carotene	2.5	0.2
β-Carotene 5,6-epoxide	2.3	0.7
β-Cryptoxanthin	5.2	1.2
Lutein	39.4	45.4
Taraxanthin	7.2	9.4
Zeaxanthin	—	6.5
Luteoxanthin	15.8	14.3
Auroxanthin	5.5	5.7
Total carotenoids [μg (g dry wt)$^{-1}$]	2340	3240
Vitamin A value	972	929

Source: From Hidaka et al. (1987).

α-cryptoxanthin (2.6%), auroxanthin (1.6%), ζ-carotene (1.6%), and neochrome (1.2%). The similarity is that in both *C. maxima* the predominant pigment was lutein followed by β-carotene, but, in contrast to pumpkins, α-carotene and its derivative α-cryptoxanthin were present at higher levels.

The carotenoid composition and vitamin A value of commercial Brazilian squashes and pumpkins was recently investigated (Arima and Rodriguez-Amaya, 1988). The carotenoids were separated by standard column chromatography. Both *C. maxima* cultivar Exposição and the hybrid Tetsukabuto are consumed when mature and were thus analyzed at this stage. The *C. pepo* cultivar Caserta, which is consumed when small and tender, was analyzed at the immature stage. The results are given in Table 73.

Discussion. The carotenoid pattern may be characterized as a modified chloroplast carotenoid pattern, although more complex and consisting of 13 to 15 carotenoids with lutein and β-carotene as the major pigments. In *C. maxima* the β-carotene level is higher than that of lutein; in the hybrid Tetsukabuto, cryptoxanthin is found as the third major pigment.

An interesting feature is the presence of α-cryptoxanthin in some Brazilian cucurbits, to which, in the present report, the structure 3′-hydroxy-α-carotene is assigned. This pigment has provitamin A activity, having the hydroxyl group attached to the ϵ-ring. The other isomer, zeinoxanthin or

Table 73. Carotenoid Distribution in Brazilian Squash and Pumpkin Cultivars*
[μg (g fresh wt)$^{-1}$]

Carotenoid Pattern	*Caserta* (C. pepo) *Immature*	*Exposição* (C. maxima) *Mature*	*Tetsukabuto Hybrid Mature*
α-Carotene	—	0.1	0.2
β-Carotene	1.4	16.6	12.9
cis-ζ-Carotene	—	0.3	—
Neurosporene	—	—	1.2
Mutatochrome	—	0.1	0.1
α-Cryptoxanthin	0.3	1.1	—
Zeinoxanthin	0.1	—	3.9
β-Cryptoxanthin	—	0.3	9.3
Lutein	4.6	10.2	19.3
cis-Lutein	tr	3.2	0.1
Zeaxanthin	—	2.4	1.3
Taraxanthin	—	1.1	1.9
cis-Taraxanthin	0.4	—	—
Flavoxanthin	tr	—	0.1
Violaxanthin	—	8.2	—
cis-Violaxanthin	0.1	—	1.3
Luteoxanthin	—	—	0.1
cis-Luteoxanthin	—	0.2	—
cis-Trihydroxy-α-carotene	0.2	—	—
Trihydroxy-α-carotene	—	0.3	0.1
Neoxanthin	tr	8.8	0.2
Total carotenoids	7.3	46.0	52.3
Vitamin A value (RE kg^{-1})	258	2875	2956

Source: From Arima and Rodriguez-Amaya (1988).
Note: tr = trace.
*Means of five samples for each vegetable collected during the year.

3-hydroxy-α-carotene, is devoid of provitamin A value (see structure on page 139). Zeinoxanthin was found in *C. pepo* and Tetsukabuto as well as in *C. moschata*. Another uncommon pigment is the polar trihydroxy-α-carotene. Additional investigations will help elucidate its exact structure. It may be identical or related to the newly described cucurbitaxanthin A and B, which were recently isolated from the fruit flesh of *C. maxima* (Matsuno et al., 1986).

The carotenoids of summer squashes and pumpkins were also analyzed by HPLC. In a study on the provitamin A content in fruits and vegetables, α-carotene, β-carotene, and β-cryptoxanthin were determined (Bureau and Bushway, 1986). Summer squashes contained 0.1 μg α-carotene and 14.8 μg β-carotene (g fresh wt)$^{-1}$.

Another HPLC method was tried for quantification of carotenes in yellow/orange vegetables (Khachik and Beecher, 1987). In the analyzed pumpkin cultivar the total carotene fraction amounted to 162.4 μg (g fresh wt)$^{-1}$. It was separated into *cis*-ζ-carotene (14.3 μg), all-*trans*-ζ-carotene (13.1 μg), α-carotene (55 μg), and β-carotene (80 μg). The numerous additional peaks of minor pigments that appeared in the HPLC chromatogram were not identified.

The carotenoids of the acorn squash (*C. pepo*) grown in New Jersey in the United States were investigated with HPLC by Khachik and Beecher (1988a) and Khachik et al. (1988). The major carotenoid pattern was the chloroplast carotenoid pattern devoid of neoxanthin. The three major carotenoids, preponderantly all-*trans* isomers, were accompanied by *cis* isomers: 15,15′-*cis*-β-carotene, neolutein A, and 9-*cis*-violaxanthin (violeoxanthin). The quantitative distribution is shown in Table 74.

Table 74. Quantitative Distribution of the Predominant Carotenoids from Saponified Extracts of Raw and Fresh Cooked Acorn Squash

	μg (100 g edible food)$^{-1}$ (wet weight/dry weight)	
	Raw	*Cooked*
Xanthophylls		
all-*trans*-Violaxanthin		
+ 13- or 15,15′-*cis*-violaxanthin	44/733	34/243
all-*trans*-Luteoxanthin	4/67	10/71
9-*cis*-Violaxanthin	7.8/130	14/100
Auroxanthin	54.5/908	122/871
di-*cis*-Violaxanthin	—	17.4/124
all-*trans*-Lutein + neolutein A or A′	38/633	66/471
Carotenes		
all-*trans*-β-Carotene		
+ 15,15′-*cis*-β-carotene	220/3667	490/3500
Total	368.3/6138	753.4/5380

Source: From Khachik and Beecher (1988a).

Discussion. Auroxanthin and luteoxanthin are presumed to be artifacts formed during the analysis. The total carotenoid content is very low, in the same range as that in immature fruits.

The carotenoids of acorn squash were also investigated without saponification so that carotenes and xanthophyll esters, as they exist in vivo, were analyzed with a specially adapted HPLC method. The qualitative distribution is given in Table 75.

Table 75. HPLC Peak Identification of the Predominant Carotenoids and Carotenol Fatty Acid Esters of Acorn Squash in the Order of Chromatographic Elution on a C-18 Reversed-Phase Column

Peak	*Component*	*Retention Time (min)*
1	all-*trans*-Violaxanthin	5.43
2	15,15′- or 13-*cis*-Violaxanthin	5.73
3	all-*trans*-Luteoxanthin	6.00
4	9-*cis*-Violaxanthin (violeoxanthin)	6.34
5	all-*trans*-Auroxanthin	6.69
6	13- or 15,15′-*cis*-Violaxanthin	6.70
7	di-*cis*-Violaxanthin	6.87
7′	di-*cis*-Violaxanthin	7.07
8	all-*trans*-Lutein	7.19
8′	Neolutein A or A′- (13- or 13′-*cis*-lutein)	7.45
9	β-apo-8-Carotenal (int std)	8.48
10	9- or 9′-*cis*-Violaxanthin monomyristate	23.90
11	all-*trans*-Violaxanthin monomyristate	24.78
12	9′- or 9-*cis*-Violaxanthin monomyristate	26.46
13	all-*trans*-β-Carotene	32.38
13′	15,15′-*cis*-β-Carotene	32.60
14	all-*trans*-Violaxanthin dilaurate	37.50
15	9-*cis*-Violaxanthin dilaurate	38.54
16	all-*trans*-Violaxanthin dimyristate	39.26
17	9-*cis*-Violaxanthin dimyristate	40.30
18	all-*trans*-Violaxanthin myristate/palmitate	40.94
19	9-*cis*-Violaxanthin myristate/palmitate	41.90
20	all-*trans*-Violaxanthin dipalmitate	42.54
21	all-*trans*-Lutein dilaurate	42.86
22	9-*cis*-Violaxanthin dipalmitate	43.50
23	all-*trans*-Lutein dimyristate	44.46
24	all-*trans*-Lutein myristate/palmitate diester	46.06

Source: From Khachik and Beecher (1988a).

Discussion. Lutein and violaxanthin, accompanied by their various *cis* isomers, were present in free and esterified form. The esters were mono- and diesters, simple or mixed. The acids were saturated fatty acids: lauric (12C), myristic (14C), and palmitic (16C). Structural elucidation of several xanthophyll esters was based on absorption, NMR, and MS spectra, and comparison with synthetic compounds. It was found that in lutein monoester the hydroxyl of the β-ring was preferentially esterified.

Numerous studies dealing with the genetic control of carotenoids in fruits of *Cucurbita* species have been carried out (Shifriss, 1981; Schaffer et al., 1984; Schaffer et al., 1986).

Distribution in Other Parts of the Plant

The carotenoids of the flowers of *C. pepo* were analyzed in petals and anthers (Szabolcs et al., 1970). The total carotenoid content of petals was 625 μg (g dry wt)$^{-1}$ and consisted of β-carotene (2.7%), β-cryptoxanthin (18.4%), lutein (8.2%), zeaxanthin (33.6%), antheraxanthin (12.8%), and violaxanthin (12.3%). In anthers the total carotenoid content amounted to 80 μg (g dry wt)$^{-1}$. Lutein, violaxanthin, luteoxanthin, and neoxanthin were identified.

In seeds of immature and mature Butternut squash the carotenoid content decreased from 73 to 35 and 77 to 30 μg (g dry wt)$^{-1}$ during winter storage (Holmes et al., 1954). Both groups of seeds also lost starch during storage. This metabolism appears to be characteristic for seeds, with carotenoid accumulation paralleling starch synthesis until just before maturation, when both begin to decrease. It occurs also in peas, chick-peas, and soybeans, as previously described.

Influence of Different Factors on Carotenoid Content of Different *Cucurbita* Species

Intercultivar Differences

The cultivar (or varietal) differences, which are genetically determined, are widely exemplified in previous reports of carotenoid distribution in pumpkins and squashes. Inherent variability of samples of the same fruits, due to factors such as stage of maturity, environmental conditions, time after harvest, and storage conditions, is also expected. This variability was illustrated in all previous reports in which carotene content ranges are given (Fitzgerald and Fellers, 1938; Holmes et al., 1948; Sweeney and Marsh, 1971a).

The extent of variation found in Brazilian squashes and pumpkins exceeds that of other fruits and vegetables. The explanation for the discrepancy could be the long harvest period for these fruits and their much longer shelf life (Arima and Rodriguez-Amaya, 1988). The total carotenoid content ranges were *C. pepo* cv. Caserta, 2.6 to 10.1; *C. maxima* cv. Exposição, 13.7 to 76.4; *C. moschata* cv. Menina Verde, 2.0 to 11.3 (immature) and 35.2 to 165.4 (mature); and hybrid Tetsukabuto 30.3 to 65.4 μg (g fresh wt)$^{-1}$, respectively.

Maturity

The color changes in the peel of the edible snakegourd fruit (*Trichosanthes cucumeroides*) during ripening were followed (Matsuno et al., 1970). The green color of the immature fruit imparted by chlorophyll changed to orange due to a massive increase of β-carotene accompanied by lycopene, the second major pigment, and by lower levels of γ-carotene. Then a change from orange to red occurred in the ripe fruit due to a twofold increase in the level of both red carotenes, lycopene and γ-carotene, whereas the β-carotene level remained unchanged.

The carotenoid analysis of the Brazilian squash *C. moschata* cv. Menina Verde, which is consumed both at immature and mature stages, permits one to follow the carotenoid changes during maturity in the edible flesh (Arima and Rodriguez-Amaya, 1988). The results are given in Table 76.

Table 76. Carotenoid Changes in the Ripening Brazilian Squash *Cucurbita moschata cv. Menina Verde* [μg (g fresh wt)$^{-1}$]

Carotenoid Pattern	*Immature*	*Mature*
α-Carotene	0.1	23.0
β-Carotene	1.5	39.3
cis-ζ-Carotene	—	6.3
α-Zeacarotene	—	2.6
Mutatochrome	tr	0.5
Aurochrome	—	0.1
δ-Carotene	—	0.2
α-Cryptoxanthin	0.2	1.5
Lutein	3.0	3.3
cis-Lutein	0.1	1.6
Taraxanthin	0.3	—
Violaxanthin	tr	0.7
cis-Violaxanthin	0.2	0.5
Total carotenoids	5.4	79.6
Vitamin A value [(μg RE kg)$^{-1}$]	27.5	862.4

Source: From Arima and Rodriguez-Amaya (1988).
Note: tr = trace.

Discussion. During ripening the cream-colored pulp turned dark orange, the result of the increased carotenogenesis from 5 to 80 μg g^{-1}. The carotenoid pattern in the immature fruit, which is a modified chloroplast carotenoid pattern devoid of neoxanthin, changed to an uncommon carotenoid pattern dominated by β- and α-carotene. Such a pattern is rarely found in fruits and is very similar to that found in banana pulp (Gross, 1987). The fruit maturation involves a massive synthesis of β- and α-carotene. As generally occurs in ripe fruits, an accumulation of the more saturated carotene precursors, especially of ζ-carotene, occurred. The epoxides mutatochrome and aurochrome, mainly those of the predominant carotenoid, were detected.

Changes During Storage, Cooking, and Processing

Storage

A characteristic for pumpkins and squashes is that the fruits can be stored much longer than other fruits, during which their development, maturation, and senescence continues, depending on the physiological stage at which they were harvested. Particularly suited for winter storage is winter squash. The carotenoid content of four cultivars of winter squashes [Blue Hubbard (*C. maxima*), Butternut (*C. moschata*), Golden Cushaw (*C. moschata*), and Delicious (*C. maxima*)] was determined as part of a study on the nutritive value of the fruits at the end of a winter storage period of 4 months (Holmes and Spelman, 1946). A high carotene was found in Butternut and Golden Cushaw squashes, 79.4 and 71.2 μg (g fresh wt)$^{-1}$ after storage. At the immature stage these fruits contained less carotene, 33.1 and 62.1 $\mu g\ g^{-1}$, respectively (Holmes et al., 1945), a fact that puzzled the investigators.

The influence of light during storage of squashes was followed (Holmes et al., 1948, 1949). The two cultivars investigated reacted differently. In Blue Hubbard squash stored in light or darkness, no consistent differences in carotene content could be observed. In mature Butternut squash stored in light and in darkness, the carotene content increased throughout the storage period and was higher in fruits stored under artificial illumination.

In another study the carotene content of immature and mature Butternut squash, in flesh and seeds, was followed during winter storage (Holmes et al., 1954). Carotene increased in immature and mature squash during a storage period of 221 days from 256 to 318 and from 286 to 318 μg (g dry wt)$^{-1}$, respectively.

In an extended study comprising six cultivars, the carotene changes were followed for a storage period of 25 weeks (Hopp et al., 1960). In all cultivars the carotene content increased during storage, but the extent of the increase was not consistent with all cultivars. On the other hand, in all cultivars the increase tended to be greatest during the first 10 weeks of storage.

During storage of Butternut squash the changes of carotenoids other than β-carotene were followed for the first time (Lewis and Merrow, 1962). The general increasing trend was observed in α- and β-carotene, being even more pronounced for the hydroxy carotenoid fraction, which increased from 9 to 21.4 μg (g fresh wt)$^{-1}$.

The carotenoids of the Japanese squash (*C. maxima*) stored at 10°C and 25°C for 3 months were analyzed by Kon and Shimba (1988). The carotenoid content and vitamin A value increased 2.4 and 2.7 times at 10°C and 4.3 and 4.8 times at 25°C. Whereas α-carotene disappeared after 2 months, lutein increased markedly during storage.

Cooking

The effects of cooking on the carotenoids of acorn squash (*C. pepo*) were investigated (Khachik et al., 1988), and the results are shown in Table 74. Following cooking, 12.4% of the total carotenoids were lost. The losses of each carotenoid were as follows: violaxanthin (including all its isomers), 23%; lutein, 26%, and β-carotene, 5%. The results indicate that β-carotene is more thermostable than hydroxy carotenoids and their epoxides.

Processing

Canning. Three winter squash cultivars, whose carotenoid distribution is shown in Table 70, were canned and their carotenoids analyzed (Lee et al., 1984). The results are shown in Table 77.

Table 77. Carotenoids and Vitamin A Value of Fresh and Canned Winter Squash [μg (100 g wet wt)$^{-1}$]

	81-568		*Golden Delicious*		*NK 530*	
Carotenoids	*Fresh*	*Canned*	*Fresh*	*Canned*	*Fresh*	*Canned*
Total carotenoids	2629	1746	2618	3076	2258	1918
Hydrocarbon	2062	1353	1873	1771	1819	1338
Monohydroxy	246	242	196	926	93	386
Polyhydroxy	188	130	283	340	173	178
Phytoene	38	35	64	21	54	31
Phytofluene	56	30	20	18	65	31
α-Carotene	9	tr	tr	tr	10	tr
β-Carotene	1745	1167	1646	1456	1742	1127
ζ-Carotene	146	33	36	25	87	31
β-Zeacarotene	32	14	13	27	23	27
γ-Carotene	33	14	74	19	32	21
Neurosporene	46	47	38	52	42	64
Lycopene	tr	tr	tr	tr	tr	tr
RE (100 g)$^{-1}$	297	197	281	246	296	192

Source: From Lee et al. (1984).
Note: tr = trace.

Discussion. In general, the carotenoid content decreased during the canning process, except for the monohydroxy fraction, which increased significantly for some unknown reason. The vitamin A value loss ranged from 12% to 34%.

Freezing and Dehydration. Changes in carotenoid distribution in frozen and freeze-dried squash immediately after processing and during storage were determined (Kon and Shimba, 1989), and the results are given in Table 78.

Discussion. The total carotenoid content did not change after blanching. A loss of 23% occurred in the frozen product, which after 3 months of storage underwent a further loss of 18%. β-Carotene was stable after freezing, but decreased during storage of the frozen product. Lutein decreased during freezing and remained stable during storage.

Freeze drying induced a loss of 25% of total carotenoids. Storage of the dehydrated product caused additional losses of 24% and 36%, respectively, during storage of 3 months at 0°C and 30°C. The loss of β-carotene after storage at 0°C was only 10%, and was 15%, 20%, and 53% after 1, 2, and 3 months at 30°C.

Spinach
(*Spinacia oleracea* L.)
Chenopodiaceae

In early reports concerned with the nutritional value of spinach, only β-carotene was determined in the leaves. The mean β-carotene was 40 μg (g fresh wt)$^{-1}$ (De Felice and Fellers, 1938; Gleim et al., 1944). The total carotene fraction of spinach leaves, 69 μg (g fresh wt)$^{-1}$, consisted of 87% all-*trans*-β-carotene and 13% of its isomer neo-β-carotene (Beadle and Zscheile, 1942).

Spinach leaves were widely used as classical material in analytical chemistry to work out methods for the separation and determination of chloroplast pigments.

The spinach chloroplast pigments, first separated by TLC, were chlorophylls *a* and *b* and the following carotenoids: β-carotene, lutein epoxide, violaxanthin lutein, antheraxanthin, neoxanthin neo A, neoxanthin, and zeaxanthin in order of increasing polarity (Hager and Meyer-Bertenrath, 1966).

The carotenoid content in μg (g fresh wt)$^{-1}$ of spinach leaves was found to be as follows: β-carotene, 40; lutein, 70; zeaxanthin, 12; violaxanthin, 10; and neoxanthin, 20 (Hirayama and Oida, 1969). With the advent of the HPLC only the four chloroplast carotenoids were separated and determined (Stransky, 1978).

As the method developed, additional carotenoids present in the spinach chloroplasts could be also separated, such as lutein epoxide (Braumann et al., 1982; Katukani et al., 1983). The thylakoids isolated from intact spinach chloroplasts contained the four chloroplast carotenoids and zeaxanthin and antheraxanthin. In the chloroplast envelopes very small quantities of vio-

Table 78. Carotenoid Changes in Frozen and Freeze-dried Squash Pulp after Processing and Storage

			Frozen		*Freeze dried at 0°C*				*Freeze dried at 30°C*		
	Raw	*Blanched*	*0 Mon.*	*3 Mon.*	*0 Mon.*	*1 Mon.*	*2 Mon.*	*3 Mon.*	*1 Mon.*	*2 Mon.*	*3 Mon.*
α-Carotene	1.0	1.2	1.0	1.1	0.9	0.5	0.7	1.1	0.5	0.7	0.9
β-Carotene	16.4	19.6	19.1	13.0	19.0	19.2	19.6	17.2	16.2	15.3	9.0
α-Cryptoxanthin	3.1	12.0	5.3	0.7	6.3	7.2	2.1	1.8	4.5	0.5	0
α-Cryptoxanthin 5,6-epoxide	1.3	2.2	0	0	0	0	0	0	0	0	0
Lutein	74.0	75.8	54.4	57.1	50.8	50.2	47.4	48.0	52.5	50.9	49.0
Zeaxanthin	4.2	4.8	4.4	4.7	2.7	2.8	3.1	1.6	2.3	2.1	1.4
Antheraxanthin	28.1	17.6	16.1	7.6	15.0	14.4	9.5	7.3	14.4	7.2	6.2
Violaxanthin	22.2	18.8	17.0	12.8	16.3	15.6	13.5	9.4	15.1	8.9	6.6
Trollein	2.3	1.0	0.9	1.0	2.1	1.0	1.6	1.0	0.7	2.1	0
Neoxanthin	2.5	2.0	1.6	1.0	2.6	2.4	1.5	1.5	2.8	1.5	1.6
Total carotenoids [μg (g fresh wt)$^{-1}$]	155.1	155.0	120.0	99.0	115.7	113.3	99.0	88.9	109.0	89.2	74.5

Source: From Kon and Shimba (1989).

laxanthin and neoxanthin were found (Grumbach, 1983a). It has also been found that β-carotene was accompanied by its 15,15′-*cis* isomer, and the main xanthophylls, except violaxanthin, were mixtures of *trans* and *cis* isomers. Besides the four chloroplast carotenoids, lutein epoxide and neochrome were also detected (Khachik et al., 1986). The results are given in Table 25.

Since spinach leaves are a good source of provitamin A, only its β-carotene content was determined. To assess more exactly the vitamin A value, *cis*-β-isomers that have a lower value were also separated and determined. The carotene fraction of spinach is a mixture of all-*trans*-β-carotene (80–82%), neo-β-carotene B (4.9–9.4%), and neo-β-carotene U (10.4–12.5%) (Sweeney and Marsh, 1971a) or all-*trans*-β-carotene (80.4%), 13-*cis*-β-carotene (8.8%), and 9-*cis*-β-carotene (10.8%) (Chandler and Schwartz, 1987). The values found for β-carotene are between 29 and 53 μg (g fresh wt)$^{-1}$ (Bureau and Bushway, 1986; Bushway, 1986; Arqués et al., 1988; Dietz et al., 1988; and Heinonen et al., 1989) (see Tables A1–A26, passim).

By HPLC, spinach leaf carotenoids were separated and determined (Kakutani et al., 1987). The total carotenoid content was 38.1 mg (100 g leaves), comprising 7.0 mg β-carotene, 19.0 mg lutein, 7.5 mg violaxanthin, and 4.6 mg neoxanthin. Similar results also obtained by HPLC were reported for spinach grown in the United States (Khachik et al., 1989) (see Table 79).

Quantitative changes of the spinach chloroplast carotenoids were followed monthly for a year (Kakutani, 1989). In winter the total carotenoid content was halved. Whereas the xanthophylls showed regular monthly fluctuation, β-carotene remained almost constant throughout the year.

Changes During Storage, Cooking, and Processing

Storage

Garden fresh spinach stored at 18.8°C to 25.5°C lost 8% of its carotene content, whereas storage for 1 week at 0°C to 4.4°C resulted in a loss of 5% (Gleim et al., 1944). Carotene changes in spinach stored under various conditions after harvest and before canning were investigated (Kelley et al., 1950). Upland grown spinach stored in the shade in baskets or in piles lost 14% and 22% of its carotenoid content after 37 h. In cold storage (−15°C) for 133 h the loss was 30%.

The influence of cultivars was also observed. Spring spinach underwent a greater loss than autumn spinach (Schlottman et al., 1961).

Cooking

Cooking did not affect the carotenoid content of spinach (Gleim et al., 1944; Weits et al., 1970). The content was 455 μg (g dry wt)$^{-1}$ in raw spinach, 443 μg (g dry wt)$^{-1}$ after 12 min of conventional cooking, and 464 μg (g dry wt)$^{-1}$ after 6 min cooking in a microwave oven (Park, 1987). The effect of different cooking methods on the individual carotenoids of spinach was also studied (Khachik et al., 1989). The results are given in Table 79.

Table 79. Carotenoid Content of Spinach

Cooking Method	*Carotenoids [mg (100 g raw food)$^{-1}$]*			
	Neoxanthin	*Violaxanthin*	*Lutein* cis + trans	*β-Carotene*
Raw	2.4	7.4	9.5	8.9
Microwave	2.3	4.8	8.7	9.1
Steaming	2.5	4.9	10.6	9.9

Source: From Khachik et al. (1989).

Discussion. It appears that violaxanthin is the most thermolabile of the pigments.

The stability of spinach leaf carotenoids during heat treatment was also investigated by Kakutani et al. (1987). The retention of each carotenoid after various heat treatments of the spinach leaves is shown in Table 80.

Table 80. Percent Retention of Spinach Chloroplast Carotenoids Subjected to Various Heat Treatments

Heat Treatment	*Carotenoids*			
	β-Carotene	*Lutein*	*Violaxanthin*	*Neoxanthin*
Raw spinach	100	100	100	100
Boiling 3 min	80	99	38	87
Boiling 10 min	86	111	23	90
Heating 2 min (150–170°C)	82	63	48	74
Heating 2 min (200–220°C)	75	51	26	58

Source: From Kakutani et al. (1987).

Discussion. The results are very similar to those in the previous table, showing that violaxanthin is the least stable, since it is the most rapidly destroyed, and is probably a source of aroma compounds. The 75% retention of spinach β-carotene during the most drastic treatment is much higher than that of pure β-carotene, which remained unchanged when boiled for 10 min but was 13% when heated at 200°C to 220°C for 2 min.

The percentages of the carotene isomers, neo-β-carotene B, all-*trans*-β-carotene; and neo-β-carotene U, which were 4.9%, 82.6%, and 12.5%, respectively, in fresh spinach, changed to 7.8%, 76.4%, and 15.8% after 5 min cooking and remained almost unchanged after 10 min cooking (Sweeney and Marsh, 1971a).

Processing

Canning. In canned spinach the isomerization of all-*trans*-β-carotene that occurs during cooking was more pronounced due to the drastic heat treatment. The relative percentages were as follows: neo-β-carotene B, 54% to 69%; all-*trans*-β-carotene, 62.5% to to 72.5%; and neo-β-carotene U, 16.3% to 29.0% (Sweeney and Marsh, 1971a). Other values reported were 8.8% for 13-*cis*-β-carotene, 80.4% for all-*trans*-β-carotene, and 10.8% for 9-*cis*-β-carotene in fresh spinach, and 15.3%, 58.4%, and 26.3% in canned spinach (Chandler and Schwartz, 1987).

Three different blanching methods did not affect the carotene content of spinach (Sistrunk et al., 1977).

Freezing. During freezing a loss of 25% of the carotene in spinach, on a moist basis, was reported (De Felice and Fellers, 1938). Quick-freezing produced a 13% loss of carotene in spinach (Gleim et al., 1944). In spinach grown in Spain and frozen an apparent increase was reported, its content being 71.2 μg (g fresh wt)$^{-1}$ β-carotene, whereas fresh spinach contained 52.2 μg (g fresh wt)$^{-1}$ (Arqués et al., 1988).

Dehydration. Air-dried and freeze-dried spinach carotene retention was 77% (Weits et al., 1970). In freeze-dried spinach carotene retention was higher after 6 months storage (88%), whereas in air-dried spinach it decreased to 47% after the same storage length.

The effect of sun drying on β-carotene content investigated by Renqvist et al. (1978) is shown in Table 81.

Carotene retention in dried spinach was highest in freeze-dried (81%) and lowest in vacuum- (72%) and microwave-dried (69%) products (Park, 1987).

Table 81. β-Carotene Content of Spinach, Fresh and after Drying [mg (kg fresh material)$^{-1}$]

Fresh spinach leaves	29
Dried leaf powder	7
Dried leaf powder, boiled with potash	8
Dried whole leaves boiled	11
Dried whole leaves, rehydrated by soaking	15
Mean value for dried spinach	10

Source: From Renqvist et al. (1978).

Tomato (*Lycopersicon esculentum* L.) Solanaceae

The red color of the tomato is due to its major carotene, lycopene, which is found at levels up to 90% of the total carotenoids. The investigation of the pigment "lycopersicon" began in 1913 by Duggar. About 15 years later the vitamin A content of tomatoes was investigated (Morgan and Smith, 1928; House et al., 1929). It was established that the pigments are carotenes, lycopene being the most abundant, accompanied by β-carotene. The structure of β-carotene and lycopene was elucidated (Karrer et al., 1930; Kuhn and Grundmann, 1932). In addition, free and esterified xanthophylls were also separated. The concentration of lycopene was 78 μg g^{-1}, and the amount of β-carotene was approximately a tenth of that of lycopene.

The total carotenoid content of tomato varies between 70 and 190 μg (g fresh wt)$^{-1}$. The carotenoids are predominantly carotenes, constituting 90% to 95% of the total carotenoids. The common red tomato also contains the colorless precursors phytoene and phytofluene (15–30%) and minor pigments such as β-carotene, ζ-carotene, γ-carotene, and neurosporene (Trombly and Porter, 1953; Kargl et al., 1960; Curl, 1961; Williams et al., 1967; Koskitalo and Ormrod, 1972; Raymundo et al., 1976; Lee and Robinson, 1980; Johjima and Ogura, 1983).

Carotenoid distribution in four red tomato cultivars is shown in Table 82.

Discussion. The major carotenoid pattern is 3—that is, large amounts of lycopene and the related partly saturated acyclic polyenes phytoene, phytofluene, ζ-carotene, and neurosporene (Goodwin, 1980).

The xanthophylls (6% of total carotenoids) were separated into 20 pigments, of which 15% were monols, 49% diols, 4% monoepoxide diols, 22%

Table 82. Carotenoid Distribution in Four Red Tomato Cultivars

	Percent of Total Carotenoids			
Carotenoid Pattern	*Common*	*Early Redchief*	*Summer Sunrise*	*New Yorker*
Phytoene	8.0	19.7	9.2	25.0
Phytofluene	3.9	8.4	7.8	5.0
β-Carotene	14.3	2.3	6.9	5.7
ζ-Carotene	0.7	2.0	1.7	1.1
γ-Carotene	2.0	0.5	0.5	0.8
Neurosporene	0.6	0.4	0.1	0.2
Lycopene	70.5	66.7	73.7	62.2
Total carotenoids [μg (g dry wt)$^{-1}$]	1140.4	950.0	1700.0	922.0

Source: Recalculated from Williams et al. (1967b), column 2; Koskitalo and Ormrod (1972), column 3; Raymundo et al. (1976), column 4; Lee and Robinson (1980), column 5.

diepoxide diols, and 11% polyols (Curl, 1961). In the New Yorker cultivar monohydroxy carotenoids constituted 2.2%, and polyhydroxy carotenoids 0.4% of the total carotenoids (Lee and Robinson, 1980). The vitamin A value of commercial tomatoes is 200 μg retinol equivalents kg^{-1} (Zakaria et al., 1979).

The carotenoids are not uniformly distributed within the fruits. The outer pericarp has the highest carotenoid concentration, and locular contents have the highest carotene content (Bauerfeind, 1981). In the cherry tomato the distribution of the pigments is different in the inner "pulp" and the outer "flesh" (Laval-Martin et al., 1975). Whereas β-carotene is equally distributed, lycopene predominates in the flesh in which the total carotenoid content is sixfold higher than in the pulp.

Besides common red tomatoes, there are numerous tomato strains and mutants of various colors, ranging from white to deep orange. In the early 1940s a comprehensive study concerning the carotenoids of *Lycopersicon* species and strains was initiated (Lincoln et al., 1943; Porter and Zscheile, 1946; Trombly and Porter, 1953). Additional pigments up to 18 carotenes, including several lycopene isomers, were separated.

The characterization of the carotenoid pattern in the tomato species and strains together with genetic data permitted the formulation of the Porter–Lincoln pathway of carotene biogenesis even before the structures of the intermediates were elucidated (Porter and Lincoln, 1950).

Some tomato strains are orange because either they do not synthesize lycopene at all or carotenes other than lycopene predominate. In beta strains, β-carotene is the major pigment (Tomes, 1963). The delta strain of reddish orange color contains δ-carotene as the main pigment (Kargl et al., 1960;

Tomes, 1963; Cabibel et al., 1981; Johjima and Ogura, 1983). Tangerine-type tomatoes do not synthesize all-*trans*-lycopene but prolycopene, a poly-*cis* isomer of lycopene. For structure see "Unique Carotenoids."

Distribution in Other Parts of the Plant

Carotenoids of cherry tomato seeds with a total carotenoid content of 23.7 μg (g fresh wt)$^{-1}$ were predominantly carotenes (98%) (Rodriguez et al., 1975). They consisted of phytofluene (1.7%), α-carotene (2.5%), β-carotene (54.4%), β-zeacarotene (1.3%), γ-carotene (3.4%), and lycopene (34.6%). In contrast to the pericarp, β-carotene was the major pigment, and lycopene was found at levels of about a tenth of its levels in commercial tomatoes.

In seeds of two red tomatoes and one tangerine tomato, lutein was the dominating pigment (Rymal and Nakayama, 1974). The seeds of the red tomato cultivars Chico Grande and Rutgers contained three major carotenoids: lutein (64.4% and 90.2%), β-carotene (21.5% and 5.3%), and lycopene (14.1% and 4.5%), respectively. The tangerine tomato Golden Jubilee seeds contained lutein (59.7%), β-carotene (22.2%), and ζ-carotene (18.1%). The carotene content of tomato leaves are given in Table A9.

Influence of Different Factors on Carotenoid Content

Changes During Ripening

The tomato is one of the most important carotenogenic fruits. Thus, the pigment changes occurring during fruit ripening were studied more intensely than other fruits. During ripening the color of the red tomato turns gradually from green to white to yellow to pink to red, as chloroplasts are transformed into chromoplasts, chlorophylls disappear, and carotenogenesis is taking over.

The total carotenoid content displays a minimum at midseason. This fact was observed for the first time in the cherry tomato (Laval-Martin et al., 1975). During the lag phase that precedes maturation, the granal system of the chloroplasts disintegrates, and chlorophylls and part of the chloroplast carotenoids disappear (Laval-Martin, 1974).

Studies on the carotenoid changes during tomato ripening began early (Kuhn and Grundmann, 1932). Changes in the content of lycopene, carotene, xanthophylls, and xanthophyll esters were followed in unripe, half-ripe, and fully ripe fruits. Concentrations of all pigments increased, particularly lycopene, which showed a 70-fold increase, and β-carotene, which showed a lower 5-fold increase. Subsequent studies were more detailed and monitored changes of individual carotenoids during maturation (Goodwin and Jamikorn, 1952; MacKinney and Jenkins, 1952).

Chlorophyll and carotenoid changes were followed in the San Marzano tomato at six ripening stages (Edwards and Reuter, 1967). During ripening, lycopene was intensely synthesized from 1 μg at the second stage to 181 $\mu g\ g^{-1}$, constituting 93% of total carotenoids. The hydroxy derivatives of lycopene (lycoxanthin and lycophyll) were detectable at the third maturity stage and found at appreciable amounts (2.2 and 1.1 $\mu g\ g^{-1}$) in the ripe fruit. The pigments phytofluene, β-carotene, and lutein 5,6-epoxide increased gradually up to 7.8, 5.6, and 2.3 $\mu g\ g^{-1}$, respectively.

Carotenoid changes occurring in ripening Homestead tomatoes investigated by Meredith and Purcell (1966) are shown in Table 83.

Table 83. Changes of Main Carotenes of Ripening Homestead Tomatoes

	Percent of Total Carotenoids					
Carotenoid Pattern	*RS 1**	*RS 2*	*RS 3*	*RS 4*	*RS 5*	*RS 6*
Phytoene	—	—	20.0	10.7	10.8	9.0
Phytofluene	—	—	2.5	2.1	2.6	2.4
β-Carotene	100	75.0	55.0	17.0	9.8	2.6
ζ-Carotene	—	—	2.5	1.4	1.6	0.6
γ-Carotene	—	8.3	10.0	4.3	2.3	0.7
Lycopene	—	16.7	10.0	64.3	72.9	84.7
Total carotenes [μg (g fresh wt)$^{-1}$]	1.2	2.4	4.0	14.0	30.6	97.7

Source: Recalculated from Meredith and Purcell (1966).
*RS = ripening stages: 1 = mature-green; 2 = breaker; 3 = turning; 4 = pink; 5 = light red; 6 = red.

In the cherry tomato, pigment evolution during growth and ripening was followed separately in the inner pulp and the outer region of the pericarp (Laval-Martin et al., 1975). The chloroplast pigments accumulate in both parts of the fruit and remain constant, except for chlorophylls and neoxanthin, which disappear. Phytofluene and lycopene appear only at the end of the ripening and almost exclusively in the external part of the fruit. Lycophyll and lycoxanthin could be detected during the entire maturation period.

Because lycopene concentration changes significantly from one maturity stage to another, it was proposed to be considered as a maturity index (Cabibel and Ferry, 1979).

In many fruits the ripening process, including carotenogenesis, continues after harvest, of which the tomato is a classical example. Vine-ripened tomatoes produce more carotenoids than those ripened in storage (Watada et al., 1976a).

Selection

Intensive studies on the genes responsible for the various colors of the numerous tomato strains were carried out (Lincoln and Porter, 1950; Jenkins and MacKinney, 1955; Tomes, 1967, 1969). An objective of the research was the development of tomato selections with a high β-carotene content. A high beta strain with orange-red flesh and a carotene content of 100 $\mu g\ g^{-1}$, of which 88% is β-carotene, was developed (Kohler et al., 1947). The cultivar Caro-Red, with an average content of 44 $\mu g\ g^{-1}$ β-carotene, was selected (Tomes and Quackenbush, 1958).

Cultural Practices

Different cultural practices, such as field, and glass and plastic greenhouses, as well as the place of the fruit on the plant affect the carotenoid content of tomatoes (Cabibel and Ferry, 1980). The field-cultured fruits were richer in carotenoids, a fact known 60 years ago.

Changes During Storage and Processing

Tomatoes are processed into juice and paste or puree, both of high color imparted by lycopene. Consequently, changes in carotenoid investigation are concerned both with β-carotene (the provitamin) and with lycopene, which determines the products' color.

Storage

In tomatoes ripened during storage, the β-carotene content varied directly with the ripeness of the fruit at harvest (Watada et al., 1976a). The average β-carotene content of ripe tomatoes harvested as mature green, breaker, and ripe was 4.0, 5.1, and 5.9 μg (g fresh wt)$^{-1}$, respectively.

Changes in β-carotene and lycopene levels, during a long storage of tomatoes were followed (Veloso de Almeida et al., 1977). The fluctuations differed among cultivars. The content of β-carotene was unchanged during the first week, but lycopene decreased more rapidly, up to 10% of the initial content at the end of the storage period.

Processing

Canning. Canned tomato juice retained over 95% of the initial carotene content during a 2-year storage at 10°C to 27°C. In whole tomatoes the retention was 84% after 18 months and decreased to 75% after 24 months (Sheft et al., 1949). During heating sterilization of tomato juice, lycopene decreased 1% to 2% after 7 min at 90°C to 100° and 17% at 130°C (Miki and Akatsu, 1970). During storage of tomato juice at room temperature for 6 months, the lycopene content remained constant and decreased by 2% when exposed to sunlight for 6 months (Miki and Akatsu, 1971).

The relative percentages of carotene in fresh tomatoes, neo-β-carotene B (13.7%), neo-β-carotene U (3.8%), and all-*trans*-β-carotene (82.5%) changed into 33.6%, 5.4%, and 61.0% in canned tomatoes (Panalaks and Murray, 1970).

Dehydration. The oxidative destruction of lycopene during the manufacture of tomato puree was studied (Monselise and Berk, 1954). The factor affecting lycopene losses during the concentration of tomato juice appeared to be the length of heating rather than the temperature.

The stability of lycopene during heating of tomato pulp was studied (Cole and Kapur, 1957b). The rate of lycopene breakdown varied according to availability of oxygen, temperature, and intensity of illumination. Heating tomato pulp at 100°C in daylight in the presence of oxygen produced a loss of 15% after 1 h and 33% after 3 h. After heating for 3 h in the presence of oxygen and in light, the loss increased with temperature: 20% at 60°C and 60% at 110°C.

Pigment and color changes as a result of heat concentration of tomato pulp were investigated in juices reconstituted from pastes of 10%, 15%, and 20% soluble solids (Noble, 1975). Total carotenoid was highest in the most concentrated pulp, resulting in losses of 11%, 22%, and 57% in lycopene and 12%, 20%, and 26% total carotenoids, relative to the unconcentrated tomato pulp.

During storage of tomato puree, both the content of β-carotene and lycopene decreased (Aczél, 1972). The rate of lycopene decrease was higher than that of β-carotene. One year of storage reduced the content of both pigments to half of the initial values, which were 190 and 380 $\mu g\ g^{-1}$. *Cis–trans* isomerization of lycopene and color stability of foam-mat-dried tomato powder during storage were studied (Lovric et al., 1970). The main cause of color fading was the oxidation of lycopene as well as *trans–cis* isomerization of lycopene. Excessive desiccation strongly promoted oxidation losses. Storage in an inert atmosphere improved color retention.

The effect of harvest maturity on carotenes in tomato pastes was studied (Liu and Luh, 1977). Tomato pastes were made of a tomato line at pink, medium red, and full red maturities. Most of the natural carotenoids were retained after processing. The total carotene content corresponded to the increasing level in the ripening fruits.

Turnip (*Brassica rapa* L.) and Swedish Turnip or Rutabaga (*B. napobrassica*) Cruciferae

Botanically there are two species belonging to the same genus as the cabbage: the turnip proper (*B. rapa*), and the Swedish turnip or rutabaga (*B. napobrassica*). The turnip is formed by the thickening of the primary root. Young turnips are used raw in salads, and cooked roots are consumed as a vegetable. They are cultivated extensively in Canada and northern Europe.

There are two kinds of rutabaga: white- and yellow-fleshed. The carotenoids of the yellow-fleshed rutabagas were investigated (Joyce, 1954). A large number of carotenes were separated by column chromatography on alumina. β-Carotene and ζ-carotene were identified, and numerous other bands, which were considered to be all-*trans*-lycopene accompanied by several poly-*cis*-lycopene isomers, were also identified. The carotenoids of the swede (*B. napobrassica*) were also investigated by Joyce (1959). The following carotenes were reported: phytoene, phytofluene, ζ-carotene, neurosporene, β-carotene, and γ-carotene. As in *B. rutabaga,* lycopene was accompanied by a number of poly-*cis*-lycopene isomers.

Two varieties of turnip, Laurentian and Wye, grown in Canada were examined for their lipids (Lepage, 1967). The Laurentian turnip is deep yellow, indicating a high carotenoid content. The major carotenoid was lycopene (41%) accompanied by γ-carotene (28.3%), cryptoxanthin (17.5%), and unidentified (12.3%). The absolute value of the total carotenoid content was not mentioned.

Yam (*Dioscorea* spp.) Dioscoreaceae

Yams have thick subterranean tubers with a high carbohydrate content. Some tropical species such as *D. sativa, D. alata, D. esculenta, D. bulbifera,* and *D. cayensis* are cultivated for food and eaten cooked.

Many edible yams have a yellow flesh. The yellow pigments of the aerial tubers of *D. bulbifera* were investigated by Martin et al. (1974). Ten varieties grown in Puerto Rico contained carotenoids that consisted exclusively of xanthophylls in free and predominantly esterified form. They were tentatively identified as cryptoxanthin, lutein, zeaxanthin, violaxanthin, neoxanthin, and auroxanthin. Chlorophylls and an anthocyanin were also detected.

The carotenoids of the most important yellow-fleshed yams of the tropics (*D. cayensis* Lam.) were investigated by the same authors (Martin and Ruberté, 1975). The carotenoid pattern was similar to that found in *D. bulbifera,* the principal carotenoids being free and esterified xanthophylls such as violaxanthin, neoxanthin, and auroxanthin. In contrast to *D. bulbifera,* the tubers of *D. cayensis* contained β-carotene, which was accompanied by small quantities of α-carotene. The β-carotene content of four varieties ranged between 4 and 14 μg (g fresh wt)$^{-1}$.

FACTORS AFFECTING CAROTENOID BIOSYNTHESIS

Phytohormones

The different effects of phytohormones, which also involve pigment changes, were previously mentioned (see p. 71). Plant carotenoids are influenced reversibly by various endogenous and exogenous plant growth regulators.

The effect of gibberellic acid (GA_3) at various concentrations on the carotene content in carrot shoots and roots was investigated (Neumann and Schwab, 1975). Application of GA_3 on the shoots of carrot plants increased shoot growth, whereas root growth was reduced. In shoots the carotene content increased from 114 to 165 μg (g dry wt)$^{-1}$, whereas in roots it decreased from 153 to 54 μg (g dry wt)$^{-1}$ at the optimum GA_3 concentration. Analogous results were obtained in carrot roots throughout the vegetative period of the first year, as a result of GA_3 treatment (Linser and Zeid, 1975). Treatment with BA had negligible effects.

Application of abscisic acid to detached Rudgers tomatoes reduced by about 50% the time to initiate ripening, as indicated by red color development (Mizrahi et al., 1978).

Differential effects of BA on the ultrastructure of chloroplasts in intact bean leaves according to their age were observed (Naito et al., 1981). First, only young leaves responded to BA; however, the late treatment efficiently prevented the last stage of leaf senescence. The total carotenoid content decreased continuously during leaf yellowing but rose again and reached the maximum concentration of mature leaves.

Ethylene, which is called the fruit-ripening hormone, is extensively used to promote ripening. The quality of ethylene-ripened tomatoes, including the carotenoid changes, was repeatedly investigated (Russo et al., 1975 and the references therein). One effect was an increased lycopene content in some cultivars. Veegan tomatoes, picked at 90% development and treated with ethylene, showed an accelerated rate of chlorophyll degradation and a rapid synthesis of carotene and lycopene (Medina et al., 1981). The same effect of ethylene was observed in various cultivars, but the reaction to ethrel differed among cultivars, New Yorker tomatoes showing the highest increase of lycopene (Elkner et al., 1984).

The effect of ethephon treatment on the carotene content of early planted carrots, cultivars Danvers 126 and Gold King, was investigated (Bewick et al., 1987). Ethephon-(2-chloroethyl)-phosphonic acid was applied as an overhead spray at various rates (280–560 g ha^{-1}) and timings. In Danvers 126 carrots, treated with the lower ethephon concentration, the carotene content per gram fresh weight increased significantly. In Gold King carrots the results indicated a significant increase of carotene per gram dry weight after treatment with both ethephon rates.

Light

Light is not essential for inducing carotenogenesis. Carotenoids are synthesized in the etioplast of plants grown in the dark, however, in smaller amounts, the pattern differing from the chloroplast carotenoids of true leaves, since xanthophyll furanoxides dominate. Upon illumination of the etiolated plants, carotenoids are rapidly synthesized, being constituents of the photosynthetic apparatus. Variation of the carotenoid composition occurs in thylakoids and envelopes under different light regimes.

Carotenoids are synthesized within fruits, tubers, and roots where only low light intensity can penetrate. For practical reasons the influence of light on color development in tomatoes was thoroughly studied. It has been shown that light augments carotenoid synthesis, with red light being more effective than white or green light (Jen, 1974). In fact, red-light-induced carotenoid synthesis and far-red light inhibited it in ripening tomatoes, thus suggesting a phytochrome-mediated response (Thomas and Jen, 1975a,b). Comparative effects of light and ethephon on the ripening of detached tomatoes indicated that carotenoid biosynthesis is light-dependent and that ethephon has no effect on the total carotenoid content (Paynter and Jen, 1976).

On the other hand, tomatoes ripened in darkness and in light both contained the same carotenoid pattern, which indicated that carotenogenesis is light-independent; it may be stimulated by light, but light is not required for its induction (Raymundo et al., 1976).

Temperature

Generally the optimal temperature for carotenogenesis in plants is relatively low. The synthesis of certain pigments has been shown to be temperature-sensitive; the temperature sensitivity varies from plant to plant.

Temperature sensitivity was studied thoroughly in connection with lycopene formation in tomato. Very early, the effect of environmental factors on tomato color was studied (Vogele, 1937). The optimal temperature for lycopene formation in the tomato is 16°C to 21°C; above 30°C it is not formed. Detached fruits, ripened at temperatures between 32°C and 38°C, became yellow; after being returned to lower temperatures (20–24°C), they developed normally (Went et al., 1942). These results were subsequently confirmed, and it was also observed that the synthesis of β-carotene is not affected by temperature (Goodwin and Jamikorn, 1952; Tomes et al., 1956; Tomes, 1963; Laval-Martin et al., 1975). In ripening tomatoes β-carotene is synthesized first, before the massive lycopene accumulation (Raymundo et al., 1970; Laval-Martin, 1975). To explain this phenomenon, Goodwin and Jamikorn suggested that a physically separated pathway of carotenoid biosynthesis is superimposed on the earlier chloroplast pathway (Goodwin and Jamikorn, 1952) or even a compartmentalization of similar enzymes (Raymundo et al., 1967).

The influence of suboptimal temperatures on color quality and pigment composition of ripening tomatoes was investigated by Koskitalo and Ormrod (1972). Fruits exposed to a diurnal regime of 18°C to 26°C had the highest pigment content, with lycopene accounting for 90% of the carotenes. At this regime the carotenoid biosynthesis rate was highest, whereas in fruits exposed to a diurnal regime of lower temperature the ripening process was delayed.

The effect of high temperature on the quality of processing tomatoes of various genotypes ripened off the vine was studied by Yakir et al. (1984a). Fruits exposed to either 27.5°C or 30°C during ripening showed a significant decrease in carotenoid concentration and color score compared with fruits kept at 25°C. The different lines responded differently to high-temperature exposure, and one was found to be insensitive. High-temperature effects on tomato fruits of various genotypes ripened on the vine in the greenhouse were also studied (Yakir et al., 1984b). The average temperature diurnal regimes were 22.8°C to 30.5°C and 16.9°C to 24.7°C. A higher lycopene content was found in all lines ripened in the high-temperature greenhouse, which also ripened at a higher rate than the control fruits. These results are explained by the fact that the fruits were exposed only part of the day to temperatures higher than 30°C. The results confirm early findings that tomatoes exposed to wide temperature fluctuations on the vine ripen normally if the night temperature is lower than 30°C (Sayre et al., 1953). The contents of phytoene and phytofluene, the first colorless precursors in the carotenoid biosynthesis, and which were determined in both recent reports, are irrelevant if not considered as part of the carotenoid pattern changes that occur during ripening (Gross, 1987).

For carrots the optimal temperature for color development was between 15°C and 21°C. Carrots sown in the spring have a carotene content three times higher than those sown in summer (Booth and Dark, 1949). The influence of temperature on the carrot carotenoid contents was investigated by growing carrots in two rooms in which temperature was the only environmental factor that differed (Banga et al., 1955). The total carotenoid content was considerably lower at 8°C than at 18°C. It seemed that in the first stages of root development the temperature does not affect the carotenoid content. There are significant differences between carrot varieties and lines in their sensitivity to growing temperatures (Bradley and Rhodes, 1969). Eleven of 39 varieties were not affected, other lines had better color in cool seasons, and others resulted in consistently fair or poor color in both growing seasons. Generally, maximal increase was obtained at temperatures between 13°C and 16°C. Quality characteristics of carrots grown under controlled temperature, relative humidity, day length, and light intensity, which simulated natural climates occurring in the United States, were investigated (Simon et al., 1982). Climate has a major influence on the total carotenoid levels, the lowest level corresponding to the lowest average temperature (10°C) and relative humidity (43%). Similarly, weather conditions strongly affect the carotene content of carrots grown in Finland, with higher temperatures being favorable (Evers, 1989).

Fertilizers

A study of the effect of fertilizers on carotene content in sweetpotatoes and carrots showed that high levels of nitrogen resulted in an increase of the carotene content, but no such effect from potassium was noted (Samuels and Landrau, 1952; Habben, 1973). Earlier results on the effect of the two fertilizers reported therein were contradictory. High doses of P_2O_5, N, and K_2O increased the carotene content of spring-sown carrots (Emura and Hosoya, 1979). Other trials on the influence of N on carrot carotenes showed that season and soil type had more effect than N application (Geissler and Rüdiger, 1980). The response of Nantes carrots to 13 different fertilizer forms applied at various rates was studied (Michalik, 1985). Nitrogen as ammonium or urea had no significant effect on the carotene content. Phosphorus as triple superphosphate at 60 kg ha^{-1} to 200 kg ha^{-1} and potassium as chloride or sulfate increased the carotene content. Magnesium had a beneficial effect. In a trial with two carrot cultivars, N was applied at 0 to 150 kg ha^{-1} in addition to basal P and K, 60 and 90 kg ha^{-1}, respectively. The highest carotenoid content was produced at 90 and 120 kg N ha^{-1}, respectively (Škrbić, 1987). A moderate positive correlation was found between N dose and carotene content (Cserni and Prohaszka, 1988). The effect of different fertilization practices on the carotene content of carrots grown in southern Finland was studied in a 2-year field experiment (Evers, 1989). PK placement with N fertirrigation had a tendency to increase the carotene content compared with other fertilization practices. The effect on the provitamin A content of early lettuce of soil fertilization and covering the culture with plastic film was studied (Rouchaud et al., 1984b). Generally, covering decreased the carotene content. At a normal dose of N, P, K, and S fertilizers the βcarotene level in lettuce was 1666 μg (100 fresh wt)$^{-1}$, whereas at a dose of N 40 times lower and lack of other fertilizers it decreased to 1421 μg (100 g fresh wt)$^{-1}$.

In spinach leaves the influence of N, P, and K fertilizers on the carotene content was studied (Sengewald, 1959). There was a positive relation between N doses and carotene, the carotene level rising even after optimal yield had been exceeded. Deficiency of K and P caused an increase in carotene content.

The highest urea dose (190 kg ha^{-1}) produced a maximum β-carotene level in spinach (Kansal et al., 1981).

Increasing amounts of K supplied to tomatoes increased the total carotenoid content. Lycopene followed a similar trend, whereas it had the opposite effect on the β-carotene levels (Trudel and Ozbun, 1971). The latter effect is due more to the precursor–product relationship between the two pigments than to the fertilizer.

Pesticides

With the beginning of pesticide application their possible effect on cultivated plants was investigated. Soil treatment with the fumigants Telone or Nemagone increased the carotene content of carrots (Wu et al., 1970). The carotene contents of Danvers and Chantenay carrots grown in soil treated with the herbicides Linuron or CIPC and that of Butternut squash grown in soil treated with Amiben and Dinoseb were significantly higher than the carotene content of control vegetables. Also, Dinoseb-treated Butternut squash contained a higher percentage of α-carotene than Amiben-treated or control squash. Herbicide treatment did not affect the carotene content of Hubbard squash or that of two spinach varieties (Sweeney and Marsh, 1971b). Treatment with chlorcholine chloride increased the yield of carotenoids in carrots (Linser and Zeid, 1975). An organophosphorus insecticide incorporated into the soil increased the carotene content to 34% in covered grown carrots. In uncovered grown carrots the carotene content was approximately 22% higher. None of the used cultural conditions modified the ratio of α-carotene to β-carotene or the ratio of *cis–trans* isomers (Rouchaud et al., 1982). In summer carrots grown on soil pretreated with a series of insecticides and herbicides, the carotene content was higher (up to 21%) than in the control, except for the herbicide Dosanex, which decreased the carotene content to 19% (Rouchaud et al., 1983).

The total carotene content (*cis* and *trans* isomers) was higher in lettuce treated with herbicides but unaffected by the fungicide treatment. The individual chloroplast xanthophyll content, including β-cryptoxanthin, was higher in the plants grown on treated soil, except for a fungicide. The results suggest that the pesticide treatment increased the violaxanthin cycle and photosynthetic activities in the lettuce (Rouchaud et al., 1984a, 1985).

Miscellaneous

The influence of the type of soil on the carotene content of carrots was emphasized early (Booth and Dark, 1949). Carotene levels were higher in carrots grown on sandy soils, moor soils, and brown and loessial chernozem (Mehwald, 1965; Habben, 1972). The effect of three different soils (muck soil, pH 7.1 and pH 5.8 and sandy clay, pH 7.7) on the carrot carotene content was less important than genotypic effects. However, the more acidic muck enhanced significantly the carotene level (Simon et al., 1982).

Different cultural methods, such as plastic greenhouse or field, as well as the place of the fruit on the plant, affect the carotenoid content of tomatoes. The highest carotene content was found in the vine-grown fruit (Cabibel and Ferry, 1980). Direct plastic covering with perforated polyethylene sheeting applied for a certain time produced a threefold increase in the carrot carotene content, whereas in uncovered culture the increase was only twofold. In lettuce the effect was opposite; covered and uncovered plants contained 11.0 and 16.6 g (g fresh wt)$^{-1}$ carotene, respectively (Benoit et al., 1984).

Titavit (titanium ascorbate) affected the dynamics of tomato fruit ripening by promoting the lycopene cyclization so that β-carotene content was about 18% to 20% higher in treated fruits (Biacs et al., 1988).

APPENDIX

Table A1. Constituents of Crude Carotene in Raw Vegetables

Vegetable	*Crude Carotene [μg × (g fresh wt)$^{-1}$]*	*Percent of Total Carotenoids*				
		Impurity	*Neo-β-carotene U*	*β-Carotene*	*Neo-β-carotene B*	*α-Carotene*
Beet green	33	4	10	76	10	—
Broccoli	12	13	1	80	6	—
Collards	59	7	10	77	6	—
Mustard greens	41	6	9	78	7	—
Spinach mustard	72	6	11	76	7	—
Spinach	46	5	12	76	7	—
Swiss chard	33	7	13	70	10	—
Turnip greens	54	4	9	79	8	—
Turnip greens, Shogoin	60	7	10	76	7	—
Squash, Ford Hook	4	42	4	45	3	6
Pumpkin, orange	20	41	2	52	2	3
Carrots	96	5	1	62	3	29
Sweetpotato	65	10	0	86	4	0

Source: From Kemmerer et al. (1945).

Table A2. Carotene Content of Some Vegetable Leaves Grown in Palestine

Plant	*Time of Sampling*	*β-Carotene [μg (g fresh wt)$^{-1}$]*
Cabbage	Nov.	15.2
Kohlrabi	Apr.	63.2
	June	64.8
Lettuce	Dec.	24.5
	May	18.9
Nettle	Dec.-March	56.4
Onion	Mar.	17.0
Parsley	Feb.-March	64.0
Purslane	Sept.	19.7
Spinach	Dec.	28.2
(leaves and stalks)	Apr.	31.6
	Aug.	11.4

Source: Recalculated from Bondi and Meyer (1946).

Table A3. Carotene Content of Vegetables Grown in Israel

Vegetable	*Carotene [μg (g fresh wt)$^{-1}$]*
Bean, kidney crystal green	tr
Bean, kidney yellow	tr
Beet leaves	81.0
Cabbage	36.0
Celery leaves	tr
Cucumber	tr
Kohlrabi	tr
Lettuce	9.6
Maize, yellow	tr
Okra	tr
Onion, green	tr
Parsley	87.8
Pea	3.2
Pepper, green	13.9
Pepper, gamba California	27.0
Pumpkin	59.1
Spinach, New Zealand	31.1
Squash	tr
Tomato	4.3

Source: From Halevy et al. (1957).
Note: tr = trace.

Table A4. β-Carotene Content of Various Vegetables Grown in Germany [μg (g fresh wt)$^{-1}$]

Vegetable	Rinno (1965)	Herrmann (1975)
Asparagus, white	tr	
Asparagus, green	2.0	2.0–6.0
Aubergine	tr	
Bean, green	2.0	1.2–6.5
Broad bean	1.7	
Broccoli	7.0	1.3–24.0
Brussels sprouts	4.0	1.5–7.0
Cabbage, green	35.0	9.0–75.8
Cabbage, white	0.2	
Carrot	60.0	60.0–110.0
Chinese cabbage	3.0	
Chives	26.0	3.0–5.0
Corn	35.0	9.0–75.0
Cucumber	0.1	
Endive		0.7–19.8
Fennel	0.3	
Garden cress	14.0	
Kohlrabi, leaf	30.0	
Kohlrabi, tuber	1.0	1.5–4.5
Leek	2.0	
Lettuce, head	10.0	1.2–23.0
Lettuce, romaine		12.0–60.0
Parsley	50.0	48.0–80.0
Pea	4.0	3.0–6.8
Pumpkin	0.3	18.0–34.0
Rhubarb	0.5	
Spinach	40.0	40.0–56.5
Spinach, New Zealand	23.0	
Sweet corn	0.2	
Tomato	3.5	1.4–23.0

Note: tr = trace.

Table A5. Composition of Leaf Carotenoids in Summer and Winter (Japan)

	Percent of Total Carotenoids					
Plants	*Carotene*	*Lutein*	*Anthera-xanthin*	*Viola-xanthin*	*Neo-xanthin*	*Total Carotenoids [μg (g fresh wt)$^{-1}$]*
In summer						
Carrot	45.5	28.2	8.5	9.1	8.7	74.4
Maize	50.0	37.0	6.3	4.2	2.5	138.9
Soya	50.6	33.6	6.3	3.3	5.4	137.6
In winter						
Radish	28.4	47.7	8.0	8.6	7.3	105.8
Lettuce	28.5	44.5	8.8	9.9	8.3	107.0
Cabbage	17.9	36.3	15.3	14.2	13.5	173.1
Leaf mustard	20.3	33.0	18.9	15.2	6.2	217.5

Source: From Takagi (1985).

Table A6. Provitamin A Content in Vegetables

	*α-Carotene**	*β-Carotene**
Asparagus	17.41	581.37
Beet greens	3.47	5027.67
Broccoli	1.61	762.89
Carrots	3789.67	7602.78
Green beans	64.44	364.23
Green pepper	33.78	217.13
Lettuce (Iceberg)	4.43	329.86
Okra	28.09	431.84
Peas	16.02	557.16
Spinach	0.00†	4312.01
Squash	12.13	177.56
Sweetpotato	0.00†	8610.86
Swiss chard	57.87	4568.46

Source: From Bureau and Bushway (1986).

*Units are μg (100 g)$^{-1}$ product, averaged over all the locations and months of analyses.

†0.00—none detected at a detection limit of 1 μg (100 g)$^{-1}$ product for all compounds.

Table A7. β-Carotene Content of Various Vegetables Grown in the United States [μg (g fresh wt)$^{-1}$]

Asparagus	1.6–4.0
Beet greens	21.8
Broccoli	4.3–7.8
Carrots*	58.4–115.0
Green pepper	6.8–10.9
Lettuce	8.3
Spinach	37.7–53.2
Squash	21.5–80.4
Sweetpotatoes	54.0–130.0

Source: From Bushway (1986).
*Also contain 31.6–51.6 μg α-carotene.

Table A8. β-Carotene Content in Various Vegetables Grown in Japan [μg (g fresh wt)$^{-1}$]

①	Collard	53.6
	Kale	11.9
	Brussels sprouts	4.1
	Cabbage, green ball	1.7
	Cabbage, red	1.3
	Cabbage, pointed	0.7
	Cabbage, Savoy	0.5
	Cabbage	0.3
	Kohlrabi, red	0.4
	Kohlrabi, green	1.0
②	Parsley	58.2
	Celery, leaves	61.1
	Onion, leaves	2.2
	Watercress	31.2
	Sweet pepper	2.5
	Broccoli	3.9
	Okura	2.8
	Green asparagus	3.0
	Celery, green	1.5
	Celery	1.1
	Celery, yellow	0.5
	Cauliflower	—
③	Malabar spinach (*Basella alba*)	39.1
	Saisin (*Brassica campestris* var. *chinensis*)	9.0
	Chinese chive	45.6
	Kousaitai (*Brassica campestris* var. *chinensis*)	7.9

(*continued*)

Table A8. Continued

	Ta-sai (*Brassica campestris* var. *narinosa*)	25.3
	Chinese chive (flower stalks)	16.3
	Garlic sprouts	6.5
	Chingentsai (*Brassica campestris* var. *chinensis*)	14.4
	Rape, flower cluster	13.8
	Komatsuna (*Brassica campestris*)	62.6
	Spinach	43.5
	Perilla, green leaves	186.6
	Hamaboufuu (*Glehnia littoralis*)	30.1
	Chinese cabbage, young	17.6
	Kyouna (*Brassica campestris japonica*)	10.5
	Mitsuba (*Cryptotaenia japonica*)	47.5
	Hanawasabi (*Wasabia japonnica,* flower stalk)	6.6
	Hatakena (*Brassica campestris* var. *oleifera*)	46.0
	Hiroshimana (*Brassica campestris* var. *pekinensis*)	36.3
	Sugukina (*Brassica rapa* var. *neo suguki*)	40.8
	Kaiwarena (Japanese radish, sprout)	7.8
	Turfed stone leek (*Allium ascalonicum*)	20.3
	Ashitaba (*Angelica keiskei*)	51.1
	Leaf mustard (*Brassica juncea*)	30.9
④	Santousai (*Brassica pekinenesis*)	8.4
	Green pepper	5.7
	Welsh onion	18.7
	Salt wort (*Salsola komarovi*)	9.8
	Turnip, leaves	46.5
	Shungiku (*Chrysanthemum coronarium*)	40.8
	Mabikina (Japanese radish, immature)	26.8
	Umaina (*Beta vulgaris,* leaves)	23.8
	Broad-leaved mustard (*Brassica juncea*)	34.8
	Carrot, leaves	19.4
	Garden peas, pods, immature	4.1
	Water dropwort (*Oenanthe javanica*)	39.5
	Chinese mustard (*Brassica campestris* var. *chinensis*)	10.6
	Vitamin-na (*Brassica campestris* var. *narinosa*)	36.9
	Cucumber	3.9
	Chinese cabbage (*Brassica campestris* var. *pekinensis*)	2.4
	Kidney beans, pods, immature	11.3
	Green peas	10.8
	Squash *Ebisu* (*Cucurbita maxima*)	26.4
	Squash *Kogiku* (*Cucurbita moschata*)	1.2
⑤	Carrot *Kintoki*	43.3
	Carrot	8.6
	Squash (*Cucurbita maxima,* red skin)	4.8

Source: From Izaki et al. (1986).

① = *Brassica* species and varieties; ② = European vegetables; ③ = Chinese vegetables; ④ = Japanese (indigenous) vegetables; ⑤ = carrots and squashes (also contain α-carotene).

Table A9. Carotene Content of Green Leaves Consumed in North Brazil [μg (g fresh wt)$^{-1}$]

	December 1984		May 1985	
Vegetable	*α-Carotene*	*β-Carotene*	*α-Carotene*	*β-Carotene*
African spinach	0.1	38.9	—	—
Bitter cress	—	49.4	—	60.0
Cassava	0.4	150.7	0.6	108.2
Cariru	0.2	12.3	2.8	30.0
Chicory	—	—	0.2	25.0
Chinese cabbage	0.3	49.2	—	—
Indian spinach	1.0	46.8	0.5	78.8
Jambo	1.0	39.2	—	—
Malabar nightshade	2.3	54.8	—	—
Okra	1.3	116.2	—	—
Orelha de macaco	0.2	33.0	0.4	58.6
Purple vinegar plant	0.3	73.2	0.3	83.8
Purslane	—	27.0	—	—
Taro	1.5	55.0	8.9	59.0
Tomato	0.2	54.8	—	—
White vinegar plant	0.2	71.3	0.8	101.8

Source: From Penteado et al. (1986).

Table A10. β-Carotene Content of Common Brazilian Leafy Vegetables

English Name	*Portuguese Name*	*Botanical Name*	*No. of Samples Analyzed*	*β-Carotene [μg (g fresh wt)$^{-1}$]*
Parsley	Salsinha	*Petroselinum hortense*	5	33.0–66.0
Roquette or rucula	Rucula	*Eruca usativa*	15	14.0–64.2
Coriander leaves	Coentro	*Coriandrum sativum*	5	39.2–53.0
Cress	Agriao	*Nasturium officinale*	5	27.0–52.0
Kale	Couve	*Brassica oleracea* var. *acephala*	10	23.6–59.0
Common chicory	Almeirao	*Cicorium intybus*	10	19.0–45.9
Endive	Chicoria	*Cicorium endivia*	8	12.0–28.8
Lettuce, unheaded curly	Alface crespa	*Lactuca sativa*	14	6.4–21.6
Lettuce, Boston	Alface lisa	*Lactuca sativa*	6	9.7–20.2
Cabbage	Repolho	*Brassica oleracea* var. *capitata*	5	0.2–1.7
Chinese cabbage	Couve-chinesa	*Brassica chinensis*	5	0.1–3.5

Source: From Ramos and Rodriquez-Amaya (1987).

Table A11. β-Carotene Content of Australian Vegetables [μg (g fresh wt)$^{-1}$]

Vegetable	*β-Carotene Equivalents**	*Vegetable*	*β-Carotene Equivalents**
Angled loofa	0.09	Endive	0.03
Artichoke, globe	0.09	Garland chrysanthemum	1.48
Artichoke, Jerusalem	0.02	Garlic	0.01
Asparagus	0.05	Ginger	0.03
Bean, broad	0.20	Gourd, wax	0.02
Bean, butter	0.09	Kohlrabi	0.03
Bean, French	0.36	Leek	0.47
Bean, purple	0.07	Lettuce, cos/mignonette	0.60
Bean, red	0.04	Lettuce, head	0.09
Bean, yard-long	0.48	Marrow	0.04
Bean sprouts	0.02	Melon, bitter	0.06
Beetroot	0.02	Okra	0.13
Broccoli	0.39	Onion	0.01
Brussels sprouts	0.18	Onion, spring	0.62
Cabbage	0.04	Parsley	4.43
Cabbage, Chinese	0.19	Parsnip	0.03
Cabbage, Chinese flowering	1.38	Pea, green	0.42
Cabbage, Chinese white	0.50	Pea, snow	0.20
Cabbage, mustard	1.64	Potato	0
Cabbage, red	0.01	Pumpkin, butternut	2.78
Capsicum, banana	0.22	Pumpkin, Golden Nugget	3.38
Capsicum, continental	0.01	Pumpkin, Queensland Blue	2.77
Capsicum, green	0.18	Radish, red	0.01
Capsicum, red	1.46	Radish, white	0.04
Carrot	10.02	Rhubarb	0.10
Cassava, white	0.04	Shallot	0.79
Cassava, yellow	0.77	Silverbeet	0.43
Cauliflower	0.01	Spinach	2.31
Celeriac	0.02	Spinach, Chinese	1.87
Celery	0.04	Spinach, water	1.22
Chicory	0.25	Squash, button	0.16
Chili, banana	0.06	Squash, scallopini	0.68
Chili, green	0.26	Sweetpotato	0.02
Chili, red	1.79	Sweetpotato, orange	6.78
Chives	2.30	Sweet corn	0.26
Chives, Chinese	3.34	Taro	0.03
Choko	0.05	Tomato	0.34
Cucumber	0.16	Watercress	2.01
Eggplant	0.44	Zucchini	0.27

Source: From Wills (1987).

*β-Carotene equivalents are calculated as μg β-carotene + 0.5 (μg α-carotene + cryptoxanthin).

Table A12. Carotene Content of Vegetables Consumed in Tanzania [μg (g fresh wt)$^{-1}$]

Vegetable	*Edible Part*	*α-Carotene*	*β-Carotene*
Amaranthus (*Amaranthus* spp)	Leaves	0.85	24.0
Cassava (*Manihot esculenta*)	Leaves	—	28.2
Cowpea (*Vigna* spp)		—	7.3
Okra (*Hibiscus esculentus*)	Pods, raw	—	1.6–2.7
Peas (*Pisum sativum*)		0.3	1.1
Pumpkin (*Cucurbita moschata*)	Pulp	11.0	11.7
leaves		—	10.5
Sweetpotato (*Ipomoea batatas*)	Yellow tuber	—	18.2

Source: From Pepping et al. (1988).

Table A13. Carotenoid Content of Thai Vegetables [μg (g fresh wt)$^{-1}$]

	Total Carotenoids	*β-Carotene*
Leaf vegetables		
Acacia insuavis Lace cha-om (local name)	61	4.2
Allium odorum Linn. (Chinese chive, onion fragrant)	109	8.3
Allium odorum Linn. (Chinese chive)	39	1.7
Amaranthus gangeticus Linn. (amaranth spineless)	93	19.6
Apium graveolens Linn. (celery)	75	7.7
Azadirachia indica Juss. var. *siamensis valeton* (neem tree)	118	9.2
Basella alba Linn. (vine spinach; Malabar nightshade)	70	5.6
Brassica chinensis Jusl. (Chinese cabbage)	72	9.7
Brassica chinensis var. *parachinensis* Tsen & Lee (flowering white cabbage)	28	2.5
Brassica juncea Czern & Coss. (pickled leaf mustard, fermented)	13	1.2
Brassica oleracea Linn. var. *acephala* DC (Chinese kale or collard)	43	2.3
Brassica oleracea Linn. var. *capitata* Linn. (cabbage)	34	4.1
Brassica sp. pak kard soi (local name)	71	9.0

(continued)

Table A13. Continued

	Total Carotenoids	β-Carotene
Careya sphaerica Roxb. kradon (local name)	93	7.5
Centella asiatica Urban. (Asiatic pennywort)	47	1.0
Coccinia indica White & Arn. (ivy gourd)	53	6.1
Coriandrum sativum Linn. (coriander)	150	19.9
Cucurbita maxima Duch. (winter squash, young leaves)	78	9.6
Daucus carota (carrot, raw)	33	18.3
Eryngium foetidum Linn. (common fennel)	85	44.4
Ipomoea aquatica Forsk. (Chinese swamp cabbage)	115	10.5
Swamp cabbage, green stalk and big leaves (wet soil)	50	8.7
Swamp cabbage, green stalk and small leaves (dry soil)	15	0.42
Lactuca sativa Linn. (lettuce, unheaded)	67	4.8
Leucaena leucocephala de Wit (lead tree)	72	1.7
Leucaena leucocephala de Wit (lead tree, tender top)	13	0.47
Mentha cordifolio Opiz (kitchen mint)	102	7.3
Momordica charantia Linn. (bitter melon leaves)	164	34.0
Neptunia oleracea Lour. (water mimosa)	103	10.6
Ocimum basilicum Linn. (sweet basil)	92	5.1
Ocimum canum Sims. (hairy basil)	63	3.5
Ocimum sanctum Linn. (holy basil)	67	4.8
Piper sarmentosum Roxb. chaa phluu (local name)	218	8.0
Tuberous vegetables		
Allium ascolonicum Linn. (onion, young complete plant)	33	2.1
Brassica oleracea Linn. var. *botrytis* Linn. (cauliflower)	0.63	0.04
Brassica pekinensis Rupr. var. *cylindrica* Tsen & Lee (celery cabbage)	4.0	1.1
Fruits		
Cucumis sativa Linn. (cucumber, not paired)	0.95	0.06
Cucurbita maxima Duch. (winter squash or pumpkin)	23.0	0.49
Hibiscus esculentus Linn. (okra or lady's finger)	8.8	0.54
Lagenaria siceraria Standl. (bottle gourd)	0.46	0.04
Luffa acutangula Roxb. (angled loofah)	1.7	0.47
Lycopersicon esculentum Mill. (tomato, ripe)	6.4	1.46
Momordica charantia Linn. (bitter melon)	9.6	0.50
Parkia speciosa Hassk. sato (local name)	8.2	0.66
Pisum sativum Linn. (sugar peas or garden peas)	8.9	1.7
Solanum melongena Linn. (eggplant, long shape)	7.9	0.87
Solanum melongena sw. (eggplant, round shape)	5.3	0.41
Solanum torvum sw. (eggplant, small and round)	8.7	0.35
Solanum xanthocarpum sw. (eggplant, oval shape)	3.9	0.10
Vigna sinensis Savi ex Hassk. (yard-long beans, green)	6.8	0.44
Zea mays (corn, yellow)	12.6	0.74

Source: From Speek et al. (1988).

Table A14. Carotenoids and Vitamin A Activity in Vegetables

	Carotenoid [μg (100 g fresh product)$^{-1}$]						
Item	*α-Carotene*	*β-Carotene*	*γ-Carotene*	*Cryptoxanthin*	*Lutein**	*Lycopene*	*RE*†
Potato							
new (Aug.)	—	3.2	—	—	13	—	0.5
old (March)	tr	7.7	—	nd	60	—	1.3
Carrot	530	7600	nd	—	300	—	1300
Rutabaga	—	tr	nd	nd	—	—	0
Turnip	tr	72	—	—	tr	—	12
Cabbage							
white	tr	66	nd	—	150	—	11
red	tr	15	—	nd	26	—	2.5
Chinese	tr	15	—	tr	40	—	2.5
Cauliflower	—	11	—	—	33	—	1.9
Broccoli	tr	1000	nd	24	1800	—	170
Brussels sprouts	—	430	—	nd	920	—	71
Spinach	—	3300	—	—	4400	—	550
Lettuce							
leaf	tr	980	—	—	1800	—	160
American	—	73	—	nd	250	—	12
Celery	—	2900	nd	nd	7200	—	490
Celeriac	—	—	nd	nd	tr	—	0
Yellow onion	—	6.9	—	—	16	—	1.2
Leek	—	1000	nd	—	1900	—	170
Radish	—	9.4	—	—	12	—	1.6
Bean	39	180	nd	—	440	—	33
Dill	—	4500	—	—	6700	—	750
Parsley	—	5600	—	—	10200	—	940
Cucumber	tr	130	—	—	470	—	22
Tomato	—	660	170	nd	100	3100	120
Pepper							
red‡	nd	2900	nd	nd	nd	nd	480
yellow	92	150	—	—	770	—	33
green	tr	240	nd	—	700	—	40

Source: From Heinonen et al. (1989).
Notes: — = not detected at a detection limit of 0.5 μg (100 g)$^{-1}$; tr = trace; nd = not determined.
*Lutein + zeaxanthin.
†Micrograms of RE (100 g)$^{-1}$ = 0.167(μg β-carotene) + 0.083(μg α-carotene + μg γ-carotene + μg cryptoxanthin).
‡Contains capsanthin, 3200 μg (100 g)$^{-1}$.

Table A15. Carotene and Chlorophyll Contents in Various Green Vegetables Grown in Japan [μg (g fresh wt)$^{-1}$]

Vegetable	*α-Carotene*	*β-Carotene*	*Total Chlorophylls*
Chingentsuai, leaves	0	90.6	1167
Komatsuna, leaves	0	85.5	1435
Perilla, leaves	1.0	72.2	1822
Spinach, leaves	0	65.7	1100
Leaf-blade, green	0	52.4	545
Spinach, whole	0	35.1	579
Komatsuna, whole	0	33.4	560
Leaf-blade, whole	0	27.9	305
Head lettuce, butter head type	0	26.2	572
Head lettuce, crisp head type, outer layer	0	21.9	602
Chingentsuai, whole	0	17.7	254
Broccoli, head	0	12.0	126
Cabbage, outer layer	0	10.1	478
Spinach, stem	0	4.9	66
Brussels sprouts	0.1	4.2	45
Sweet pepper, green	0	3.9	103
Komatsuna, stem	0	3.2	39
Kidney beans, pods, immature	0.2	2.3	71
Leaf-blade, yellow-green	0	2.5	15
Garden peas, pods, immature	0	2.2	64
Green asparagus, stem	0	2.1	28
Cabbage, inner layer	0	1.5	34
Head lettuce, crisp head, inner layer	0	1.4	0
Chingentsuai, stem	0	0.6	3

Source: From Kon (1989).

Table A16. Carotene Content in Processed Vegetables

	Constituents of Crude Carotene (% of Carotene Fraction)				
Food and Treatment	*Crude Carotene ($\mu g\ g^{-1}$)*	*Neo-β-Carotene U*	*β-Carotene*	*Neo-β-Carotene B*	*α-Carotene*
Leafy vegetables, boiled					
Beet greens, fresh, 20 min	81	13	67	16	0
Collards, 20 min	44	10	55	26	0
Mustard greens, 20 min	56	15	65	16	0
Mustard greens, 30 min	80	20	59	16	0
Spinach, 20 min	52	12	58	26	0
Swiss chard, 20 min	36	15	66	11	0
Turnip greens, fresh, 30 min	107	0	75	17	3
Turnip greens, 20 min	66	16	62	17	0
Leafy vegetables, dehydrated					
Beet greens	518	12	75	10	0
Collards	432	11	51	13	0
Swiss chard	383	10	76	8	0
Turnip greens	533	13	72	10	0
Leafy vegetables, canned					
Spinach	54	15	61	11	0
Spinach	33	24	38	8	0
Spinach	58	17	53	19	0
Spinach	70	17	45	10	0
Turnip greens	40	26	51	10	0
Turnip greens	74	23	49	16	0
Squash and pumpkin, boiled					
Squash, Fordhook, 30 min	3	7	40	11	3
Squash, Fordhook, 20 min	3	6	31	6	8
Pumpkin, orange shelled, 30 min	7	4	35	8	4
Pumpkin, canned					
No. 1	38	5	47	8	28
No. 2	49	9	39	9	25
No. 3	21	10	26	13	21
Sweetpotatoes, baked					
70 min	46	0	67	16	0
60 min	55	0	62	16	0
90 min	43	2	68	15	0
90 min	57	0	74	13	0
90 min	44	0	68	20	0

(continued)

Table A16. Continued

Food and Treatment	Constituents of Crude Carotene (% of Carotene Fraction)				
	Crude Carotene ($\mu g\ g^{-1}$)	*Neo-β-Carotene U*	*β-Carotene*	*Neo-β-Carotene B*	*α-Carotene*
Sweetpotatoes, dehydrated					
No. 1	131	0	76	11	0
No. 2	126	0	74	13	0
No. 3	143	6	71	13	0
No. 4	134	3	72	15	0
No. 5	132	7	64	16	0
No. 6	160	0	78	13	0
No. 7	156	0	81	10	0
No. 8	88	27	35	21	0
Carrots, boiled					
30 min	121	2	50	11	34
20 min	111	0	49	11	32
30 min	103	5	55	15	23
30 min	118	0	44	9	41
Carrots, dehydrated					
No. 1	1048	1	67	3	26
No. 2	698	0	63	2	29
No. 3	750	3	63	4	23
No. 4	923	0	64	5	13
No. 5	855	0	69	7	21

Source: From Kemmerer et al. (1945).

Table A17. Carotene Content in Canned Vegetables ($\mu g\ g^{-1}$)

Asparagus	
green	3.1
white	0.3
Beans	
green cut	1.8
lima green	0.8
Carrots	73.4
Corn	
white whole kernel	0.2
yellow whole kernel	0.9
Peas, sweet, wrinkled	2.6
Peppers, sweet red	19.9
Pimentos	13.8
Sweetpotatoes	59.4
Spinach	32.9
Tomatoes	5.8
Tomato juice	5.1
Turnip greens	26.4

Source: From Clifcorn (1948).

Table A18. β-Carotene in Frozen Vegetables ($\mu g\ g^{-1}$)

Asparagus	
cuts and tips	5.1
spears	4.7
Beans	
cut wax	0.6
green cut	3.5
French style	3.2
lima baby	1.3
Fordhook	1.4
Broccoli	
chopped	15.5
spears	11.3
Brussels sprouts	3.4
Cauliflower	0.2
Collard greens	40.9
Corn, cut	0.5
Kale (leaf and chopped)	48.9
Mustard greens	35.9
Okra, cut	2.9
Peas	
black-eyed	1.0
green sweet	4.1
Potatoes, mashed	0.2
Spinach	42.2
Squash	
winter	23.5
yellow crookneck	0.9
Turnip greens	41.3

Source: Reprinted with permission, from Burger et al. (1956); copyright © 1956 American Chemical Society.

Table A19. Stereoisomers of β-Carotene in Fresh and Frozen Green Vegetables as Affected by Cooking Time

Vegetable Sample	*Cooking Time (min)*	*Neo-β-carotene E (%)*	*Neo-β-carotene B (%)*	*all-*trans *β-Carotene (%)*	*Neo-β-carotene U (%)*	*Biologic Value (%)*	*Xanthophylls (% total carotenoids)*	*Total Carotene as β-carotene [μg (100 g)$^{-1}$]*	*Effective Carotene [μg (100 g)$^{-1}$]*
Broccoli									
Fresh	0	—	5.7	81.0	13.3	89.1	54	1080	960
	15	—	4.7	74.5	20.8	84.9			920
	25	—	5.4	70.4	24.2	82.5			890
Frozen									
heads	0	—	6.6	83.8	9.6	90.9	68	950	860
	8	—	8.6	78.2	13.2	87.8			830
	15	1.0	7.0	76.4	15.6	86.5			820
stems	0	—	5.4	81.1	13.5	89.1	70	270	240
	8	1.2	11.2	73.0	14.6	85.1			230
	15	0.7	9.0	73.5	16.8	85.0			230
Brussels sprouts									
Fresh	0	—	6.8	82.5	10.7	90.2	74	480	320
	10	—	6.1	75.3	18.6	85.6			310
	20	—	6.2	70.6	23.2	82.7			300
Frozen	0	—	5.7	82.0	12.3	89.7	71	370	330
	10	—	5.7	74.6	19.7	85.1			310
	20	—	4.4	70.4	25.2	82.3			300
Collards									
Fresh	0	—	6.5	80.9	12.6	89.1	70	5400	4810
	7	—	7.3	77.5	15.2	87.2			4710
	15	—	6.7	77.5	15.8	87.1			4700

Frozen	0	—	7.8	81.0	11.2	89.4	77	5510	4930
	15	—	8.3	78.3	13.4	87.8			4840
	25	—	9.0	76.4	14.6	86.7			4780
Kale									
Fresh	0	—	3.8	85.7	10.5	91.7	65	4720	4330
	20	—	5.4	77.0	17.6	86.6			4090
	30	—	6.6	70.6	22.8	82.8			3910
Frozen	0	—	9.6	80.6	9.8	89.4	73	2840	2540
	20	—	8.7	75.5	15.8	86.1			2540
	30	—	12.7	70.3	17.0	83.5			2380
Spinach									
Fresh	0	—	4.9	82.6	12.5	90.0	59	3300	2970
	5	—	7.8	76.4	15.8	86.5			2860
	10	—	7.7	76.9	15.4	86.9			2860
Fresh	0	—	9.4	80.2	10.4	89.2	69	4840	4320
Fresh	0	—	9.5	79.9	10.6	88.9	65	3770	3350
Frozen	0	—	5.6	84.2	10.2	91.1	48	4940	4510
	5	0.3	8.0	80.5	11.2	89.2			4410
	10	0.4	8.2	78.6	12.8	88.1			4360
Beet greens, fresh	0	—	7.0	80.7	12.3	89.1	67	2560	2240
	10	—	5.8	77.4	16.8	86.9			2190
	20	—	6.8	74.8	18.4	85.4			2150
Endive, fresh	0	—	5.2	80.8	14.0	88.9	65	960	850
	7	—	6.5	76.5	17.0	86.4			830
	15	—	4.8	76.5	18.7	86.1			830

Source: Reprinted from J.P. Sweeney and A.C. Marsh (1971), Effect of processing on provitamin A in vegetables, *J. Am. Diet. Assoc.*, vol. 59, pp. 238–243.

Table A20. Stereoisomers of β-Carotene in Fresh and Frozen Red and Yellow Vegetables as Affected by Cooking Time

Vegetable	*Cultivar*	*Cooking Time (min)*	*Neo-α-carotene B (%)*	*all-trans-α-Carotene (%)*	*Neo-α-carotene U (%)*	*Neo-β-carotene E (%)*	*Neo-β-carotene B (%)*	*all-trans-β-Carotene (%)*	*Neo-β-carotene U (%)*	*Biologic Value (%)*	*Xanthophylls (% total carotenoids)*	*Total Carotene as β-carotene [μg (100 g)⁻¹]*	*Effective Carotenes [μg (100 g)⁻¹]*
Carrots													
Fresh	—	0	1.0	21.9	—	—	2.9	72.6	1.6	86.5	18	15180	13130
		15	1.1	22.4	0.4	—	9.0	65.9	1.2	83.3			12640
		25	3.9	19.0	0.5	—	10.2	65.4	1.0	82.0			12450
Fresh	Danvers	0	1.2	38.6	0.7	—	5.3	52.4	1.8	76.7		14160	10860
	Chantenay	0	1.8	37.4	0.9	—	3.2	54.6	1.1	76.9		6270	4820
Frozen	—	0	0.4	29.0	0.8	—	5.0	63.0	1.8	82.1	7	11090	9070
		10	1.8	27.5	0.5	0.9	12.9	54.2	2.2	77.3			8570
		15	2.2	25.2	0.8	0.8	15.9	51.5	3.6	75.7			8390
Squash													
Fresh		0	0.6	19.0	0.6	—	4.2	71.6	4.0	85.6	61	5700	4880
	Butternut	20	2.9	16.1	1.0	—	14.9	60.5	4.6	79.2			4510
		40	4.8	14.0	0.9	—	21.2	53.5	5.6	75.1			4280
Fresh	Butternut	0	1.2	32.5	1.2	—	6.6	53.1	5.4	76.3	66	2060	1570
	Hubbard	0	—	—	—	—	8.6	84.2	7.2	91.5	83	820	740

Frozen	—	0	—	4.0	—	—	20.8	71.0	4.2	85.7	85	2780	2380
		10	—	3.7	—	—	23.2	68.1	5.0	84.3	85		2340
Sweetpotatoes													
Fresh		0	—	—	—	—	0.1	96.7	3.2	97.9	12	7620	7470
		20	—	—	—	—	24.9	69.7	5.4	85.0			6480
		40	—	—	—	5.2	23.6	66.0	5.2	83.3			6350
Frozen	—	0	—	—	—	—	13.3	83.1	3.6	91.5	12	6220	5690
		20	—	—	—	3.6	16.0	76.2	4.2	88.2			5460
		40	—	—	—	3.7	18.9	74.2	3.2	87.4			5430
Pumpkin, fresh		0	—	3.1	—	—	5.6	86.0	5.3	92.6	71	1100	1020
		10	—	3.0	—	—	16.2	75.3	5.5	87.6			970
		20	—	3.0	—	—	14.3	75.5	7.2	87.4			970
Red peppers, fresh	—	0	—	2.6	—	—	6.3	83.5	7.6	91.1	85	2280	2080
		5	—	2.3	—	—	24.8	64.6	8.3	82.1			1870
		10	—	3.1	—	—	31.6	58.3	7.0	79.4			1810

Source: Reprinted from J.P. Sweeney and A.C. Marsh (1971), Effect of processing on provitamin A in vegetables, *J. Am. Diet. Assoc.*, vol. 59, pp. 238–243.

Table A21. Carotene Stereoisomers in Canned or Pressure-Cooked Foods

Vegetable	*Treatment*	*Neo-α-carotene B (%)*	*all-trans-α-Carotene (%)*	*Neo-α-carotene U (%)*	*Neo-β-carotene (%)*	*Neo-β-carotene B (%)*	*all-trans-β-Carotene (%)*	*Neo-β-carotene U (%)*	*Neo-β-carotene V (%)*	*Biologic Value (%)*	*Xanthophylls (% total carotenoids)*	*Total Carotene as β-carotene [μg (100 g)$^{-1}$]*	*Effective Carotenes [μg (100 g)$^{-1}$]*
Carrots													
Brand 1	Canned	4.7	29.5	0.3	2.8	4.5	55.9	2.3	—	76.4		8100	6180
Brand 2	Canned	1.8	22.8	0.9	1.7	3.6	65.2	4.0	—	82.0	5	7800	6400
Brand 3	Canned	8.2	24.2	0.7	4.8	13.9	43.8	4.4	—	69.6		8370	5830
Carrots	15 lb, 5 min	0.8	30.0	0.3	2.3	12.4	50.4	3.8	—	75.6		14460	10920
Squash	Canned	—	1.8	—	—	15.4	73.2	9.6	—	85.9	75	940	810
Squash	15 lb, 5 min	—	4.0	—	—	23.8	67.0	5.2	—	83.7		2920	2450
Spinach													
Brand 1	Canned	—	—	—	—	5.4	72.5	21.6	6.5	86.1	53	5200	4360
Brand 2	Canned	—	—	—	1.6	6.9	62.5	29.0	—	78.0	57	4430	3460
Spinach	15 lb, 5 min	—	—	—	2.1	9.6	72.0	16.3	—	84.4		4960	4180
Pumpkin													
Brand 1	Canned	5.3	35.1	3.3	2.0	6.2	43.2	4.2	0.7	68.9	20	9880	6800
Brand 2	Canned	—	3.3	—	—	20.5	62.4	13.8	—	80.2	62	580	460
Sweetpotatoes													
Brand 1	Canned	—	—	—	3.3	10.2	79.5	7.0	—	89.4	17	4640	4150
Brand 2	Canned	—	—	—	5.7	6.2	82.9	5.2	—	91.2	17	6600	6020
Kale													
Brand 1	Canned	—	—	—	—	6.4	69.1	24.5	—	81.8	48	3810	3110
Brand 2	Canned	—	—	—	—	5.2	63.9	30.9	—	78.4	54	3930	3080

Source: Reprinted from J.P. Sweeney and A.C. Marsh (1971), Effect of processing on provitamin A in vegetables, *J. Am. Diet. Assoc.*, vol. 59, pp. 238–243.

Table A22. Carotene Content of Raw and Cooked Leaf Vegetables Grown in Pakistan

Vegetables	*Raw* [μg (100 g)$^{-1}$]	*Cooked* [μg (cooked wt)$^{-1}$] *Fried*	*Pressure*	*Traditional*
Amaranthus (*Amaranthus viridis*)	5183	4417	5033	4353
Bathua (*Chenopodium album*)	3097	2619	2988	2498
Beetroot (*Beta vulgaris*)	5266	4848	4792	4460
Cabbage (*Brassica oleracea* var. *capitata*)	1229	1088	1092	932
Fenugreek (*Trigonella foenum graecum*)	2458	2189	2207	2073
Knol Kholl (*Brassica oleracea* var. *caulorapa*)	4251	3864	3992	3542
Radish (*Raphanus sativus*)	4981	4602	4359	3948
Spinach (*Spinacia oleracea*)	5222	4394	4937	3569

Source: From Sood and Bhat (1974).

Table A23. β-Carotene Content of Tropical Vegetables, Raw and After Cooking ($\mu g\ kg^{-1}$ fresh material)

Method of Preparation	*Vegetables* *Cassava* (Manihot utilissima)	*Ceylon spinach* (Basella alba)	*Waterleaf* (Talinum triangulare)	*Taro* (Colocasia esculenta)
Raw (whole or ground)	174	19	21	78
Boiled, 15–120 min	168	19	31	77
Boiled, with potash	162	19	31	79
Stewed, 25–30 ml oil, 100 ml water	173	20	—	78
Fried 100–150 ml oil	—	—	29	—

Source: From Renqvist et al. (1978).

Table A24. Vitamin A Value* in Uncooked and Cooked Vegetables Grown in Pakistan

Vegetables	*Uncooked*	*Cooked*†	*Percent Loss of Vitamin A Value*	*Residue in Cooking Water*
Bitter gourd (*Momordica charantia*)	28.4	22.1	22	3.8
Brinjal (*Solanum melongena*)	16.4	10.2	38	2.2
Cabbage (*Brassica oleracea capitata*)	1.6	0.9	44	na
Colocasia (*Colocasia esculenta*)	13.6	9.1	33	na
Carrot, red part (*Daucus carota*)	48.3	19.6	59	5.3
Carrot, white part (*Daucus carota*)	43.7	29.1	33	4.0
Cauliflower (*Brassica oleracea botrytis*)	0.3	0.2	33	na
Colbash (*Legenaria vulgaris*)	15.8	10.5	34	2.0
Gand gobhi‡ (*Brassica oleracea caulorapa*)	1.0	0.5	50	na
Green turnip (*Brassica napus*)	3.3	2.2	33	na
Lady's fingers (*Hibiscus esculentus*)	1.6	0.8	50	na
Large chilies (*Capsicum annuum*)	20.2	15.3	24	4.5
Peas (*Pisum sativum*)	2.6	1.6	39	1.0
Spinach (*Spinacia oleracea*)	33.3	25.0	25	3.1
Squash (*Citrullus fistulosus*)	19.1	9.5	50	3.0
Vegetable sponge (*Luffa acutangula*)	3.9	3.5	10	na
Yellow turnip (*Brassica napus*)	2.7	2.2	19	na

Source: From Nagra and Khan (1988).

*The figures for vitamin A value are all expressed as micrograms RE per gram of the original uncooked vegetable.

†Includes the residue of carotene present in the cooking water.

‡Local name.

Table A25. Influence of Processing on Total Carotenoids and β-Carotene Contents and Vitamin A Activity of Thailand Vegetable Products

Vegetable	*Min*	*Total Carotenoid*	*β-Carotene*	*Vitamin A Activity*
		Carotenoids as percentage of fresh after cooking		
Bitter melon leaves	5	91	89	91
Coriander	4	93	85	91
Sweet basil	2	89	68	88
Celery	4	81	84	81
Vine spinach	8	66	89	68
Lettuce, unheaded	4	89	77	88
Hairy basil	2	91	76	90
Cha-om (local name)	4	99	69	90
Swamp cabbage	4	94	50	91
Average ± SD, $n = 9$		88 ± 10	76 ± 13	86 ± 8
		Carotenoids as percentage of fresh after frying		
Chinese swamp cabbage	4	92	76	88
Chinese cabbage	4	99	84	97
Chinese kale	4	67	24	59
Chinese chive	4	66	19	57
Cabbage (*B. oleracea* var. *capitata* Linn.)	4	90	57	82
Sugar peas	4	65	92	70
Average ± SD, $n = 6$		80 ± 16	59 ± 31	76 ± 16
	Day	Carotenoids as percentage of fresh after fermenting		
Chinese chive	1.5	69	95	70
Whole onion plant	1.5	98	78	97
Pickled leaf mustard	1.5	84	67	82
Egg plant, round shape	1.5	35	22	34
Average ± SD, $n = 4$		72 ± 27	66 ± 31	71 ± 27
	Day	Carotenoids as percentage of fresh after sun drying		
Neem tree	2	64	53	63
Neem tree, duplicate	2	68	56	66
Amaranth	2	49	50	49
Sweet basil	2	45	39	45
Average ± SD, $n = 4$		57 ± 11	50 ± 7	56 ± 10

Source: From Speek et al. (1988).

Table A26. Carotenoids and Vitamin A Value in Vegetable Products [μg (100 g fresh product)$^{-1}$]

Item	*α-Carotene*	*β-Carotene*	*γ-Carotene*	*Crypto-xanthin*	*Lutein**	*Lycopene*	*RE*†
Peas, frozen	tr	360	—	9.2	1700	—	61
Corn, frozen	50	51	nd	nd	730	—	13
Red beet, pickled	—	tr	—	—	4.4	—	0
Pumpkin, pickled	tr	490	nd	nd	630	—	81
Outdoor cucumber, pickled	—	180	—	nd	510	—	29
Tomato ketchup	—	5000	nd	nd	210	9900	830

Source: Reprinted with permission, from Heinonen et al. (1989); copyright © 1989 American Chemical Society.

Notes: — = not detected at a detection limit of 0.5 μg (100 g)$^{-1}$; tr = trace; nd = not determined.

*Lutein + zeaxanthin.

†Micrograms of RE (100 g)$^{-1}$ = 0.167(μg β-carotene) + 0.083(μg α-carotene + μg γ-carotene + μg cryptoxanthin).

REFERENCES

Abdul-Baki, A.A. (1971). Changes in chlorophylls and carotene contents of green and bleached lima bean seeds during development and maturation. *J. Am. Soc. Hortic. Sci.* **96**, 576–580.

Abu-Shakra, S., Mirza, S., and Tannous, R. (1970). Chemical composition and amino acid content of chickpea seeds at different stages of development. *J. Sci. Food Agric.* **21**, 91–93.

Aczél, A. (1971a). Data on the color of green peas. *Lebensm. Wiss. Technol.* **4**, 167.

Aczél, A. (1971b). Dünnschichtchromatographische Untersuchungen über die Farbveränderungen in grünen Erbsen während der Konservierung. *Ind. Obst-Gemüseverwert.* **56**, 335–337.

Aczél, A. (1972). Influence of chemical processes on the colour of labeled tomatoes during storage. II. Changes in carotenoids during storage. *Ind. Obst-Gemüseverwert.* **57**, 436–440.

Aczél, A. (1973). The color change of green peas during technological operations. *Chem. Mikrobiol. Technol. Lebensm.* **2**, 23–25.

Aczél, A. (1986). Application of overpressured layer chromatography in red pepper analysis. Study of the carotenoids responsible for the red color in ground red pepper. *J. High Resolution Chromatogr.* **9**, 407–408.

Aczél, A. (1988). Application of overpressured layer chromatography in red pepper analysis. *Acta Aliment.* **17**, 37–41.

Adams, C.F. (1975). *Nutritive Value of American Foods.* U.S. Dept. of Agriculture, Handbook 456, Washington, D.C.

Aharoni, N., and Lieberman, M. (1979). Ethylene as a regulator of senescence in tobacco leaf discs. *Plant Physiol.* **64**, 801–804.

Aharoni, N., and Richmond, A.E. (1978). Endogenous gibberellin and abscisic acid content as related to senescence of detached lettuce leaves. *Plant Physiol.* **62**, 224–228.

Aharoni, N., Anderson, J.D., and Lieberman, M. (1979a). Production and action of ethylene in senescing leaf discs. Effects of IAA, kinetin, silver ion and CO_2. *Plant Physiol.* **64**, 805–809.

Aharoni, N., Lieberman, M., and Sisler, H.D. (1979b). Patterns of ethylene production in senescing leaves. *Plant Physiol.* **64**, 796–800.

Alban, C., Joyard, J., and Douce, R. (1988). Preparation and characterization of envelope membranes from nongreen plastids. *Plant Physiol.* **88,** 709–717.

Almeida, L.B., and Penteado, M.V.C. (1987). Carotenoids with provitamin A activity of carrots (*Daucus carota* L.) commercialized in São Paulo, Brazil. *Rev. Farm. Bioquim. Univ. São Paulo* **23,** 133–141.

Almeida, L.B., and Penteado, M.V.C. (1988). Carotenoids and provitamin A value of white fleshed Brazilian sweet potatoes (*Ipomoea batatas* Lam.). *J. Food Compos. Anal.* **1,** 341–352.

Almeida, L.B., Penteado, M.V.C., Britton, G., Übelhart, P., Acemoglu, M., and Eugster, C.H. (1988). Isolation and absolute configuration of β,β-carotene diepoxide. *Helv. Chim. Acta* **71,** 31–32.

Almeida, L.B., Penteado, M.V.C., Simpson, K.L., Britton, G., Acemoglu, M., and Eugster, C.H. (1986). Isolation and characterization of (5*R*, 6*S*, 5′*R*, 8′*R*,)- and (5*R*, 6*S*, 5′*R*, 8′*S*)-luteochrome from Brazilian sweet potatoes (*Ipomoea batatas* Lam.). *Helv. Chim. Acta* **69,** 1554–1558.

Altman, L.J., Ash, L., Kowerski, R.C., Epstein, W.W., Larsen, B.R., Rilling, H.C., Muscio, F., and Gregonis, D.E. (1972). Prephytoene pyrophosphate, a new intermediate in the biosynthesis of carotenoids. *J. Am. Chem. Soc.* **94,** 3257–3259.

Amir-Shapira, D. (1986). Studies on Breakdown of Chlorophylls in Plant Tissues During Senescence and Stress Conditions. Thesis, The Hebrew University of Jerusalem.

Amir-Shapira, D., Goldschmidt, E.E., and Altman, A. (1986). Autolysis of chlorophyll in aqueous and detergent suspensions of chloroplast fragments. *Plant Sci.* **43,** 201–206.

Amir-Shapira, D., Goldschmidt, E.E., and Altman, A. (1987). Chlorophyll catabolism in senescing plant tissues: *In vivo* breakdown intermediates suggest different degradative pathways for citrus fruit and parsley leaves. *Proc. Natl. Acad. Sci. USA* **84,** 1901–1905.

Anderson, J.M., Waldron, J.C., and Thorne, S.W. (1978). Chlorophyll-protein complexes of spinach and barley thylakoids. *FEBS Lett.* **92,** 227–233.

André, E., and Hou, K. (1932). Sur les lipoxydases des grains de *Glycine soja* (Sieb.) et de *Phaseolus vulgaris* (L.). *C.R. Acad. Sci.* **195,** 172–174.

Anstis, P.J.P., and Northcote, D.H. (1973). Development of chloroplasts from amyloplasts in potato tuber discs. *New Phytol.* **72,** 449–463.

AOAC (1984). *Official Methods of Analysis,* 14th ed. Association of Official Analytical Chemists, Arlington, VA.

Aparicio-Cuesta, P., Rivas-Gonzalo, J.C., Santos-Buelga, C., and García-Moreno, C. (1989). Quality of frozen beans (*Phaseolus vulgaris*) subjected to different storage conditions. *J. Sci. Food Agric.* **48,** 249–259.

Ardao, C., and Vennesland, B. (1960). Chlorophyllase activity of spinach chloroplasts. *Plant Physiol.* **35,** 368–371.

Arens, D., Seilmeier, W., Weber, F., Kloos, G., and Grosch, W. (1973). Purification and properties of a carotene co-oxidizing lipoxygenase from peas. *Biochem. Biophys. Acta* **327,** 295–305.

Arima, H.K., and Rodriguez-Amaya, D.B. (1988). Carotenoid composition and vitamin A value of commercial Brazilian squashes and pumpkins. *J. Micronutr. Anal.* **4,** 177–191.

Arnold, V., and Fletcher, R.A. (1986). Stimulation of chlorophyll synthesis by benzyladenine and potassium in excised and intact cucumber cotyledons. *Physiol. Plant.* **68,** 169–174.

Arnon, D.I. (1949). Copper enzymes in isolated chloroplasts. Polyphenoloxidase in *Beta vulgaris. Plant Physiol.* **24,** 1–15.

Aronoff, S. (1966). *The Chlorophyll.* L.P. Vernon and G.R. Seely, eds. Academic Press, London.

Aronoff, S. (1981). Chlorophyllide *b. Biochem. Biophys. Res. Commun.* **102,** 108–112.

Arqués, M.J., Font, G., and Farre, R. (1988). Determination of thiamin, riboflavin and β-carotene in fresh and frozen vegetables. *Anal. Bromatol.* **XL-1,** 181–186.

Arya, S.S., Natesan, V., Parihar, D.B., and Vijayaraghavan, P.K. (1979). Stability of carotenoids in dehydrated carrots. *J. Food Technol.* **14,** 579–586.

Ashikawa, I., Miyata, A., Koike, H., Inoue, Y., and Koyama, Y. (1986). Light-induced structural change of β-carotene in thylakoid membranes. *Biochemistry* **25,** 6154–6160.

Axelrod, B., Cheesbrough, T.M., and Laakso, S. (1981). Lipoxygenase from soybeans. In *Methods in Enzymology,* vol. 71, I.M. Lowenstein, ed. Academic Press, New York, pp. 441–451.

Ayers, J.E., Fishwick, M.J., Land, D.G., and Swain, T. (1964). Off-flavor of dehydrated carrot stored in oxygen. *Nature* **202,** 81–82.

Bacon, M.F. (1965). Separation of chlorophylls *a* and *b* and related compounds by thin-layer chromatography on cellulose. *J. Chromatogr.* **17,** 322–326.

Bacon, M.F., and Holden, M. (1967). Changes in chlorophyll resulting from various chemical and physical treatments of leaves and leaf extracts. *Phytochemistry* **6,** 193–210.

Badenhuizen, N.P., and Salema, R. (1976). Observation of the development of chloroamyloplasts. *Rev. Biol.* **6,** 139–155.

Baloch, A.K., Buckle, K.A., and Edwards, R.A. (1977a). Effect of processing variables on the quality of dehydrated carrot. *J. Food Technol.* **12,** 285–293.

Baloch, A.K., Buckle, K.A., and Edwards, R.A. (1977b). Separation of carrot carotenoids on Hyflo Super Cel–magnesium oxide–calcium sulfate thin layers. *J. Chromatogr.* **139,** 149–155.

Baloch, A.K., Buckle, K.A., and Edwards, R.A. (1977c). Stability of β-carotene in model systems containing sulphite. *J. Food Technol.* **12,** 309–316.

Baloch, A.K., Buckle, K.A., and Edwards, R.A. (1987). Effect of sulphur dioxide and blanching on the stability of carotenoids of dehydrated carrot. *J. Sci. Food Agric.* **40,** 179–187.

Banga, O., and de Bruyn, J.W. (1954). Selection of carrots for carotene content. *Euphytica* **3,** 203–211.

Banga, O., de Bruyn, J.W., and Smeets, L. (1955). Selection of carrots for carotene content. II. Subnormal content at low temperature. *Euphytica* **4,** 183–189.

Baranyai, M., Kajtár, J., Bujtás, Gy., and Szabolcs, J. (1977). The configuration of capsanthin epoxides. *Acta. Chim. Acad. Sci. Hung.* **94,** 67–74.

Baranyai, M., Matus, Z., and Szabolcs, J. (1982). Determination, by HPLC, of carotenoids in paprika products. *Acta Aliment.* **11,** 309–323.

Barber, M.S., Jackman, C.M., Warren, C.K., and Weedon, B.C.L. (1960). The structures of paprika ketones. *Proc. Chem. Soc.* 19–20.

Barnes, E.C. (1936). Effects of some environmental factors on growth and colour of carrots. *Cornell Agric. Exp. Sta. Memoir* **186,** 1–36.

Barton, R. (1966). Fine structure of mesophyll cells in senescing leaves of Phaseolus. *Planta* **71,** 314–325.

Basiouny, F.M., and Biswas, P.K. (1981). Ascorbic acid, pigments and mineral element content associated with growth and development of pimiento pepper. *Proc. Fla. State Hortic. Soc.* **94,** 268–269.

Bassi, R., Machold, O., and Simpson, D. (1985). Chlorophyll proteins of two photosystem I preparations from maize. *Carlsberg Res. Commun.* **50,** 145–162.

Batal, K.M., Heaton, E.K., Granberry, D.M., and Beuchat, L.R. (1982). Effects of N^6-benzyladenine and storage temperature on shelf-life and quality of raw and cooked broccoli (*Brassica oleracea* L var. *italica*). *J. Food Sci.* **47,** 1675–1678.

Bates, R.P., and Matthews, R.F. (1975). Ascorbic acid and β-carotene in soybeans as influenced by maturity, sprouting, processing and storage. *Proc. Fla. State Hortic. Soc.* **88,** 266–271.

Battersby, A.R., Fookes, C.J.R., Matcham, G.W.J., and Pandy, P.S. (1981). Biosynthesis of natural porphyrins: specificity and action of cosynthetase on isomeric hydroxymethylbilanes. *Angew. Chem.* **93,** 290–292.

Battle, A.M. del C., Llambias, E.B.C., Wider de Xifra, E.A., and Tigier, H.A. (1975). Porphyrin biosynthesis in the soybean callus tissue system. XV. The effect of growth conditions. *Int. J. Biochem.* **6,** 591–606.

Bauernfeind, J.C. (1972). Carotenoid vitamin A precursors and analogs in foods and feeds. *J. Agric. Food Chem.* **20,** 456–473.

Bauernfeind, J.C., ed. (1981). *Carotenoids as Colorants and Vitamin A Precursors.* Academic Press, New York.

Bazzaz, N., and Rebeiz, C.A. (1979). Chloroplast culture. V. Spectrofluorometric determination of chlorophyll(ide) *a* and *b* and pheophytin (or pheophorbide) *a* and *b* in unsegregated pigment mixtures. *Photochem. Photobiol.* **30,** 709–721.

Beadle, B.W., and Zscheile, F.P. (1942). Studies on the carotenoids. II. The isomerization of β-carotene and its relation to carotene analysis. *J. Biol. Chem.* **144,** 21–33.

Beale, S.I. (1978). δ-Aminolevulinic acid in plants: its biosynthesis, regulation and role in plastid development. *Ann.Rev. Plant Physiol.* **29,** 95–120.

Beale, S.I., and Castelfranco, P.A. (1974). The biosynthesis of δ-aminolevulinic acid in plants. II. Formation of ^{14}C-δ-aminolevulinic acid from labeled precursors in greening plant tissues. *Plant Physiol.* **53,** 297–303.

Belanger, F.C., and Rebeiz, C.A. (1979). Chloroplast biogenesis. XXVII. Detection of novel chlorophyll and chlorophyll precursors in higher plants. *Biochem. Biophys. Res. Commun.* **88,** 365–372.

Belanger, F.C., and Rebeiz, C.A. (1980). Chloroplast biogenesis. Detection of divinyl protochlorophyllide in higher plants. *J. Biol. Chem.* **255,** 1266–1272.

Belanger, F.C., and Rebeiz, C.A. (1982). Chloroplast biogenesis. Detection of monovinyl magnesium-protoporphyrin monoester and other monovinyl magnesium-porphyrins in higher plants. *J. Biol. Chem.* **257,** 1360–1371.

Ben-Aziz, A., Britton, G., and Goodwin, T.W. (1973). Carotene epoxides of *Lycopersicon esculentum. Phytochemistry* **12,** 2759–2764.

Ben-Aziz, A., Grossman, S., Ascarelli, I., and Budowski, P. (1971). Carotene bleaching activities of lipoxygenase and heme proteins as studied by a direct spectrophotometric method. *Phytochemistry* **10,** 1445–1452.

Bendix, G.H., Henry, R.E., and Strodz, N.H. (1952). Preservation of green color in canned vegetables. U.S. Patent 2,589,037.

Benedict, H.M., and Swidler (1961). Non-destructive method for estimating chlorophyll content of leaves. *Science* **133,** 2015–2016.

Benoit, F., Ceustermans, C., Rouchaud, J., and Vlassak, K. (1984). Influence of direct plastic covering upon the quality of carrots and lettuce. *Acta Hortic.* **154,** 321–328.

Ben-Shaul, Y., and Naftali, Y. (1969). The development and ultrastructure of lycopene bodies in chromoplasts of *Lycopersicum esculentum. Protoplasma* **67,** 333–344.

Ben-Shaul, Y., Treffry, T., and Klein, S. (1968). Fine structure studies of carotene body development. *J. Microscopie* **7,** 265–274.

Benz, Z., and Rüdiger, W. (1981). Chlorophyll biosynthesis: various chlorophyllides as exogenous substrates for chlorophyll synthetase. *Z. Naturforsch.* **36c,** 51–57.

Berger, and Bergmann, L. (1967). Light-colour and differentiation of plastids in storage tissue of *Solanum tuberosum* L. *Z. Pflanzenphysiol.* **56,** 439–445.

Berry, D.L., Singh, B., and Salunkhe, O.K. (1972). Incorporation of β-hydroxy-β-methylglutaric acid into carotenoids and other soluble compounds in maize seedlings. *Plant Cell Physiol.* **13,** 157–165.

Berset, C., and Caniaux, P. (1983). Relationship between color evaluation and chlorophyllian pigment content in dried parsley leaves. *J. Food Sci.* **48,** 1854–1857.

Bewick, T.A., Binning, L.K., and Simon, P.W. (1987). Effect of ethephon on the carotene content of early planted carrots. *Acta Hortic.* **201,** 125–131.

Biacs, P.A., Daood, H.G., Czinkotai, B., Hajdu, F., and Kiss-Kutz, N. (1988). Effect of Titavit treatment on the dynamics of tomato fruit ripeness. *Acta Hortic.* **220,** 433–438.

Biacs, P.A., Daood, H.G., Pavisa, A., and Hajdu, F. (1989). Studies on the carotenoid pigments of paprika (*Capsicum annuum* L. var. S2-20). *J. Agric. Food Chem.* **37,** 350–353.

Biale, J.B. (1942). Preliminary studies on modified air storage of the Fuerte avocado fruit. *Proc. Am. Soc. Hortic. Sci.* **41,** 113–118.

Bianchi, A., and Stegwee, D. (1979). Porphobilinogenase in greening leaves of *Phaseolus vulgaris. Z. Pflanzenphysiol.* **91,** 377–382.

Bickel, H., and Schulz, G. (1976). Biosynthesis of plastoquinone and β-carotene in isolated chloroplasts. *Phytochemistry* **15,** 1253–1255.

Bickoff, E.M., Atkins, M.E., Bailey, G.F., and Stitt, F. (1949). Stereoisomeric analysis of β-carotene. *JAOAC* **32,** 766–774.

Bills, C.E., and McDonald, F.G. (1932). The carotene content of ten varieties of carrots. *Science* **76,** 108.

Bjørnland, T., Borch, G., and Liaaen-Jensen, S. (1984). Configurational studies on red algae carotenoids. *Phytochemistry* **23,** 1711–1715.

Blain, J.A. (1970). Carotene-bleaching activity in plant tissue extracts. *J. Sci. Food Agric.* **21,** 35–38.

Blain, J.A., Patterson, S.D.E., and Pearce, M. (1968). Carotene bleaching activity of tomato extracts. *J. Sci. Food Agric.* **19,** 1713–1717.

Blair, J.S. (1940). Color stabilization of green vegetables. U.S. Patent 2,189,774; 2,189,003.

Blessin, C.W., Brecher, J.D., and Dimler, R.J. (1963a). Carotenoids of corn and sorgum. V. Distribution of xanthophylls and carotenes in hand dissected and dry-milled fractions of yellow dent corn. *Cereal Chem.* **40,** 582–586.

Blessin, C.W., Brecher, J.D., Dimler, R.J., Grogan, C.O., and Campbell, C.M. (1963b). Carotenoids of corn and sorghum. III. Variation in xanthophylls and carotenes in hybrid, inbred and exotic corn lines. *Cereal Chem.* **40,** 436–442.

Block, M.A., Joyard, J., and Douce, R. (1980). Site of synthesis of geranylgeraniol in intact spinach chloroplasts. *Biochim. Biophys. Acta* **631,** 210–219.

Boardman, N.K., and Thorne, S.W. (1971). Sensitive fluorescence method for the determination of chlorophyll *a*/chlorophyll *b* ratio. *Biochim. Biophys. Acta* **253,** 228–231.

Bodea, C., and Nicoară, E. (1957). The constitution of physoxanthin. *Ann. Chem.* **609,** 181–185.

Bodea, C., Nicoară, E., and Gross, J. (1954). On the preparation of α- and β-carotene from vegetable material. *Stud. Cerc. Stiint.* **1,** 73–81.

Bodea, C., Nicoară, E., and Mecea, E. (1957). Autoxidation of carotenoids. II. Mechanism of formation of epoxides and furanoxides of carotene and of xanthophyll in the process of autoxidation of α- and β-carotene. *St. Cerc. Chem.* **1,** 17–25.

Bogorad, L. (1958a). The enzymatic synthesis of porphyrins from porphobilinogen. I. Uroporphyrin I. *J. Biol. Chem.* **233,** 501–509.

Bogorad, L. (1958b). The enzymatic synthesis of porphyrins from porphobilinogen. II. Uroporphyrin III. *J. Biol. Chem.* **233,** 510–515.

Bogorad, L. (1976). Chlorophyll biosynthesis. In *Chemistry and Biochemistry of Plant Pigments,* 2nd ed., vol. 1, T.W.Goodwin, ed. Academic Press, New York, pp. 64–148.

Bond, C.P. (1967). The isolation of leaf components. III. The separation, identification and estimation of carotenoids and some quinones. *J. Sci. Food Agric.* **18,** 161–163.

Bondi, A., and Meyer, H. (1946). Carotene in Palestinian crops. *J. Agric. Sci.* **36,** 1–5.

Booth, V.H. (1951a). Distribution of carotenoids in different parts of the carrot. *J. Sci. Food Agric.* **2,** 350–352.

Booth, V.H. (1951b). Chromogenesis in stored carrots. *J. Sci. Food Agric.* **2,** 353–358.

Booth, V.H. (1956). α-Carotene in the leaves of the carrot plant. *J. Sci. Food Agric.* **7,** 386–389.

Booth, V.H. (1957). Carotene in the leaves of the carrot. *J. Sci. Food Agric.* **8,** 371–376.

Booth, V.H., and Dark, S.O.S. (1949). The influence of environment and maturity on total carotenoids in carrots. *J. Agric. Sci.* **39,** 226–236.

Borchgrevink, N.C., and Charley, H. (1966). Color of cooked carrots related to carotene content. *J. Am. Diet. Assoc.* **49,** 116–121.

Böttcher, H. (1986). Quality changes during the storage of green beans (*Phaseolus vulgaris* L. *nanus Juslen. Aschers.*). Part 2. Nutritional value. *Nahrung* **30**, 921-939.

Böttcher, H. (1987). Quality changes of green peas (*Pisum sativum* L.) during storage. Part 2. Nutritional quality. *Nahrung* **31**, 723-738.

Böttcher, H. (1988). Quality changes during the storage of head lettuce (*Lactuca sativa* L. var. *capitata* L.) Part 2. Nutritional value. *Nahrung* **32**, 27-36.

Bradley, G.A., and Rhodes, B.B. (1969). Carotenes, xanthophylls and color in carrot varieties and lines as affected by growing temperatures. *J. Am. Soc. Hortic. Sci.* **94**, 63-65.

Bradley, G., and Smittle, D. (1965). Carrot quality as affected by variety, planting and harvest dates. *Proc. Am. Soc. Hortic. Sci.* **86**, 397-405.

Bradley, G.A., Smittle, D.A., Kaltan, A.A., and Sistrunk, W.A. (1967). Planting date, irrigation, harvest sequence and varietal effect on carrot yields and quality. *Proc. Am. Soc. Hortic. Sci.* **90**, 223-234.

Braithwaite, G.D., and Goodwin, T.W. (1960a). Studies in carotenogenesis. XXV. The incorporation of [1-^{14}C]acetate [2-^{14}C]acetate and $^{14}CO_2$ into lycopene by tomato slices. *Biochem. J.* **76**, 1-5.

Braithwaite, G.D., and Goodwin, T.W. (1960b). Studics in carotenogenesis. XXVII. Incorporation of [2-^{14}C]acetate, DL[2-^{14}C]mevalonate and $^{14}CO_2$ into carrot root preparations. *Biochem. J.* **76**, 194-197.

Bramley, P.M. (1985). The *in vitro* biosynthesis of carotenoids. In *Advances in Lipid Research,* vol. 21, R. Paoletti and D. Kritchevsky, eds. Academic Press, New York, pp. 243-279.

Braumann, Th., and Grimme, H.L. (1981). Reversed-phase high-performance liquid chromatography of chlorophylls and carotenoids. *Biochim. Biophys. Acta* **637**, 8-17.

Braumann, Th., Weber, G., and Grimme, L.H. (1982). Carotenoid and chlorophyll composition of light-harvesting and reaction center proteins of the thylakoid membrane. *Photochem. Photobiophys.* **4**, 1-8.

Britton, G. (1976). Carotenoids and polyterpenoids. *Terpenoids Steroids* **6**, 144-168.

Britton, G. (1979). Carotenoid biosynthesis—a target for herbicide activity. *Z. Naturforsch.* **34c**, 979-985.

Britton, G. (1982). Carotenoid biosynthesis in higher plants. *Physiol. Vég.* **20**, 735-755.

Britton, G. (1983). *The Biochemistry of Natural Pigments.* Cambridge University Press, Cambridge.

Britton, G. (1985). General carotenoid methods. In *Methods in Enzymology. Steroids and Isoprenoids,* vol. 111B, J.H. Law and H.C. Rilling, eds. Academic Press, New York, pp. 113-149.

Britton, G. (1989). Carotenoids and polyterpenoids. *Nat. Prod. Rep.* **6**, 359-392.

Britton, G., and Goodwin, T.W. (1969). The occurrence of phytoene 1,2-oxide and related carotenoids in tomatoes. *Phytochemistry* **8**, 2257-2258.

Britton, G., and Goodwin, T.W. (1971). Biosynthesis of carotenoids. In *Methods in Enzymology,* vol. 18C, D.B. McCormick and Wright, eds. Academic Press, New York, pp. 654-701.

Britton, G., and Goodwin, T.W. (1975). Carotene epoxides from the Delta tomato mutant. *Phytochemistry* **14**, 2530-2532.

Britton, G., and Goodwin, T.W., eds. (1982). *Carotenoid Chemistry and Biochemistry,* Proc. 6th IUPAC Int. Symp. on Carotenoids. Pergamon Press, Oxford.

Brooker, J.D., and Russell, D.W. (1975). Subcellular localization of 3-hydroxy-3-methyl-glutaryl coenzyme A reductase in *Pisum sativum* seedlings. *Arch. Biochem. Biophys.* **167**, 730–737.

Brown, G.B. (1947). The effect of maturity and storage on the carotene content of carrot varieties. *Proc. Am. Soc. Hortic. Sci.* **50**, 347–352.

Brown, G.B. (1949). The effect of winter storage on the carotene content of carrot varieties. *Proc. Am. Soc. Hortic. Sci.* **54**, 304–306.

Brown, E.D., Micozzi, M.S., Craft, N.E., Bieri, J.G., Beecher, G., Edwards, B.K., Rose, A., Taylor, P.R., and Smith, J.C., Jr. (1989). Plasma carotenoids in normal men after a single ingestion of vegetables or purified β-carotene. *Am. J. Clin. Nutr.* **49**, 1258–1265.

Brunson, A.M., and Quackenbush, F.W. (1962). Breeding corn with provitamin A content in the grain. *Crop Sci.* **2**, 344–347.

Brunstetter, B.C., and Wiseman, H.G. (1947). Carotenoid pigments in tubers of the Katahdin variety of Irish potato. *Plant Physiol.* **22**, 421–437.

Bruyant, P., and Kannangara, C.G. (1987). Biosynthesis of δ-aminolevulinate in greening barley leaves. VIII. Purification and characterization of the glutamate tRNA ligase. *Carlsberg Res. Commun.* **52**, 99–109.

Buckle, K.A., and Edwards, R.A. (1969). Chlorophyll degradation products from processed pea puree. *Phytochemistry* **8**, 1901–1906.

Buckle, K.A., and Edwards, R.A. (1970a). Chlorophyll degradation and lipid oxidation in frozen unblanched peas. *J. Sci. Food Agric.* **21**, 307–312.

Buckle, K.A., and Edwards, R.A. (1970b). Chlorophyll color and pH changes in H.T.S.T. processed green pea puree. *J. Food Technol.* **5**, 173–186.

Buckle, K.A., and Rahman, F.M.M. (1979). Separation of chlorophyll and carotenoid pigments of capsicum cultivars. *J. Chromatogr.* **171**, 385–391.

Budowski, P., Ascarelli, J., Gross, J., and Nir, I. (1963). Provitamin A_2 from lutein. *Science* **142**, 969–971.

Budowski, P., and Bondi, A. (1960). Autoxidation of carotene and vitamin A. Influence of fat and antioxidants. *Arch. Biochem. Biophys.* **89**, 66–73.

Budowski, P., and Gross, J. (1965). Conversion of carotenoids to 3-dehydroretinol (vitamin A_2) in the mouse. *Nature* **206**, 1254–1255.

Buggy, M.J., Britton, G., and Goodwin, T.W. (1969). Stereochemistry of biosynthesis by isolated chloroplasts. *Biochem. J.* **114**, 641–643.

Buggy, M.J., Britton, G., and Goodwin, T.W. (1974). Terpenoid biosynthesis by chloroplasts isolated in organic solvents. *Phytochemistry* **13**, 125–129.

Buishand, J.G., and Gabelman, W.H. (1979). Investigations on the inheritance of color and carotenoid content in phloem and xylem of carrot roots (*Daucus carota* L.). *Euphytica* **23**, 611–632.

Bureau, J.L., and Bushway, R.J. (1986). HPLC determination of carotenoids in fruits and vegetables in the United States. *J. Food Sci.* **51**, 128–130.

Burger, M., Hein, L.W., Teply, L.J., Derse, P.H., and Krieger, C.H. (1956). Nutrients in frozen foods. Vitamin, mineral and proximate composition of frozen fruits, juices and vegetables. *J. Agric. Food Chem.* **4**, 418–425.

Burke, S., and Aronoff, S. (1980). Semiquantitative paper chromatography of chlorophylls *a* and *b*. *Anal. Biochem.* **101**, 103–106.

Burke, J., and Aronoff, S. (1981). High-performance liquid chromatography of the plastid pigments and some of their immediate derivatives. *Anal. Biochem.* **114,** 367–370.

Burton, G.W., and Ingold, K.U. (1984). β-Carotene: an unusual type of antioxidant. *Science* **224,** 569–573.

Bushway, R.J. (1985). Separation of carotenoids in fruits and vegetables by high performance liquid chromatography. *J. Liq. Chromatogr.* **8,** 1527–1547.

Bushway, R.J. (1986). Determination of α- and β-carotene in some raw fruits and vegetables by high-performance liquid chromatography. *J. Agric. Food Chem.* **34,** 409–412.

Bushway, R.J., and Wilson, A.M. (1982). Determination of α- and β-carotene in fruit and vegetables by high-performance liquid chromatography. *Can. Inst. Food Sci. Technol.* **15,** 165–169.

Cabibel, M., and Ferry, P. (1980). Evolution of carotenoids in tomatoes as a function of ripeness levels and cultural conditions. *Ann. Technol. Agric.* **29,** 27–45.

Cabibel, M., Lapize, F., and Ferry, P. (1981). Determination of carotenes in tomatoes by high performance liquid chromatography. Application to some varieties of tomatoes. *Sci. Aliment.* **I**(4), 489–500.

Caldwell, J.S., Brunstetter, B.C., Culpepper, C.W., and Ezell, B.D. (1945). Causes and control of discoloration in dehydration of white potatoes. *The Canner* **100,** 13, 14, 15.

Camara, B. (1980). *In vitro* conversion of violaxanthin to capsorubin by a chromoplast enriched fraction of *Capsicum* fruits. *Biochem. Biophys. Res. Commun.* **93,** 113–117.

Camara, B., Bardat, F., and Monéger, R. (1982). Nature du premier precurseur en C_{40} des caroténoïdes dans les chromoplastes de fruit de poivron (*Capsicum annuum* L.). *C.R. Acad. Sci.* (Ser. 3) **294,** 549–551.

Camara, B., and Brangeon, J. (1981). Carotenoid metabolism during chloroplast to chromoplast transformation in *Capsicum annuum* fruit. *Planta* **151,** 359–364.

Camara, B., and Dogbo, O. (1986). Demonstration and solubilization of lycopene cyclase from Capsicum chromoplast membranes. *Plant Physiol.* **80,** 172–174.

Camara, B., Dogbo, O., Bardat, F., and Monéger, R. (1985a). Conversion enzymatique du pyrophosphate de préphytoène en caroténoïdes par des plastes isolés. *C.R. Acad. Sci. Paris* (Ser. 3) **301,** 749–752.

Camara, B., Dogbo, O., d'Harlingue, A., Kleinig, H., and Monéger, R. (1985b). Metabolism of plastid terpenoids: lycopene cyclization by *Capsicum* chromoplast membranes. *Biochim. Biophys. Acta* **836,** 262–266.

Camara, B., and Monéger, R. (1978). Free and esterified carotenoids in green and red fruits of *Capsicum annuum. Phytochemistry* **17,** 91–93.

Camara, B., and Monéger, R. (1981). Carotenoid biosynthesis: *in vivo* conversion of antheraxanthin to capsanthin by a chromoplast enriched fraction of *Capsicum* fruits. *Biochem. Biophys. Res. Commun.* **99,** 1117–1122.

Camara, B., and Monéger, R. (1982). Biosynthetic capabilities and localization of enzymatic activities in carotenoid metabolism of *Capsicum annuum* isolated chromoplasts. *Physiol. Vég.* **20,** 757–773.

Cameron, E.J., Pilcher, R.W., and Clifcorn, L.E. (1949). Nutrient retention during canned food production. *Am. J. Publ. Health* **39,** 756–763.

Campbell, H. (1937). Undesirable color change in frozen peas stored at insufficiently low temperature. *Food Res.* **2,** 55–57.

Carballido, A., Lorente, V., and Centeno, M. (1971). Composition of maize. *Anal. Bromatol.* **23**, 291–309.

Carey, E.E., and Rebeiz, C.A. (1985). Chloroplast biogenesis. 49. Differences among angiosperms in the biosynthesis and accumulation of monovinyl and divinyl protochlorophyllide during photoperiodic greening. *Plant Physiol.* **79**, 1–6.

Carnevale, J., Cole, P.E.R., and Crank, G. (1980). Photocatalyzed oxidation of paprika pigments. *J. Agric. Food Chem.* **28**, 953–956.

Castelfranco, P.A., and Beale, S.I. (1981). Chlorophyll biosynthesis. In *The Biochemistry of Plants,* vol. 8, M.D. Hatch and N.K. Boardman, eds. Academic Press, New York, pp. 375–421.

Castelfranco, P.A., and Beale, S.I. (1983). Chlorophyll biosynthesis: recent advances and areas of current interest. *Ann. Rev. Plant Physiol.* **34**, 241–278.

Castelfranco, P.A., Weinstein, J.D., Schwarcz, S., Pardo, A.D., and Wezelman, B.E. (1979). The Mg insertion step in chlorophyll biosynthesis. *Arch. Biochem. Biophys.* **192**, 592–598.

Chandler, L.A., and Schwartz, S.J. (1987). HPLC separation of cis-trans carotene isomers in fresh and processed fruits of vegetables. *J. Food Sci.* **52**, 669–672.

Chandler, L.A., and Schwartz, S.J. (1988). Isomerization and losses of *trans-β*-carotene in sweet potatoes as affected by processing treatments. *J. Agric. Food Chem.* **36**, 129–133.

Charlton, J.M., Treharne, K.J., and Goodwin, T.W. (1967). Incorporation of [2-^{14}C]mevalonic acid into phytoene by isolated chloroplasts. *Biochem. J.* **105**, 205–212.

Chen, T.C., and Miller, G.W. (1974). Purification and characterization of uroporphyrinogen decarboxylase from tobacco leaves. *Plant Cell Physiol.* **15**, 993–1005.

Chereskin, B.M., Wong, Y.-S., and Castelfranco, D.A. (1982). *In vitro* synthesis of the chlorophyll isocyclic ring. Transformation of magnesium-protoporphyrin IX and magnesium-protoporphyrin monomethylester into magnesium-2,4-divinyl pheoporphyrin a_5. *Plant Physiol.* **70**, 987–993.

Chichester, C.O., and Nakayama, T.O. (1965). Pigment changes in senescent and stored tissue. In *Chemistry and Biochemistry of Plant Pigments,* T.W. Goodwin, ed. Academic Press, London, pp. 439–458.

Cholnoky, L. (1937). The pigments of Hungarian paprika and their vitamin A effects. *Exp. Comm.* **40**, 173–186.

Cholnoky, L. (1939). Determination of color components of paprika spice. *Z. Untersuch. Lebensm.* **78**, 157–161.

Cholnoky, L., Györgyfy, K., Nagy, E., and Pánczél, M. (1955). Investigations of carotenoid pigments. I. Pigments of red paprika. *Acta Chim. Acad. Sci. Hung.* **6**, 143–171.

Cholnoky, L., and Szabolcs, J. (1960). Bemerkungen zur Struktur des Capsorubins. *Experientia* **16**, 483–484.

Cholnoky, L., Szabolcs, J., Cooper, R.D.G., and Weedon, B.C.L. (1963). The structure of cryptocapsin. *Tetrahedron Lett.* **19**, 1257–1259.

Cholnoky, L., Szabolcs, J., and Nagy, E. (1958a). Carotenoid coloring matters. IV. α-Cryptoxanthin. *Ann. Chem.* **616**, 207–218.

Cholnoky, L., Györgyfy, K., Nagy, E., and Pánczél, M. (1958b). Investigations of carotenoid pigments. III. Pigments of the yellow paprika. *Acta Chim. Acad. Sci. Hung.* **16,** 227–246.

Chou, H.-E., and Breene, W.M. (1972). Oxidative decoloration of β-carotene in low-moisture model systems. *J. Food Sci.* **37,** 66–68.

Christ, J., and Dye, M. (1929). The association of vitamin A with greenness in plant tissue. II. The vitamin A content of asparagus. *J. Biol. Chem.* **81,** 525–531.

Clifcorn, L.E. (1948). Factors influencing vitamin content of canned foods. *Adv. Food Res.* **1,** 39–104.

Clough, J.M., and Pattenden, G. (1979). Naturally occurring poly-*cis*-carotenoids. Stereochemistry of poly-*cis*-lycopene and its congeners in "Tangerine" tomato fruits. *J.C.S. Chem. Commun.* 616–619.

Clough, J.M., and Pattenden, G. (1983). Stereochemical assignment of prolycopene and other poly-*Z*-isomeric carotenoids in fruits of the Tangerine tomato *Lycopersicon esculentum* var. "Tangella." *J. Chem. Soc. Perkin Trans. I,* 3011–3018.

Clydesdale, F.M., Fleischmann, D.L., and Francis, F.J. (1970). Maintenance of color in processed green vegetables. *Food Prod. Dev.* **4,** 127, 130, 134, 136, 138.

Clydesdale, F.M., and Francis, F.J. (1968). Chlorophyll changes in thermally processed spinach as influenced by enzyme conversion and pH adjustment. *Food Technol.* **22,** 135–138.

Cohen, B., Grossman, S., Klein, B.P., and Pinsky, A. (1985). Pigment bleaching of soybean lipoxygenase type-2 and the effect of specific chemical modifications. *Biochim. Biophys. Acta* **837,** 279–287.

Cole, E.R., and Kapur, N.S. (1957a). The stability of lycopene. I. Degradation by oxygen. *J. Sci. Food Agric.* **8,** 360–365.

Cole, E.R., and Kapur, N.S. (1957b). The stability of lycopene. II. Oxidation during heating of tomato pulps. *J. Sci. Food Agric.* **8,** 366–368.

Cook, B.B., Gunning, B., and Uchimoto, D. (1961). Variations in nutritive value of frozen green baby Lima beans as a result of methods of processing and cooking. *J. Agric. Food Chem.* **9,** 316–321.

Cooper, R.D.G., Jackman, L.M., and Weedon, B.C.L. (1962). Stereochemistry of capsorubin and synthesis of its optically inactive epimers. *Proc. Chem. Soc.,* 215.

Cordner, H.B., Reder, R., and Odell, G. (1959). Carotene and ascorbic acid content in improved sweet potato variants. *J. Agric. Food Chem.* **7,** 53–54.

Costes, C. (1966). Biosynthèse du phytol des chlorophylles et du squelette tetraterpenique des caroténoïdes dans les feuilles vertes. *Phytochemistry* **5,** 311–324.

Costes, C., Burghoffer, C., Carrayol, E., Ducet, G., and Diano, M. (1976). Occurrence of carotenoids in non-plastidial materials from potato tuber cells. *Plant. Sci. Lett.* **6,** 153–159.

Costes, C., Burghoffer, C., Joyard, J., Block, M., and Douce, R. (1979). Occurrence and biosynthesis of violaxanthin in isolated spinach chloroplast envelope. *FEBS Lett.* **103,** 17–21.

Cserni, I., and Prohaszka, K. (1988). The effect of N supply on the nitrate, sugar and carotene content of carrots. *Acta Hortic.* **220,** 303–307.

Curl, A.L. (1953). Application of countercurrent distribution to Valencia orange juice carotenoids. *J. Agric. Food Chem.* **1,** 456–460.

Curl, L.A. (1956). On the structure of hydroxy-α-carotene from orange juice. *Food Res.* **21,** 689–693.

Curl, A.L. (1961). The xanthophylls of tomatoes. *J. Food Sci.* **26,** 106–111.

Curl, A.L. (1962). The carotenoids of red bell peppers. *J. Agric. Food Chem.* **10,** 504–509.

Curl, A.L. (1964). The carotenoids of green bell peppers. *J. Agric. Food Chem.* **12,** 522–524.

Czuba, M., and Ormrod, D.P. (1974). Effects of cadmium and zinc on ozone-induced phytotoxicity in cress and lettuce. *Can. J. Bot.* **52,** 645–649.

Czygan, F.-C., ed. (1980). *Pigments in Plants,* 2nd ed. Fischer, Stuttgart.

Daley, R.J., Gray, C.B.J., and Brown, S.R. (1973). Reversed-phase thin-layer chromatography of chlorophyll derivatives. *J. Chromatogr.* **76,** 175–183.

Daood, H.G., and Biacs, P.A. (1986). Evidence for the presence of lipoxygenase and hydroperoxyde-decomposing enzyme in red pepper seeds. *Acta Aliment.* **15,** 307–318.

Daood, H.G., Biacs, P.A., Hoschke, A., Harkay-Vinkler, M., and Hajdu, F. (1987). Separation and identification of tomato fruit pigments by TLC and HPLC. *Acta Aliment.* **16,** 339–350.

Daood, H.G., Czinkotai, B., Agoston, H., and Biacs, P. (1989). High performance liquid chromatography of chlorophyll and carotenoids from vegetables. *J. Chromatogr.* **472,** 296–302.

Dark, S.O.S., and Booth, V.H. (1946). Total carotenoids in carrots. *J. Agric. Sci.* **36,** 192–198.

Davies, B.H. (1976). Carotenoids. In *Chemistry and Biochemistry of Plant Pigments,* vol. 2, T.W. Goodwin, ed. Academic Press, London, pp. 38–165.

Davies, B.H., Matthews, S., and Kirk, J.T.O. (1970). The nature and biosynthesis of the carotenoids of different colour varieties of *Capsicum annuum. Phytochemistry* **9,** 797–805.

Dean, L.L. (1968). Progress with persistent-green color and green-coat in snap beans (*Phaseolus vulgaris* L.) for commercial processing. *HortScience* **3,**177–178.

Dedolph, R.R., Wittwer, S.H., and Tuli,V. (1961). Senescence inhibition and respiration. *Science* **134,** 1075.

De Felice, D., and Fellers, C.R. (1938). Carotene content of spinach. *Proc. Am. Soc. Hortic. Sci.* **35,** 728–733.

Dei, M. (1983). Benzyladenine-induced stimulation of chlorophyll formation in attached cotyledons of etiolated cucumber seedlings. *Plant Sci. Lett.* **30,** 251–257.

de la Mar, R., and Francis, F.J. (1969). Carotenoid degradation in bleached paprika. *J. Food Sci.* **34,** 287–290.

de la Torre Boronat, M.C., and Farré Rovira, R. (1975). Los carotenoides del pimenton. *Anal. Bromatol.* **XXVII-2,** 149–196.

Della Monica, E.S., and McDowell, P.E. (1965). Comparison of β-carotene content of dried carrots prepared by three dehydration processes. *Food Technol.* **19,** 141–143.

De Ritter, E., and Purcell, A.E. (1981). Carotenoid analytical methods. In *Carotenoids as Colorants and Vitamin A Precursors,* J.C. Bauernfeind, ed. Academic Press, New York, pp. 815–882.

Dicks, J.W., and Friend, J. (1967). The oxidation of carotenoids by mitochondria from sugar beet leaves. 2. Crocin oxidation by an unsaturated fat oxydase system in Triton extracts of mitochondria. *Phytochemistry* **6,** 1193–1202.

Diehl, H.C., and Berry, J.A. (1933). Relation of scalding practice and storage temperature to quality retention in frozen-pack peas. *Proc. Am. Soc. Hort. Sci.* **30,** 496–498.

Dietrich, W.C. (1958). Determination of the conversion of chlorophyll to pheophytin. *Food Technol.* **12,** 428–431.

Dietrich, W.C., Lindquist, F.E., Miers, J.C., Bohart, G.S., Neumann, H.J., and Talburt, W.F. (1957). Objective tests to measure adverse changes in frozen vegetables. *Food Technol.* **11,** 109–113.

Dietrich, W.C., and Neumann, H.J. (1965). Blanching Brussels sprouts. *Food Technol.* **19,** 150–153.

Dietz, J.M., Sri Kantha, S., and Erdman, J.W., Jr. (1988). Reversed-phase HPLC analysis of α- and β-carotenes from selected raw and cooked vegetables. *Plant Foods Hum. Nutr.* **38,** 333–341.

Di Mascio, P., Kaiser, S., and Sies, H. (1989). Lycopene as the most efficient biological singlet oxygen quencher. *Arch. Biochem. Biophys.* **279,** 532–538.

Dobreanu, M., Doru, G., and Ciofu, R. (1983). The chemical composition of some vegetable species in relation to the time and stage of harvesting. *Lucr. St. I.A.N.B.* (Ser. B) **XXVI,** 43–47.

Dogbo, O., Bardat, F., and Camara, B. (1984). Terpenoid metabolism in plastids: activity, localization and substrate specificity of chlorophyll synthetase in *Capsicum annuum* plastids. *Physiol. Vég.* **22,**75–82.

Dogbo, O., Bardat, F., Laferrière, A., Quennemet, J., Brangeon, J., and Camara, B. (1987). Metabolism of plastid terpenoids. I. Biosynthesis of phytoene in plastid stroma isolated from higher plants. *Plant Sci.* **49,** 89–101.

Dostal, H.C., and Leopold, A.C. (1967). Gibberellin delays ripening of tomatoes. *Science* **158,** 1579–1580.

Douce, R., Holtz, R.B., and Benson, A.A. (1973). Isolation and properties of the envelope of spinach chloroplasts. *J. Biol. Chem.* **248,** 7215–7222.

Drake, S.R., and Carmichael, D.M. (1986). Frozen vegetable quality as influenced by high temperature short time (HTST) steam blanching. *J. Food Sci.* **51,** 1378–1379.

Drake, S.R., Spayd, S.E., and Thompson, J.B. (1981). The influence of blanch and freezing methods on the quality of selected vegetables. *J. Food Quality* **4,** 271–277.

Dua, P.N., Day, E.J., Hill, J.E., and Grogan, C.O. (1967). Utilization of xanthophylls from natural sources by the chick. *J. Agric. Food Chem.* **15,** 324–328.

Duggan, J.X., and Rebeiz, C.A. (1982). Chloroplast biogenesis. 38: quantitative detection of a chlorophyllide *b* pool in higher plants. *Biochim. Biophys. Acta* **679,** 248–260.

Duggar, B.M. (1913). Lycopersicon: the red pigment of the tomato and the effects of conditions upon its development. *Washington Univ. Stud.* **1,** 22–45.

Dupont, J., and Siegenthaler, P.A. (1986). A parallel study of pigment bleaching and cytochrome breakdown during aging of thylakoid membranes. *Plant Cell Physiol.* **27,** 473–484.

Dutton, H.J., Bailey, G.F., and Kohake, E.(1943). Dehydrated spinach changes in color and pigments during processing and storage. *Ind. Eng. Chem.* **35,** 1173–1177.

Duvekot, W.S., van der Meer, O.P., and Apeland, J. (1960). Tests on the storage life of vegetables and soft fruit under different external conditions. Inst. Res. Storage Processing Hort. Products. Wageningen Ann. Report.

Edwards, C.G., and Lee, C.Y. (1986). Measurement of provitamin A carotenoids in fresh and canned carrots and green peas. *J. Food Sci.* **51,** 534–535.

Edwards, R.A., and Reuter, F.H. (1967). Pigment changes during the maturation of tomato fruits. *Food Technol. Aust.* **19,** 352–357.

Egger, K. (1964). Vergleichende Untersuchung der Xanthophyllveresterung in Blüten, Früchten und Herbstlaub. *Ber. Dtsch. Bot. Ges.* **77,** 145–150.

Egger, K. (1968). Die Farbwachse roter Paprikaschoten. *Ber. Dtsch. Bot. Ges.* **81,** 153–156.

Egger, K., and Schwenker, U. (1966). The esters of lutein in autumn leaves. *Z. Pflanzenphysiol.* **54,** 407–416.

Egle, K. (1944). Untersuchungen über die Resistenz der Plastidenfarbstoffe. *Bot. Arch.* **44,** 93–148.

Egle, K. (1960). Biologischer Chlorophyllabbau. In *Handbuch der Pflanzenphysiologie,* vol. V/1, Ruhland, ed. Springer, Berlin, pp. 354–386.

Eheart, M.S., and Odland, D. (1973). Use of ammonium compounds for chlorophyll retention in frozen green vegetables. *J. Food Sci.* **38,** 202–205.

El-Difrawi, E.A., and Hudson, B.J.F. (1979). Identification and estimation of carotenoids in the seeds of four *Lupinus* species. *J. Sci. Food Agric.* **30,** 1168–1670.

Elkins, E.R. (1979). Nutrient content of raw and canned green beans, peaches and sweet potatoes. *Food Technol.* **33,** 66–70.

Elkner, K., Bakowski, J., and Babik, J. (1984). The effect of ethrel on the chemical composition of tomatoes cultivated for processing. *Biul. Warzywniczy* **27,** 531–553.

Ellsworth, R.K., and Nowak, C.A. (1974). A gas chromatographic-mass spectrophotometric analysis of esterifying alcohols of pumpkin seed protochlorophylls. *Anal. Biochem.* **57,** 534–546.

Ellsworth, R.K., Perkins, H.J., Detwiller, S.P., and Liu, K. (1970). On the enzymatic conversion of ^{14}C-labeled chlorophyll *a* to ^{14}C-labeled chlorophyll *b. Biochim. Biophys. Acta* **223,** 275–280.

El-Tinay, A.H., and Chichester, C.O. (1970). Oxidation of β-carotene. Site of initial attack. *J. Org. Chem.* **35,** 2290–2293.

Emura, K., and Hosoya, T. (1979). Effect of fertilizer on the quality and yield of spring sown carrots. *Bul. Saitama Hortic. Exp. St.* **8,** 13–23.

Englert, G., Brown, B.O., Moss, G.P., Weedon, B.C.L., Britton, G., Goodwin, T.W., Simpson, K.L., and Williams, R.S.H. (1979). Prolycopene, a tetra-*cis*-carotene with two hindered *cis* double bonds. *J.C.S. Chem. Commun.,* 545–547.

Eskin, M.A.N., Grossman, S., and Pinsky, A. (1977). Biochemistry of lipoxygenase in relation to food quality. *Critic. Rev. Food Sci. Nutr.* **9,** 1–40.

Eskins, K., and Harris, L. (1981). High-performance liquid chromatography of etioplast pigments in red kidney bean leaves. *Photochem. Photobiol.* **33,** 131–133.

Eskins, K., Scholfield, C.R., and Dutton, H.J. (1977). High performance liquid chromatography of plant pigments. *J. Chromatogr.* **135,** 217–220.

Evans, N., Games, D.E., Jackson, A.H., and Mallin, S.A. (1975). Applications of high-performance liquid chromatography and field desorption mass spectrometry in studies of natural porphyrins and chlorophyll derivatives. *J. Chromatogr.* **115,** 325–333.

Evers, A.-M. (1989). Effects of different fertilization practices on the carotene content of carrot. *J. Agric. Sci. Finland* **61,** 7–14.

Ezell, B.D., and Wilcox, M.S. (1946). The ratio of carotene to carotenoid pigments in sweet potato varieties. *Science* **103,** 193–194.

Ezell, B.D., and Wilcox, M.S. (1948). Effect of variety and storage on carotene and total carotenoid pigments in sweet potatoes. *Food Res.* **13,** 203–212.

Ezell, B.D., and Wilcox, M.S. (1952). Influence of storage temperature on carotene, total carotenoids and ascorbic acid content of sweet potatoes. *Plant Physiol.* **27,** 81–94.

Ezell, B.D., and Wilcox, M.S. (1958). Variation in carotene content of sweet potatoes. *J. Agric. Food Chem.* **6,** 61–65.

Ezell, B.D., Wilcox, M.S., and Crowder, J.N. (1952). Pre- and post-harvest changes in carotene, total carotenoids and ascorbic acid content of sweet potatoes. *Plant Physiol.* **27,** 355–369.

Ezell, B.D., Wilcox, M.S., and Demaree, K.D. (1956). Physiological and biochemical effects of storage humidity on sweet potatoes. *J. Agric. Food Chem.* **4,** 640–644.

Faigle, J.W., Müller, H., von Philipsborn, W., and Karrer, P. (1964). The configuration of natural capsanthin and capsorubin. *Helv. Chem. Acta* **47,** 741–745.

Falconer, M.E., Fishwick, M.J., Land, D.G., and Sayer, E.R. (1964). Carotene oxidation and off-flavour development in dehydrated carrot. *J. Sci. Food Agric.* **15,** 897–901.

Falkowski, P.G., and Sucher, J. (1981). Rapid quantitative separation of chlorophylls and their degradation products by high performance liquid chromatography. *J. Chromatogr.* **213,** 349–351.

FAO/WHO Joint Expert Committee on Food Additives (1967). WHO Tech. Rep. Ser. 362.

Farhangi, M., and Valadon, G.L.R. (1981). Effect of acidified processing and storage on carotenoids (provitamin A) and vitamin C in mung bean sprouts. *J. Food Sci.* **48,** 1464–1466.

Fellers, C.R., Young, R.E., Isham, P.D., and Clague, J.A. (1934). Effect of fertilization, freezing, cooking and canning on the vitamin C and A content of asparagus. *Proc. Am. Soc. Hortic. Sci.* **31,** 145–151.

Feltwell, J.S.E., and Valadon, L.R.G. (1972). Carotenoids of *Pieris brassicae* and of its food plants. *J. Insect Physiol.* **18,** 2203–2215.

Feltwell, J.S.E., and Valadon, L.R.G. (1974). Carotenoid changes in *Brassica oleracea* var *capitata* L. with age in relation to the large White Butterfly, *Pieris brassicae* L. *J. Agric. Sci. Camb.* **83,** 19–26.

Firn, R.D., and Friend, J. (1972). Enzymatic production of the plant growth inhibitor xanthoxin. *Planta* **103,** 263–266.

Fischbach, H. (1943). Microdeterminations for organically combined metals in pigment of okra. *J. Assoc. Off. Agric. Chem.* **26,** 139–143.

Fischbach, H., and Newburger, S.H. (1943). Spectrophotometric study of the green color in okra. *J. Assoc. Off. Agric. Chem.* **26,** 134–139.

Fisher, C., and Kocis, G.A. (1987). Separation of paprika pigments by HPLC. *J. Agric. Food Chem.* **35,** 55–57.

Fishwick, M.J., and Wright, A.J. (1980). Isolation and characterization of amyloplast envelope membranes from *Solanum tuberosum. Phytochemistry* **19,** 55–59.

Fitzgerald, G.A., and Fellers, C.R. (1938). Carotene and ascorbic acid content of fresh market and commercially frozen fruits and vegetables. *Food Res.* **3,** 109–120.

Fletcher, R.A., and McCullagh, D. (1971). Benzyladenine as a regulator of chlorophyll synthesis in cucumber cotyledons. *Can. J. Bot.* **49,** 2197–2201.

Flynn, L.M., Hibbard, A.D., Hogan, A.G., and Murneek, A.E. (1946). Effect of maturity on nutrients of snap beans. *J. Am. Diet. Assoc.* **22,** 415–419.

Fordham, J.R., Wells, C.E., and Chen, L.H. (1975). Sprouting of seeds and nutrient composition of seeds and sprouts. *J. Food Sci.* **40,** 552–556.

Forni, E., Ghezzi, M., and Polesello, A. (1988). HPLC separation and fluorimetric estimation of chlorophylls and pheophytins in fresh and frozen peas. *Chromatographia* **26,** 120–124.

Francis, G.W., and Isaksen (1989). Droplet counter current chromatography of the carotenoids of parsley (*Petroselinum crispum*). *Chromatographia* **272,** 549–551.

Francis, J.E., and Smith, A.G. (1984). Assay of mouse liver uroporphyrinogen decarboxylase by reverse phase high performance liquid chromatography. *Anal. Biochem.* **138,** 404–410.

Fraps, G.S. (1933). Losses of vitamin A in drying fresh raw carrots and sweet potatoes and canned spinach. *J. Agric. Res.* **47,** 539–541.

Fraps, G.S., and Kemmerer, A.R. (1941). Determination of carotene and cryptoxanthin in yellow corn. *Ind. Eng. Chem. Anal.* **13,** 806–809.

Frey-Wyssling, A., and Schwegler, F. (1965). Ultrastructure of the chromoplasts in the carrot root. *J. Ultrastruct. Res.* **13,** 543–559.

Friend, J. (1958). The coupled oxidation of β-carotene by a linoleate-lipoxidase system and by autoxydizing linoleate. *Chem. Ind.* **18,** 597–598.

Friend, J., and Nakayama, T.O.M. (1959). Destruction of carotenoids in isolated chloroplasts. *Nature* **184,** B.A. 66.

Fritz, D., and Habben, J. (1975). Determination of ripeness of carrots (*Daucus carota* L.). *Acta Hortic.* **52,** 231–238.

Frydman, R.B., and Frydman, B. (1970). Purification and properties of porphobilinogen deaminase from wheat germ. *Arch. Biochem. Biophys.* **136,** 193–202.

Fuesler, T.P., Hanomoto, C.M., and Castelfranco, P.A. (1982). Separation of Mg-protoporphyrin IX and Mg-protoporphyrin IX monomethyl ester synthesized de novo by developing cucumber etioplasts. *Plant Physiol.* **69,** 421–423.

Fuesler, T.P., Wong, Y.-S., and Castelfranco, P.A. (1984). Localization of Mg-chelatase and Mg-protoporphyrin IX monomethyl ester (oxidative) cyclase within isolated developing cucumber chloroplasts. *Plant Physiol.* **75,** 662–664.

Fuesler, T.P., Wright, L.A., and Castelfranco, P.A. (1981). Properties of magnesium chelatase in greening etioplasts. *Plant Physiol.* **67,** 246–249.

Ganguly, J., and Sastry, P.S. (1985). Mechanism of conversion of β-carotene into vitamin A—central cleavage versus random cleavage. *Wld. Rev. Nutr. Diet.* **45,** 198–220.

Garrido-Fernández, J., and Mínguez Mosquera, M.I. (1983). Chlorophylls *a* and *b* and their derivatives. Methods of extraction, isolation and determination in plants. *Grasas y Aceites* **34,** 199–205.

Gassman, M.L., Pluscec, J., and Bogorad, L. (1966). δ-Aminolevulinic acid transaminase from *Chlorella* and *Phaseolus. Plant Physiol.* **41:** xiv.

Geissler, T., and Rüdiger, K. (1980). Appropriate mineral fertilizer application—a prerequisite for higher yields and good quality in carrots. *Gartenbau.* **27,** 168–169.

Gepstein, S., and Thimann, K.V. (1981). The role of ethylene in the senescence of oat leaves. *Plant Physiol.* **68,** 349–354.

Gershoni, J.M., and Ohad, I. (1980). The use of an internal standard for semiquantitative analysis of low temperature (77°K) fluorescence of photosynthetic cells. *Anal. Biochem.* **104,** 315–320.

Gibson, K.D., Neuberger, A., and Scott, J.J. (1955). The purification and properties of δ-aminolevulinic acid dehydrase. *Biochem. J.* **61,** 618–629.

Gieseker, L.E. (1949). Art of processing and maintaining color of green vegetables. U.S. Patent 2,437,747.

Gill, H.S., and Kataria, A.S. (1974). Some biochemical studies in European and Asiatic varieties of carrot (*Daucus carota*). *Current Sci.* **43,** 184–185.

Gilpin, G.L., Sweeney, J.P., Chapman, V.J., and Eisen, J.N. (1959). Effect of cooking methods on broccoli. II. Palatability. *J. Am. Diet. Assoc.* **35,** 359–363.

Glass, R.W., and Simpson, K.L. (1976). The isolation of γ-carotene and poly-*cis*-γ-carotene from the Tangerine tomato. *Phytochemistry* **15,** 1077–1078.

Gleim, E.G., Tressler, D.K., and Fenton, F. (1944). Ascorbic acid, thiamin, riboflavin and carotene contents of asparagus and spinach in the fresh, stored and frozen states both before and after cooking. *Food. Res.* **9,** 471–490.

Glover, J. (1960). The conversion of β-carotene into vitamin A. *Vitams. Horm.* **18,** 371–386.

Glover, J., and Redfearn, E.R. (1954). The mechanism of transformation of β-carotene into vitamin A *in vivo. Biochem. J.* **58,** XV.

Gnaegi, F., Perraudin, G., and Schopfer, J.-F. (1977). Conservation et qualite des carottes indigenes. *Rev. Suisse Vitic. Arboric. Hortic.* **9,** 173–177.

Goedherr, J.C. (1979). Carotenoids in the photosynthetic apparatus. *Ber. Deutsch. Bot. Ges.* **92,** 436.

Gold, H.J., and Weckel, H.G. (1959). Degradation of chlorophyll to pheophytin during sterilization of canned green peas by heat. *Food Technol.* **13,** 281–286.

Goldman, M., Horev, B., and Saguy, I. (1983). Decolorization of β-carotene in model systems stimulating dehydrated foods. Mechanism and kinetic principles. *J. Food Sci.* **48,** 751–754.

Goodwin, T.W. (1952). *The Comparative Biochemistry of the Carotenoids.* Chapman and Hall, London.

Goodwin, T.W. (1958a). Studies on carotenogenesis. The changes in carotenoid and chlorophyll pigments in the leaves of deciduous trees during autumn necrosis. *Biochem. J.* **68,** 503–511.

Goodwin, T.W. (1958b). Studies in carotenogenesis. XXV. The incorporation of $^{14}CO_2$,[2-^{14}C]acetate and [2-^{14}C]mevalonate into β-carotene by illuminated etiolated maize seedlings. *Biochem. J.* **70,** 612–617.

Goodwin, T.W., ed. (1976). *Chemistry and Biochemistry of Plant Pigments,* vols. 1, 2, 2nd ed. Academic Press, London.

Goodwin, T.W. (1980). *The Biochemistry of the Carotenoids,* vol. 1, 2nd ed. Chapman and Hall, London.

Goodwin, T.W., and Jamikorn, M. (1952). Biosynthesis of carotenoids in ripening tomatoes. *Nature* **170,** 104–105.

Goodwin, T.W., and Phagpolngarm, S. (1960). Studies on carotenogenesis. The effect of illumination on carotenoid synthesis in French bean (*Phaseolus vulgaris*) seedlings. *Biochem. J.* **76,** 197–199.

Gorchein, A. (1972). Magnesium protoporphyrin chelatase activity in *Rhodopseudomonas spheroides. Biochem. J.* **127,** 97–106.

Gough, S.P., and Kannangara, C.G. (1977). Synthesis of δ-aminolevulinate by a chloroplast stroma preparation from greening barley leaves. *Carlsberg Res. Commun.* **42,** 459–464.

Granick, S. (1950). Magnesium vinyl pheoporphyrin a_5, another intermediate in the biological synthesis of chlorophyll. *J. Biol. Chem.* **183,** 713–730.

Granick, S. (1954). Enzymatic conversion of δ-aminolevulinic acid to porphobilinogen. *Science* **120,** 1105–1106.

Grant, J.L., Kramer, V.J., Ding, R., and Kispert, L.D. (1988). Carotenoid cation radicals: electrochemical, optical EPR study. *J. Am. Chem. Soc.* **110,** 2151–2157.

Gray, J.C., and Kekwick, R.G.O. (1973). Mevalonate kinase in green leaves and etiolated cotyledons of the French bean *Phaseolus vulgaris. Biochem. J.* **133,** 335–347.

Gregory, G.K., Chen, T.-S., and Philip, T. (1987). Quantitative analysis of carotenoids and carotenoid esters in fruits by HPLC: red bell peppers. *J. Food Sci.* **52,** 1071–1073.

Griffiths, W.T. (1978). Reconstitution of chlorophyllide formation by isolated etioplast membranes. *Biochem. J.* **186,** 267–278.

Grilli, M. (1965). Origin and development of the chromoplasts in the pumpkin *Cucurbita pepo* L. cv Small sugar 1. Origin of the chromoplasts from amyloplasts. *Caryologia* **18,** 409–433.

Gröschel, E.C., Nelson, A.I., and Steinberg, M.P. (1966). Changes in color and other characteristics of green beans stored in controlled refrigerated atmospheres. *J. Food. Sci.* **31,** 488–496.

Grogan, C.O., and Blessin, C.W. (1968). Characterization of major carotenoids in yellow maize lines of differing pigment concentration. *Crop. Sci.* **8,** 730–732.

Grosch, W., Weber, F., and Fischer, K.H. (1977). Bleaching of carotenoid by the enzyme lipoxygenase. *Ann. Technol. Agric.* **26,** 133–137.

Gross, J. (1967). Provitamins A_2. Thesis, The Hebrew University of Jerusalem.

Gross, J. (1979). The pigments of three hybrid varieties of broccoli (*Brassica oleracea* var. *italica*). *Gartenbauwiss.* **44,** 213–216.

Gross, J. (1980). A rapid separation of citrus carotenoids by thin-layer chromatography. *Chromatographia* **13,** 572–573.

Gross, J. (1985). Carotenoid pigments in the developing cherry (*Prunus avium*) v. "Doenissen's Gelbe." *Gartenbauwiss.* **50,** 88–90.

Gross, J. (1987). *Pigments in Fruits.* Academic Press, London.

Gross, J., and Budowski, P. (1966). Conversion of carotenoids into vitamin A_1 and A_2 in two species of freshwater fish. *Biochem. J.* **101,** 747–754.

Gross, J., Carmon, M., Lifshitz, A., and Costes, C. (1976). Carotenoids of banana pulp, peel and leaves. *Lebensm. Wiss. Technol.* **9,** 211–214.

Gross, J., Carmon, M., Lifshitz, A., and Sklarz, B. (1975). Structural elucidation of some orange juice carotenoids. *Phytochemistry* **14,** 249-252.

Gross, J., and Costes, C. (1976). Nicotine effect on carotenogenesis in α-carotene containing banana leaves. *Physiol. Vég.* **14,** 427–435.

Gross, J., Gabai, M., Lifshitz, A., and Sklarz, B. (1973). Carotenoids in pulp, peel and leaves of *Persea americana. Phytochemistry* **12,** 2259-2263.

Grossmann, S., and Leshem, Y.Y. (1978). Lowering of endogenous lipoxygenase activity in *Pisum sativum* foliage by cytokinin as related to senescence. *Physiol. Plant.* **43,** 359-362.

Grumbach, K.H. (1983a). Distribution of chlorophylls, carotenoids and quinones in chloroplasts of higher plants. *Z. Naturforsch. Biosci.* **38c,** 996–1002.

Grumbach, K.H. (1983b). Interconversion of carotenoids and quinones after onset of photosynthesis in chloroplasts of higher plants. *Z. Naturforsch. Biosci.* **38c,** 393–398.

Grumbach, K.H. (1984). Does the chloroplast envelope contain carotenoids and quinones in vitro? *Physiol. Plant.* **60,** 180–186.

Grumbach, K.H., and Forn, B. (1980). Chloroplast autonomy in acetyl-coenzyme A-formation and terpenoid biosynthesis. *Z. Naturforsch.* **35c,** 645–648.

Guerrant, N.B., and O'Hara, M.B. (1953). Vitamin retention in peas and lima beans after blanching, freezing, processing in tin and in glass, after storage and after cooking. *Food Technol.* **7,** 473–477.

Guerrant, N.B., Vavich, M.G., Fardio, O.B., Dutcher, R.A., and Stern, R.M. (1946). The nutritive value of canned foods. Changes in the vitamin content of foods during canning. *J. Nutr.* **32,** 435–438.

Gupta, V.K., and Mukherjee, D. (1981). Effect of pre- and postharvest applications of morphactin on the post-harvest behaviour of okra. *Sci. Hortic.* **15,** 107–115.

Gupte, S.M., El-Bisi, H.M., and Francis, F.J. (1964). Kinetics of thermal degradation of chlorophyll in spinach puree. *J. Food Sci.* **29,** 379–382.

Gupte, S.M., and Francis, F.J. (1964). Effect of pH adjustment and high-temperature short time processing on color and pigment retention in spinach puree. *Food Technol.* **18,** 141–144.

Haas, L.W., and Bohn, R.M. (1934). Bleaching bread dough. U.S. Patent 1,957,333-337.

Habben, J. (1972). Influence of some environmental factors on yields and quality of carrots (*Daucus carota* L.). *Gartenbauwiss.* **37,** 345–359.

Habben, J. (1973). Einfluss der Stickstoff- und Kaliumdüngung auf Ertrag und Qualität der Möhre (*Daucus carota.* L.). *Landwirtsch. Forsch.* **26,** 156–172.

Hager, A. (1975). The reversible light-induced conversion of xanthophylls in the chloroplast. *Ber. Deutsch. Bot. Ges.* **88,** 27–44.

Hager, A., and Bertenrath, T. (1962). Verteilungschromatographische Trennung von Chlorophyllen und Carotinoiden grüner Pflanzen an Dünnschichten. *Planta* **58,** 564–568.

Hager, A., and Meyer-Bertenrath, T. (1966). Extraction and quantitative determination of carotenoids and chlorophylls of leaves, algae and isolated chloroplasts with the aid of thin-layer chromatography. *Planta* **69,** 198–217.

Hager, A., and Perz, H. (1970). Veränderung der Lichtabsorption eines Carotinoids in Enzym (De-Epoxidase)-substrat (Violaxanthin)-Komplex. *Planta* **93,** 314–322.

Haisman, D.R., and Clarke, M.W. (1975). The interfacial factor in the heat induced conversion of chlorophyll to pheophytin in green leaves. *J. Sci. Food. Agric.* **26,** 1111–1126.

Halevy, S., Koth, H., and Guggenheim, K. (1957). The vitamin and mineral content of fruits and vegetables grown in Israel. *Brit. J. Nutr.* **11,** 409–413.

Halevy, A.A., and Wittwer, S.H. (1966). Effect of growth retardants on longevity of vegetables, mushrooms and cut flowers. *Proc. Am. Soc. Hortic. Sci.* **88,** 582–590.

Hall, T.C. (1968). Protein, aminoacid and chlorophyll metabolism during ontogeny of snap beans. *J. Am. Soc. Hortic. Sci.* **93,** 379–387.

Hansen, E. (1945). Variations in the carotene content of carrots. *Proc. Am. Soc. Hortic. Sci.* **46,** 355–358.

Hansen, S., and Maret, W. (1988). Retinal is not formed *in vitro* by enzymatic central cleavage of β-carotene. *Biochemistry* **27,** 200–206.

Harper, R.H., and Zscheile, F.P. (1945). Carotenoid content of carrot varieties and strains. *Food Res.* **10,** 84–97.

Hart, G.J., and Battersby, A.R. (1985). Purification and properties of uroporphyrinogen III synthase (co-synthase) from *Euglena gracilis. Biochem. J.* **232,** 151–160.

Hart, G.J., Leeper, F.J., and Battersby, A.R. (1984). Modification of hydroxymethylbilane synthase (porphobilinogen deaminase) by pyridoxal 5′-phosphate. *Biochem. J.* **222,** 93–102.

Harvey, B.M.R., Lu, B.C., and Fletcher, R.R. (1974). Benzyladenine accelerates chlorophyll differentiation and stimulates photosynthetic enzyme activity in cucumber cotyledons. *Can. J. Bot.* **52,** 2581–2586.

Hasegawa, K., MacMillan, J.D., Maxwell, W.A., and Chichester, C.O. (1973). Photosensitized bleaching of β-carotene with light at 632.8 nm from a continuous wave gas laser. *Photochem. Photobiol.* **9,** 165–169.

Hayakawa, K., and Timbers, G.E. (1977). Influence of heat treatment on the quality of vegetables: changes in visual green color. *J. Food Sci.* **42,** 778–781.

Hayashi, S., and Noguchi, T. (1983). Response to light of a specific aminotransferase for δ-aminolevulinate formation in higher plants. *J. Biol. Chem.* **258,** 13,693–13,696.

Hayden, F.R., Heinze, P.H., and Wade, B.L. (1948). Vitamin content of snap beans grown in South Carolina. *Food Res.* **13,** 143–161.

Heinonen, M.I., Ollilainen, V., Linkola, E.K., Varo, P.T., and Koivistoinen, P.E. (1989). Carotenoids in Finnish foods: vegetables, fruits and berries. *J. Agric. Food Chem.* **37,** 655–659.

Heinze, P.H., Hayden, F.R., and Wade, B.L. (1947). Vitamin studies of varieties and strains of peas. *Plant Physiol.* **22,** 548–560.

Hendry, G.A.F., Houghton, J.D., and Brown, S.B. (1987). Tansley review No. 11: The degradation of chlorophyll—a biological enigma. *New Phytol.* **107,** 255–302.

Hendry, G.A.F., and Stobart, A.K. (1986). Chlorophyll turnover in greening barley. *Phytochemistry* **25,** 2735–2737.

Herrmann, K. (1975). Ueber den Carotin-(Provitamin A-) Gehalt der Gemüse- und Obstarten. *Ernähr.-Umschau* **22,** 45–49; 75–77.

Hidaka, T., Anno, T., and Nakatsu, S. (1987). The composition and vitamin A value of the carotenoids of pumpkins of different colors. *J. Food Biochem.* **11,** 59–68.

Hidaka, T., Katsuki, S., Nagata, Y., and Nakatsu, S. (1986). Partial purification and properties of pumpkin lipoxygenase with carotene-bleaching activity. *J. Food Biochem.* **10,** 55–73.

Hildebrand, D.F., and Hymowitz, Th. (1982). Carotene and chlorophyll bleaching by soybeans with and without seed lipoxygenase-1. *J. Agric. Food Chem.* **30,** 705–708.

Hill, H.M., Calderwood, S.K., and Rogers, L.J. (1971). Conversion of lycopene to β-carotene by plastids isolated from higher plants. *Phytochemistry* **10,** 2051–2058.

Hirayama, O., and Oida, H. (1969). Lipids and lipoprotein complexes in photosynthetic tissues. VI. Changes of lipid and pigment compositions in spinach leaves during their storage. *J. Agric. Chem. Soc.* **43,** 423–428.

Hoffman, J.C., and Kanapaux, M.S. (1955). Relation of visual color rating to chlorophyll contents of snap bean pods. *Proc. Am. Soc. Hortic. Sci.* **66,** 339–349.

Hohler, A., Nitz, S., and Drawert, F. (1988). Über die Bildung und die sensorischen Eigenschaften flüchtiger Nor-Carotinoide 1. Photooxidation von β-Carotin in *N,N*-Dimethylformamid. *Chem. Mikrobiol. Technol. Lebensm.* **11,** 115–126.

Hojilla, M.P., Garcia, V.V., and Raymundo, L.C. (1985). Thermal degradation of β-carotene in carrot juice. *ASEAN Food J.* **1,** 157–161.

Holden, M. (1961). The breakdown of chlorophyll by chlorophyllase. *Biochem. J.* **78,** 359–364.

Holden, M. (1965). Chlorophyll bleaching by legume seeds. *J. Sci. Food Agric.* **16,** 312–325.

Holden, M. (1970a). Lipoxidase activity of leaves. *Phytochemistry* **9,** 507–512.

Holden, M. (1970b). Chlorophyll bleaching systems in leaves. *Phytochemistry* **9,** 1771–1777.

Holden, M. (1976). Chlorophylls. In *Chemistry and Biochemistry of Plant Pigments,* 2nd ed., vol. 2, T.W. Goodwin, ed. Academic Press, London, pp. 1–37.

Holmes, A.D., Smith, C.T., and Lachman, W.H. (1948). Variation in composition of winter squashes. *Food Res.* **13,** 123–127.

Holmes, A.D., and Spelman, A.F. (1946). Composition of squashes after winter storage. *Food Res.* **11,** 345–350.

Holmes, A.D., Spelman, A.F., and Jones, C.P. (1945). Ascorbic acid, carotene, chlorophyll, riboflavin and water content of summer squash. *Food Res.* **10,** 489–496.

Holmes, A.D., Spelman, A.F., Rogers, C.J., and Lachman, W.H. (1948). Influence of light during storage on composition of Blue Hubbard squash. *Food Res.* **13,** 304–307.

Holmes, A.D., Spelman, A.F., and Wetherbee, R.T. (1949). Comparison of light vs. darkness for storing Butternut squashes. *Food Res.* **3,** 269–271.

Holmes, A.D., Spelman, A.F., and Wetherbee, R.T. (1954). Effect of storage on Butternut squash and its seeds. *J. Am. Diet. Assoc.* **30,** 138–141.

Hopp, R.J., Merrow, S.B., and Elbert, E.M. (1960). Varietal differences and storage changes in β-carotene content of six varieties of winter squashes. *Proc. Am. Soc. Hortic. Sci.* **76,** 568–576.

Houghen, C.L., Meller, E., and Gassman, M.L. (1982). Mg-protoporphyrin monoester destruction by extracts of etiolated Red Kidney bean leaves. *Plant Sci. Lett.* **24,** 289–296.

House, M.C., Nelson, P.M., and Haber, E.S. (1929). The vitamin A, B and C content of artificially versus naturally ripened tomatoes. *J. Biol. Chem.* **31,** 495–504.

Houssier, C., and Sauer, K. (1969). Optical properties of the protochlorophyll pigments. I. Isolation, characterization and infrared spectra. *Biochim. Biophys. Acta* **172,** 476–491.

Howes, C.D. (1974). Nicotine inhibition of carotenoid cyclization in *Cucurbita ficifolia* cotyledons. *Phytochemistry* **13,** 1469–1471.

Hsieh, Y.-P., and Karel, M. (1983). Rapid extraction and determination of α- and β-carotenes in foods. *J. Chromatogr.* **259,** 515–518.

Huang, D.-D., Wang, W.-Y., Gough, S.P., and Kannangara, C.G. (1984). δ-Aminolevulinic acid synthesizing enzymes need an RNA moiety for activity. *Science* **225,** 1482–1484.

Huault, C., and Bruyant, P. (1984). Affinity chromatography of δ-aminolevulinic dehydratase from radish seedlings. *Physiol. Vég.* **22,** 209–213.

Huff, A. (1982). Peroxidase-catalized oxidation of chlorophyll by hydrogen peroxide. *Phytochemistry* **21,** 261–266.

Hulme, A.C., ed. (1970/71). *The Biochemistry of Fruits and Their Products,* vols. 1, 2. Academic Press, London.

Humphrey, A.M. (1980). Chlorophyll. *Food Chem.* **5,** 57–67.

Hunter, R.F., and Williams, N.E. (1945). Chemical conversion of β-carotene into vitamin A. *J. Chem. Soc.* 554–556.

Hynninen, P.H. (1973). Chlorophyll. IV. Preparation and purification of some derivatives of chlorophylls *a* and *b*. *Acta Chim. Scand.* **27,** 1771–1780.

Hynninen, P.H. (1981). Mechanism of the allomerization of chlorophyll: inhibition of the allomerization by carotenoid pigments. *Z. Naturforsch.* **36b,** 1010–1016.

Hynninen, P.H., and Ellfolk, N. (1973). Chlorophylls. I. Separation and isolation of chlorophylls *a* and *b* by multiple liquid-liquid partition. *Acta Chim. Scand.* **27,** 1463–1477.

Ikediobi, C.O., and Snyder, H.E. (1977). Cooxidation of β-carotene by an isoenzyme of soybean lipoxygenase. *J. Agric. Food Chem.* **25,** 124–127.

Ikuno, Y., Maoka, T., Shimizu, M., Komori, T., and Matsuno, T. (1985). Direct diastereomeric resolution of carotenoids. II. All ten stereoisomers of tunaxanthin (ε,ε-carotene-3,3′-diol). *J. Chromatogr.* **328,** 387–391.

Imamura, M., and Shimizu, S. (1974). Metabolism of chlorophyll in higher plants. IV. Relationship between fatty acid oxidation and chlorophyll bleaching in plant extracts. *Plant Cell Physiol.* **15,** 187–190.

Inada, Y., and Shibata, K. (1960). Vinyl pheoporphyrin a_5 in the inner seed coats of pumpkins. *Plant Cell Physiol.* **1,** 311–316.

Inskeep, W.P., and Bloom, P.R. (1985). Extinction coefficients of chlorophyll *a* and *b* in *N,N*-dimethylformamide and 80% acetone. *Plant Physiol.* **77,** 483–485.

Iriyama, K., Shiraki, M., and Yoshiura, M. (1979). An improved method for extraction, partial purification, separation and isolation of chlorophyll from spinach leaves. *J. Liq. Chromatogr.* **2,** 255–276.

Iriyama, K., and Yoshiura, M. (1979). Separation of chlorophyll *a* and chlorophyll *b* by column chromatography with Sephadex LH-20 or powdered sugar. *J. Chromatogr.* **177,** 154–156.

Iriyama, K., Yoshiura, M., and Shiraki, M. (1978). Micromethod for the qualitative

and quantitative analysis of photosynthetic pigments using high-performance liquid chromatography. *J. Chromatogr.* **154,** 302–305.
Isler, O., ed. (1971a). *Carotenoids.* Birkhäuser, Basel.
Isler, O., ed. (1971b). *Tentative Rules of the Nomenclature of Carotenoids in Carotenoids.* IUPAC, pp. 851–864.
Isoe, S., BeHyeon, S., and Sakan, T. (1969). Photo-oxygenation of carotenoids. I. The formation of dihydroactinidiolide and β-ionone from β-carotene. *Tetrahedron Lett.,* 279–281.
Ito, R., Oba, K., and Uritani, I. (1979). Mechanism for the induction of 3-hydroxy-3-methylglutaryl coenzyme A reductase in $HgCl_2$-treated sweet potato root tissue. *Plant Cell Physiol.* **20,** 867–874.
Ito, Y., Shinohara, R., Ohno, Y., Fukaya, Y., Shima, Y., and Otani, M. (1988). Interrelationship between serum carotenoid concentration and serum lipid peroxides level in healthy adults. *Jpn. J. Clin. Chem.* **17,** 71–76.
IUPAC (1967). Main lectures presented at the First International Symposium on Carotenoids other than Vitamin A. *Pure Appl. Chem.* **14,** 215–278.
IUPAC (1969). Main lectures presented at the Second International Symposium on Carotenoids other than Vitamin A. *Pure Appl. Chem.* **20,** 365–553.
IUPAC (1973). Plenary lectures presented at the Third International Symposium on Carotenoids other than Vitamin A. *Pure Appl. Chem.* **35,** 1–130.
IUPAC (1974). *Nomenclature of Carotenoids.* Butterworth, London.
IUPAC (1976). Main lectures presented at the Fourth International Symposium on Carotenoids. *Pure Appl. Chem.* **47,** 97–243.
IUPAC (1979). Main lectures presented at the Fifth International Symposium on Carotenoids. *Pure Appl. Chem.* **51,** 435–675.
IUPAC (1985). Main lectures presented at the Seventh International Symposium on Carotenoids. *Pure Appl. Chem.* **57,** 639–821.
Iwahori, S., and Lyons, J.M. (1970). Maturation and quality of tomatoes with preharvest treatments of 2-chloroethylphosphonic acid. *J. Am. Hortic. Soc.* **95,** 88–91.
Iwanzik, W., Tevini, M., Stute, R., and Hilbert, R. (1983). Carotinoidgehalt und -zusammensetzung verschiedener deutscher Kartoffelsorten und deren Bedeutung fuer die Fleischfarbe der Knolle. *Potato Res.* **26,** 149–162.
Izaki, Y., Yoshida, K., Hidaka, K., and Toda, K. (1986). Chlorophylls, carotenes and tocopherols in green vegetables and their relationships. *J. Jpn. Soc. Nutr. Food Sci.* **39,** 485–493.
Jackson, A.H. (1976). Structure, properties and distribution of chlorophylls. In *Chemistry and Biochemistry of Plant Pigments,* 2nd ed., vol. 1, T.W. Goodwin, ed. Academic Press, London, pp. 1–63.
Jacobs, J.M., and Jacobs, N.J. (1984). Protoporphyrinogen oxidation, an enzymatic step in heme and chlorophyll synthesis: Partial characterization of the reaction in plant organelles and comparison with mammalian and bacterial system. *Arch. Biochem. Biophys.* **229,** 312–319.
Jacobs, J.M., and Jacobs, N.J. (1987). Oxidation of protoporphyrinogen to protoporphyrin, a step in chlorophyll and haem biosynthesis. *Biochem. J.* **244,** 219–224.
Jacobs, J.M., Jacobs, N.J., and De Maggio, A.E. (1982). Protoporphyrinogen oxi-

dation in chloroplasts and plant mitochondria, a step in heme and chlorophyll synthesis. *Arch. Biochem. Biophys.* **218,** 233–239.

Jahn, O.L. (1975). Comparison of instrumental methods for measuring ripening changes of intact tomato fruit. *J. Am. Soc. Hortic. Sci.* **100,** 688–691.

Jeffrey, S.W., Douce, R., and Benson, A.A. (1974). Carotenoid transformations in the chloroplast envelope. *Proc. Natl. Acad. Sci. USA.* **71,** 807–810.

Jen, J.J. (1974). Influence of spectral quality of light on pigment systems of ripening tomatoes. *J. Food Sci.* **39,** 907–910.

Jen, J.J., and MacKinney, G. (1970a). On the photodecomposition of chlorophyll *in vitro.* I. Reaction rates. *Photochem. Photobiol.* **11,** 297–302.

Jen, J.J., and MacKinney, G. (1970b). On the photodecomposition of chlorophyll *in vitro.* II. Intermediates and breakdown products. *Photochem. Photobiol.* **11,** 303–308.

Jenkins, J.A., and MacKinney, G. (1955). Carotenoids of the apricot tomato and its hybrids with yellow and tangerine. *Genetics* **40,** 715–720.

Johjima, T., and Ogura, H. (1983). Analysis of tomato carotenoids by thin-layer chromatography and a *cis*- form γ-carotene newly identified in tangerine tomato. *J. Jpn. Soc. Hortic. Sci.* **52,** 202–209.

Jones, B.L., and Porter, J.W. (1985). Enzymatic synthesis of phytoene. In *Methods in Enzymology,* vol. 110A, J.H. Law and H.C. Rilling, eds., pp. 209–220.

Jones, B.L., and Porter, J.W. (1986). Biosynthesis of carotenes in higher plants. *CRC Crit. Rev. Plant Sci.* **3,** 295–324.

Jones, I.D., Butler, L.S., Gibbs, E., and White, R.C. (1972a). An evaluation of reversed-phase partition for thin-layer chromatographic identification of chlorophylls and derivatives. *J. Chromatogr.* **70,** 87–98.

Jones, I.D., Butler, L.S., Gibbs, E., and White, R.C. (1972b). Thin-layer chromatographic identification of zinc and copper complexes of pheophytins and pheophorbides in mixtures with chlorophylls and derivatives. *J. Chromatogr.* **70,** 206–209.

Jones, I.D., White, R.C., and Gibbs, E. (1961). The formation of pheophorbides during brine preservations of cucumbers. *Food Technol.* **15,** 172–175.

Jones, I.D., White, R.C., and Gibbs, E. (1962). Some pigment changes in cucumber brining and brine storage. *Food Technol.* **16,** 96–102.

Jones, I.D., White, R.C., and Gibbs, E. (1963). Influence of blanching or brining treatments on the formation of chlorophyllides, pheophytins and pheophorbides in green plant tissue. *J. Food Sci.* **28,** 437–439.

Jones, I.D., White, R.C., Gibbs, E., and Butler, L.S. (1977a). Estimation of zinc pheophytins, chlorophylls and pheophytins in mixtures in diethyl ether or 80% acetone by spectrophotometry and fluorometry. *J. Agric. Food Chem.* **25,** 146–149.

Jones, I.D., White, R.C., Gibbs, E., Butler, L.S., and Nelson, L.A. (1977b). Experimental formation of zinc and copper complexes of chlorophyll derivatives in vegetable tissue by thermal processing. *J. Agric. Food Chem.* **25,** 149–153.

Jones, O.T.G. (1966). A protein-protochlorophyll complex obtained from inner seed coats of *Cucurbita pepo. Biochem. J.* **101,** 153–160.

Jordan, P.M., and Seehra, J.S. (1979). The biosynthesis of uroporphyrinogen III: order of assembly of the four porphobilinogen molecules in the formation of the tetrapyrrole ring. *FEBS Lett.* **104,** 364–366.

Jordan, P.M., and Seehra, J.S. (1980). Mechanism of action of δ-aminolevulinic acid dehydratase: stepwise order of addition of the two molecules of δ-aminolevulinic acid in the enzymatic synthesis of porphobilinogen. *J. Chem. Soc. Chem. Commun.* 240–242.

Joyce, A.E. (1954). Some polyenes of *Brassica rutabaga. Nature* **173,** 311–312.

Joyce, A.E. (1959). Carotenoids of *Brassica napus. J. Sci. Food. Agric.* **10,** 342–348.

Jungalwala, F.B., and Porter, J.W. (1965). The configuration of phytoene. *Arch. Biochem. Biophys.* **110,** 291–299.

Jungalwala, F.B., and Porter, J.W. (1967). Biosynthesis of phytoene from isopentenyl and farnesyl pyrophosphate by partially purified tomato enzyme system. *Arch. Biochem. Biophys.* **119,** 209–219.

Kader, A.A., and Morris, L.L. (1978). Tomato fruit color measured with an Agtron E5-W reflectance spectrophotometer. *HortScience* **13,** 577–578.

Kader, A.A., Morris, L.L., and Maxie, E.C. (1966). Effect of growth regulating substances on the ripening and shelf-life of tomatoes. *HortScience* **1,** 90–91.

Kakutani, Y. (1989). Changes of contents of carotenoids and chlorophyll *a* and *b* depending on season. *Bull. Kobe Women's Univ.* **22,** 195–203.

Kakutani, Y., Hayashi, R., and Kiku, N. (1987). Stability of spinach under heat-treatment. *Bull. Kobe Women's Univ.* **20,** 175–181.

Kakutani, Y., and Kiku, N. (1988). Improvement of the separation and quantitative analysis of carotenoids by HPLC. *Bull. Kobe Women's Univ.* **21,** 273–278.

Kakutani, Y., Ogava, S., and Hamana, K. (1983). Quantitative analysis of carotenoids by HPLC. *Bull Kobe Women's Univ.* **16,** 17–23.

Kaminski, E., Wasowicz, E., Zawirska, R., and Wower, M. (1986). The effect of drying and storage of dried carrots on sensory characteristics and volatile constituents. *Nahrung* **30,** 819–828.

Kannangara, C.G., Gough, S.P., Oliver, R.P., and Rasmussen, S.K. (1984). Biosynthesis of δ-aminolevulinate in greening barley leaves. VI. Activation of glutamate by ligation to RNA. *Carlsberg Res. Commun.* **49,** 417–437.

Kanner, J., Mendel, H., and Budowski, P. (1976). Carotene oxidizing factors in red pepper fruits (*Capsicum annuum* L.): ascorbic acid. *J. Food Sci.* **41,** 183–185.

Kanner, J., Mendel, H., and Budowski, P. (1977). Carotene oxidizing factors in red pepper fruits (*Capsicum annuum* L.): peroxidase activity. *J. Food Sci.* **42,** 1549–1551.

Kansal, B.D., Singh, B., Bajaj, K.L., and Kaur, G. (1981). Effect of different levels of nitrogen and farm yard manure on yield and quality of spinach (*Spinacia oleracea* L.). *Qual. Plant. Pl. Foods Hum. Nutr.* **31,** 163–170.

Kargl, T.E., Quackenbush, F.W., and Tomes, M.L. (1960). The carotene polyene system in a strain of tomatoes high in delta-carotene and its comparison with eight other tomato strains. *Proc. Am. Soc. Hortic. Sci.* **75,** 574–578.

Karrer, P., Eugster, C.H., and Faust, N. (1950). The occurrence of carotenoids in the pollen and anthers of various plants. *Helv. Chim. Acta* **33,** 300–301.

Karrer, P., Helfenstein, A., Wehrli, H., and Wettstein, A. (1930). Pflanzenfarbstoffe. XXV. Über die Konstitution des Lycopins und Carotins. *Helv. Chim. Acta* **14,** 154–162.

Karrer, P., and Jucker, E. (1945). Oxides of β-carotene: β-carotene monoepoxide, β-carotene diepoxide, mutatochrome, aurochrome, luteochrome. *Helv. Chim. Acta* **28,** 427–436.

Karrer, P., and Jucker, E. (1948). *Carotinoide.* Birkhäuser, Basel.

Karrer, P., and Morf, R. (1931). Zur Konstitution der zweiten Carotinform (α-Carotin). *Helv. Chim. Acta* **14,** 833–838.

Karrer, P., Salomon, H., and Wehrli, H. (1929). Pflanzenfarbstoffe XIV. Über einen Carotinoidfarbstoff aus Mais: Zeaxanthin. *Helv. Chim. Acta* **12,** 790–792.

Kasemir, H. (1983). Action of light on chlorophyll(ide) appearance. *Photochem. Photobiol.* **37,** 701–708.

Kasim, M. (1967a). Beitrag zur qualitativen Analyse der Carotinoide von Kartoffeln. *Nahrung* **11,** 405–409.

Kasim, M. (1967b). Untersuchungen über die Zusammensetzung der Kartoffel-Carotinoide. *Nahrung* **11,** 411–415.

Kato, M., and Shimizu, S. (1987). Chlorophyll metabolism in higher plants. VII. Chlorophyll degradation in senescing tobacco leaves: phenolic-dependent peroxidative degradation. *Can. J. Bot.* **65,** 729–736.

Katz, J.J., Norman, G.D., Svec, W.A., and Strain, H.H. (1968). Chlorophyll diastereoisomers. The nature of chlorophyll *a′* and *b′* and evidence for bacteriochlorophyll epimers from proton magnetic resonance studies. *J. Am. Chem. Soc.* **90,** 6841–6845.

Kaufman, J., and Ringel, S.M. (1961). Test of growth regulators to retard yellowing and abscission of cauliflower. *Proc. Am. Soc. Hortic. Sci.* **78,** 347–352.

Kaur, B., and Manjrekar, S.P. (1975). Effect of dehydration on the stability of chlorophyll and β-carotene content of green leafy vegetables available in northern India. *J. Food Sci. Technol. (India)* **12,** 321–323.

Kelley, L., Bennet, B.B., Rafferty, J.P., and Paul, P. (1950). The nutritive value of canned foods. V. Changes in carotene content of vegetables during storage prior to canning. *Food Technol.* **4,** 269–272.

Kemmerer, A.R., and Fraps, G.S. (1945). Constituents of crude carotene of carrots. *Food Res.* **10,** 457–460.

Kemmerer, A.R., Fraps, G.S., and Meinke, W.W. (1945). Constituents of the crude carotene of certain human foods. *Food Res.* **10,** 66–71.

Ketsa, S., and Uthaiburt, J. (1987). Chemical composition of white and green spears of asparagus (*A. officinalis* L.). *ASEAN Food J.* **3,** 77–78.

Khachik, F., and Beecher, G.R. (1987). Application of a C-45-β-carotene as an internal standard for the quantification of carotenoids in yellow/orange vegetables by liquid chromatography. *J. Agric. Food Chem.* **35,** 732–738.

Khachik, F., and Beecher, G.R. (1988a). Separation and identification of carotenoids and carotenol fatty acid esters in some squash products by liquid chromatography. 1. Quantification of carotenoids and related esters by HPLC. *J. Agric. Food Chem.* **36,** 929–937.

Khachik, F., and Beecher, G.R. (1988b). Separation of carotenol fatty acid esters by high performance liquid chromatography. *J. Chromatogr.* **449,** 119–133.

Khachik, F., Beecher, G.R., and Lusby, W.R. (1988). Separation and identification of carotenoids and carotenol fatty acid esters in some squash products by liquid chromatography. 2. Isolation and characterization of carotenoids and related esters. *J. Agric. Food Chem.* **36,** 938–946.

Khachik, F., Beecher, G.R., and Whittaker, N.F. (1986). Separation, identification and quantification of the major carotenoid and chlorophyll constituents in extracts of several green vegetables by liquid chromatography. *J. Agric. Food Chem.* **34,** 603–616.

Khachik, F., Goli, M.B., Beecher, G.R., Holden, J.M., Lusby, W.R., Tenorio, M.D., and Barrera, M.R. (1989). The effect of food preparation on qualitative and quantitative distribution of major carotenoid constituents of several green vegetables. *Abst. AGFD 39, 198th Natl. Am. Chem. Soc. Meeting.*

Khan, N.A., Khalil, J.K., and Khan, I. (1975). Changes in the carotene content of carrots during maturity, processing and storage under different temperature conditions. *Pak. J. Sci. Res.* **27,** 112–114.

Kim, H., Simpson, K.L., and Gerber, L.E. (1988). Serum carotenoids and retinol of human subjects consuming carrot juice. *Nutr. Res.* **8,** 1119–1127.

Kimbrough, W.D., Fieger, E.A., and Lewis, H.F. (1946). Effect of date of planting and time of harvest on the carotene content of sweet potatoes of the Porto Rico variety. *Proc. Am. Soc. Hortic. Sci.* **47,** 400–402.

Kirk, J.T.O., and Tilney-Basset, R.A.E. (1978). *The Plastids. Their Chemistry, Structure, Growth and Inheritance,* 2nd ed. Elsevier, Amsterdam.

Klein, B.P., Grossman, S., King, D., Cohen, B.S., and Pinsky, A. (1984). Pigment bleaching, carbonyl production and antioxidant effects during the anaerobic lipoxygenase reaction. *Biochim. Biophys. Acta* **793,** 72–79.

Klein, B.P., and Perry, A.K. (1982). Ascorbic acid and vitamin A activity in selected vegetables from different geographical areas of the United States. *J. Food Sci.* **47,** 941–945, 948.

Kobayashi, K., Tsurumizu, A., Toyoda, M., and Sayto, Y. (1989). Contents of chlorophylls, β-carotene and pesticide residues in butter head lettuce produced by various cultivation methods. *Nippon Shokuhin Kogyo Gakkaishi* **36,** 676–681.

Kohler, G.W., Lincoln, R.E., Porter, J.W., Zscheile, F.P., Caldwell, R.M., Harper, R.H., and Silver, W. (1947). Selection and breeding for high β-carotene content (provitamin A) in tomato. *Bot. Gaz.* **109,** 219–225.

Kon, M. (1989). The color of green vegetables and fruits and the contents of carotene and chlorophyll in them. *Kaseijaku Kenkya* **35,** 99–105.

Kon, M., and Shimba, R. (1988). Accumulation of carotenoids in squash. *Nippon Kasei Gakkaishi* **39,** 1059–1064.

Kon, M., and Shimba, R. (1989). Changes in carotenoid composition during preparation and storage of frozen and freeze-dried squash. *Nippon Shokuhin Kogyo Gakkaishi* **36,** 619–624.

Koskitalo, L.N., and Ormrod, D.P. (1972). Effects of sub-optimal ripening temperatures on the color quality and pigment composition on tomato fruit. *J. Food Sci.* **37,** 56–59.

Koyama, X., Hosomi, M., Miyata, A., Hashimoto, H., and Reames, S.A. (1988). Supplementary and revised assignment of the peaks of the 7,9-,9,9′,13,13′,9,13′-di-*cis* and 9,9′, 13-tri-*cis* isomers of β-carotene in high-performance liquid chromatography using a column of calcium hydroxide. *J. Chromatogr.* **439,** 417–422.

Kozup, J., and Sistrunk, W.A. (1982). Quality attributes of fermented and acidified green beans. *J. Food Sci.* **47,** 1001–1005.

Kramer, A., Haut, I.C., Scott, L.E., and Ide, L.E. (1949). Objective methods for measuring quality factors of raw, canned and frozen asparagus. *J. Am. Soc. Hortic. Sci.* **53,** 411–425.

Kramer, A., and Smith, H.R. (1946). Preliminary investigations on measurement of color in canned foods. *Food Res.* **11,** 14–31.

Krinsky, N.I. (1971). Function. In *Carotenoids,* O. Isler, ed. Birkhäuser Verlag, Basel, pp. 669–716.

Krinsky, N.I. (1979). Carotenoid protection against oxidation. *Pure Appl. Chem.* **51,** 649–660.

Krinsky, N.I. (1989). Beta-carotene: Functions. In *New Protective Roles for Selected Nutrients,* G.A. Spiller and J. Scala, eds. Alan R. Liss, New York, pp. 1–15.

Krinsky, N.I., and Deneke, S.M. (1982). Interaction of oxygen and oxy-radicals with carotenoids. *J. Natl. Cancer Inst.* **69,** 205–210.

Krinsky, N.I., and Welankiwar, S. (1984). Assay of carotenoids. *Methods. Enzymol.* **105,** 155–162.

Krossing, G. (1940). Versuche zur Localisation einiger Fermente in den verschiedenen Zellbestandteilen der Spinatblätter. *Biochem. Z.* **305,** 359–373.

Kufner, R.B. (1980). The biological degradation of chlorophyll in senescent tissues. In *Pigments in Plants,* 2nd ed., E.-C. Czygan, ed. Fischer, Stuttgart, pp. 308–313.

Kuhn, R., and Brockmann, H. (1932). Bestimmung von Carotinoiden. *Hoppe-Seyler's Z. Physiol. Chem.* **206,** 41–64.

Kuhn, R., and Grundmann, C. (1932). Die Konstitution des Lycopins. *Ber. Deutsch. Chem. Ges.* **65,** 1880–1889.

Kuhn, R., and Grundmann, C. (1934). Kripto-xanthin aus gelbem Mais (Über das Vitamin des Wachstums). *Ber. Deutsch. Chem. Ges.* **67,** 593–595.

Kuhn, R., and Lederer, E. (1931). Ueber α- und β-Carotin. *Hoppe-Seyler's Z. Physiol. Chem.* **200,** 246–254.

Kushwaha, S.C., Suzue, G., Subbarayan, C., and Porter, J.W. (1970). The conversion of phytoene ^{14}C to acyclic, monocyclic and dicyclic carotenes and the conversion of lycopene 15,15′-^{3}H to mono- and dicyclic carotenes by soluble enzyme systems obtained from plastids of tomato fruits. *J. Biol. Chem.* **245,** 4708–4717.

Lakshmanan, M.R., Chansang, N., and Olson, J. (1972). Purification and properties of carotene 15,15′-dioxygenase of rabbit intestine. *J. Lipid. Res.* **13,** 477–482.

Lakshmanan, M.R., Mychkovsky, I., and Attlesey, M. (1989). Enzymatic conversion of all-*trans*-β-carotene to retinal by a cytosolic enzyme from rabbit and rat intestinal mucosa. *Proc. Natl. Acad. Sci. USA* **86,** 9124–9128.

Laferriere, L., and Gabelman, W.H. (1968). Inheritance of color, total carotenoids, α-carotene and β-carotene in carrots, *Daucus carota* L. *Proc. Am. Soc. Hortic. Sci.* **93,** 408–418.

Lajolo, F., and Lanfer Marquez, U.M. (1982). Chlorophyll degradation in a spinach system at low and intermediate water activities. *J. Food. Sci.* **47,** 1995–1998, 2003.

Lajolo, F., Tannenbaum, S.R., and Labuza, T.P. (1971). Reaction at limited water concentration. 2. Chlorophyll degradation. *J. Food. Sci.* **36,** 850–853.

Lambers, J.W.J., and Terpstra, W. (1985). Inactivation of chlorophyllase by negatively charged plant membrane lipids. *Biochim. Biophys. Acta* **831,** 225–235.

Lambers, J.W.J., Velthuis, H.W., and Terpstra, W. (1985). Molecular mass estimation of chlorophyllase in situ by radiation inactivation analysis. Studies on the composition of the isolated enzyme. *Biochim. Biophys. Acta* **831,** 213–224.

Lampitt, L.H., and Goldenberg, N. (1940). The composition of the potato. *Chem. & Ind.* **18,** 748–761.

Lanier, J.J., and Sistrunk, W.A. (1979). Influence of cooking method on quality attributes and vitamin content of sweet potatoes. *J. Food Sci.* **44,** 374–376.

Larsen, E.C. (1949). Investigations on cause and prevention of greening in potato tubers. *Idaho Agric. Exp. Sta. Bull.* No. 16.

Larsen, C.E. (1950). Chlorophyll formation in potato tubers as affected by temperature and time. *Science* **111,** 206–207.

Lauber, J.J., Taylor, G.A., and Drinkwater, W.O. (1967). The use of tristimulus colorimetry for the estimation of carotenoid content of raw sweet potato roots. *Proc. Am. Soc. Hortic. Sci.* **91,** 472–477.

Laval-Martin, D. (1974). La maturation du fruit de tomate "cerise": mise en évidence, par cryodécapage de l'évolution des chloroplastes en deux types de chromoplastes. *Protoplasma* **82,** 33–59.

Laval-Martin, D., Quennemet, J., and Monéger, R. (1975). Pigment evolution in *Lycopersicon esculentum* fruits during growth and ripening. *Phytochemistry* **14,** 2357–2362.

Lease, J.G., and Lease, E.J. (1956a). Factors affecting the retention of red colour in peppers. *Food Technol.* **10,** 368–373.

Lease, J.G., and Lease, E.J. (1956b). Effect of fat-soluble antioxidants on the stability of the red color in peppers. *Food Technol.* **10,** 403–405.

Lease, J.G., and Lease, E.J. (1962). Effect of drying conditions on initial color, color retention and pungency of red peppers. *Food Technol.* **16,** 104–106.

Lederer, E. (1934). *Les Caroténoïdes des plantes.* Hermann, Paris.

Lee, C.Y. (1986). Changes in carotenoid content of carrots during growth and post-harvest storage. *Food Chem.* **20,** 285–293.

Lee, C.Y., Massey, L.M., Jr., and van Burren, J.P. (1982). Effects of post-harvest handling and processing on vitamin contents of peas. *J. Food Sci.* **47,** 961–964.

Lee, C.Y., McCoon, P.E., and LeBowitz, J.M. (1981). Vitamin A value of sweet corn. *J. Agric. Food Chem.* **29,** 1294–1295.

Lee, C.Y., and Robinson, R.W. (1980). Influence of the crimson gene on vitamin A content of tomato. *HortScience* **15,** 260–261.

Lee, C.Y., Simpson, K.L., and Gerber, L. (1989). Vegetables as a major vitamin A source in our diet. *Food Life Sci. Bull.* **126,** 1–11.

Lee, C.Y., Smith, N.L., and Robinson, R.W. (1984). Carotenoids and vitamin A value of fresh and canned winter squashes. *Nutr. Rep. Int.* **29,** 129–133.

Lee, F.A. (1958). The blanching process. *Adv. Food Res.* **8,** 63–109.

Lee, F.A., Gortner, W.A., and Whitcombe, J. (1946). Effect of freezing rate of vegetables. *Ind. Eng. Chem.* **38,** 341–346.

Lee, W.G., and Ammerman, G.R. (1974). Carotene stereoisomerization in sweet potatoes as affected by rotating and still retort canning processes. *J. Food Sci.* **39,** 1188–1190.

Leeper, F.J. (1985). The biosynthesis of porphyrins, chlorophylls and vitamin B_{12}. *Nat. Prod. Rep.* **2,** 19–47; 561–580.

Lepage, H. (1968). The lipid components of white potato tubers (*Solanum tuberosum*). *Lipids* **3,** 477–481.

Lepage, M. (1967). Identification and composition of turnip root lipids. *Lipids* **2,** 244–254.

Leshem, Y.Y. (1984). Interaction of cytokinins with lipid-associated oxygen free radicals during senescence: a prospective mode of cytokinin action. *Can. J. Bot.* **62,** 2943–2949.

Lewis, E.P., and Merrow, S.B. (1962). Influence on the estimation of β-carotene by other carotenoids in Butternut squashes at harvest and during storage. *J. Agric. Food Chem.* **10,** 53–56.

Liaaen-Jensen, S. (1971). Isolation, reactions. In *Carotenoids,* O. Isler, ed. Birkhaeuser, Basel, pp. 61–188.

Liaaen-Jensen, S. (1985). Carotenoids of lower plants—recent progress. *Pure Appl. Chem.* **57,** 649–658.

Lichtenthaler, H.K. (1971). The unequal synthesis of the lipophilic plastid quinones in sun and shade leaves of *Fagus silvatica* L. *Z. Naturforsch.* **26b,** 832–841.

Lichtenthaler, H.K. (1984). Lipids. In *CRC Handbook of Chromatography,* vol. 2, H.K. Mangold, ed. Boca Raton, Fla., p. 115.

Lichtenthaler, H.K. (1987). Chlorophylls and carotenoids: pigments of photosynthetic biomembranes. *Methods Enzymol.* **148,** 350–382.

Lichtenthaler, H.K., Burkhardt, G., Kuhn, G., and Prenzel, U. (1981). Light induced accumulation and stability of chlorophylls and chlorophyll-proteins during chloroplast development in radish seedlings. *Z. Naturforsch.* **36c,** 421–430.

Lichtenthaler, H.K., and Grumbach, K.H. (1974). Kinetic of lipoquinone and pigment synthesis in green *Hordeum* seedlings during an artificial day-night rhythm with a prolonged dark phase. *Z. Naturforsch.* **29c,** 533–540.

Lichtenthaler, H.K., and Wellburn, A.R. (1983). Determination of total carotenoids and chlorophylls *a* and *b* of leaf extracts in different solvents. *Biochem. Soc. Trans.* **603,** 591–592.

Lieberman, M., and Hardenburg, R.E. (1954). Effect of modified atmosphere on respiration and yellowing of broccoli at 75°F. *Proc. Am. Soc. Hortic. Sci.* **63,** 409–414.

Lincoln, R.E., and Porter, J.W. (1950). Inheritance of β-carotene in tomatoes. *Genetics* **35,** 206–211.

Lincoln, R.E., Zscheile, F.P., Porter, J.W., Kohler, G.W., and Caldwell, R.M. (1943). Provitamin A and vitamin C in the genus *Lycopersicon. Bot. Gaz.* **105,** 113–115.

Linser, H., and Zeid, F.A. (1975). Pure protein, chlorophyll, carotene and carbohydrates of *Daucus carota* throughout the vegetative period of first years as influenced by growth regulators (CCC, GA and BA). *Z. Pflanzenern. Bodenk.* **2,** 181–196.

Lipton, W.J., and Ceponis, M.J. (1962). Retardation of senescence and stimulation of oxygen consumption in head lettuce treated with N^6-benzyladenine. *Proc. Am. Soc. Hortic. Sci.* **87,** 379–384.

Liu, Y.K., and Luh, B.S. (1977). Effect of harvest maturity on carotenoids in pastes made from VF-145-7879 tomatoes. *J. Food Sci.* **42,** 216–220.

Ljubešić, N. (1968). The fine structure of chloroplasts during the yellowing and regreening of leaves. *Protoplasma* **66,** 369–379.

Ljubešić, N. (1972). Ultrastructural changes of plastids during the yellowing of the fruit of *Cucurbita pepo* var. *pyriformis. Acta Bot. Croat.* **31,** 47–53.

Ljubešić, N. (1977). The formation of chromoplasts in fruits of *Cucurbita maxima* Duch. *"turbaniformis." Bot. Gaz.* **138,** 286–290.

Lohr, J.B., and Friedmann, H.C. (1976). New pathway for δ-aminolevulinic acid biosynthesis: formation from α-ketoglutaric acid by two partially purified plant enzymes. *Biochem. Biophys. Res. Commun.* **69,** 908–913.
Loomis, W.D., and Battaile, J. (1963). Biosynthesis of terpene. III. Mevalonic kinase from higher plants. *Biochem. Biophys. Acta* **67,** 54–63.
Lovric, T., Sablek, Z., and Bošković, M. (1970). *Cis-trans* isomerisation of lycopene and color stability of foam-mat dried tomato powder during storage. *J. Sci. Food Agric.* **21,** 641–647.
Lüthy, B., Martinoia, E., Matile, Ph., and Thomas, H. (1984). Thylakoid-associated "chlorophyll oxidase": distinction from lipoxygenase. *J. Plant Physiol.* **113,** 423–434.
Lütke-Brinkhaus, F., and Kleinig, H. (1987). Carotenoid and chlorophyll biosynthesis in isolated plastids from mustard seedling cotyledons (*Sinapsis alba* L.) during etioplast-chloroplast conversion. *Planta* **170,** 121–129.
Lütke-Brinkhaus, F., Liedvogel, B., Kreuz, K., and Kleinig, H. (1982). Phytoene synthase and phytoene dehydrogenase associated with envelope membranes from spinach chloroplasts. *Planta* **156,** 176–180.
Lütz, C., Roper, U., Beer, N.S., and Griffiths, T. (1981). Subetioplast localization of the enzyme NADPH-protochlorophyllide oxidoreductase. *Eur. J. Biochem.* **118,** 347–353.
Lynch, L.J., Mitchell, R.S., and Casimir, D.J. (1959). Preservation of green peas. *Adv. Food Res.* **9,** 61–151.
Lyons, J.M., and Rappaport, L. (1959). Effect of temperature on respiration and quality of Brussels sprouts during storage. *Proc. Am. Soc. Hortic. Sci.* **73,** 361–366.
Lyons, J.M., and Rappaport, L. (1962). Effect of controlled atmospheres on storage quality of Brussels sprouts. *Proc. Am. Soc. Hort. Sci.* **81,** 324–331.
MacKinney, G. (1935). Leaf carotenes. *J. Biol. Chem.* **111,** 75–84.
MacKinney, G. (1941). Absorption of light by chlorophyll solutions. *J. Biol. Chem.* **140,** 315–322.
MacKinney, G., Aronoff, S., and Bornstein, B.T. (1942). Some assays of provitamin A carotenoids. *Ind. Eng. Chem. Anal.* **14,** 391–395.
MacKinney, G., and Jenkins, J.A. (1952). Carotenoid differences in tomatoes. *Proc. Natl. Acad. Sci. USA* **38,** 48–52.
MacKinney, G., and Weast, C.A. (1940). Color changes in green vegetables, frozen-pack peas and string beans. *Ind. Eng. Chem.* **33,** 392–395.
MacLean, D.C., and Dedolph, R.R. (1964). Phytokinins and senescence in broccoli. *Am. J. Bot.* **51,** 618–621.
MacLean, D.C., Dedolph, R.R., and Wittwer, S.H. (1963). Respiratory responses of broccoli (*Brassica oleracea* var. *italica*) to pre- and post-harvest treatments with N^6-benzyladenine. *Proc. Am. Soc. Hortic. Sci.* **83,** 484–487.
MacLeod, F.L., Armstrong, M.R., Heap, M.E. and Tolbert, L.A. (1935). The vitamin A content of 5 varieties of sweet potato. *J. Agric. Res.* **50,** 181–187.
MacNair, V. (1956). Effects of storage and cooking on carotene and ascorbic acid content of some sweet potatoes grown in North West Arkansas. *A.K. Agric. Exp. Sta. Bull.* 574.
Märki-Fischer, E., and Eugster, C.H. (1985). Karpoxanthin and 6-epikarpoxanthin. *Helv. Chim. Acta* **68,** 1704–1707.

Malchev, E., Ioncheva, N., Tanchev, S., and Kalpakchieva, K. (1982). Quantitative changes in carotenoids during the storage of dried red pepper and red pepper powder. *Nahrung* **26,** 415–420.

Markham, M.C., and Liaaen-Jensen, S. (1968). Carotenoids of higher plants. I. The structures of lycoxanthen and lycophyll. *Phytochemistry* **7,** 839–844.

Marquard, R.D., and Tipton, T.L. (1987). Relationship between extractable chlorophyll and an in situ method to estimate leaf greenness. *HortScience* **22,** 1327.

Martin, F.W. (1983). The carotenoid pigments of white-fleshed sweet potatoes. *J. Agric. Univ. P.R.* **67,** 494–500.

Martin, F.W., and Ruberté, R. (1975). Carotenoid pigments of *Dioscorea cayenensis. Ann. Appl. Biol.* **80,** 317–322.

Martin, F.W., Telek, L., and Ruberté, R. (1974). Yellow pigments of *Dioscorea bulbifera. J. Agric. Food Chem.* **22,** 335–337.

Martin, M.E., Sweeney, J.P., and Gilpin, G.L. (1960). Broccoli nutrients. Factors affecting the ascorbic acid and carotene content of broccoli. *J. Agric. Food Chem.* **8,** 387–390.

Martinoia, E., Dalling, M.J., and Matile, Ph. (1982). Catabolism of chlorophyll: demonstration of chloroplast-localized peroxidative and oxidative activities. *Z. Pflanzenphysiol.* **107,** 269–279.

Marty, C., and Berset, C. (1986). Degradation of *trans-β*-carotene during heating in sealed glass tubes and extrusion cooking. *J. Food Sci.* **51,** 698–702.

Mathews, M.M. (1964). Effect of β-carotene against lethal photosensitization by hematoporphyrin. *Nature* **203,** 1092.

Mathews-Roth, M.M. (1982). Anti-tumor activity of beta-carotene, canthaxanthin and phytoene. *Oncology* **39,** 33–37.

Mathews-Roth, M.M. (1985). Carotenoids and cancer prevention—experimental and epidemiological studies. *Pure Appl. Chem.* **57,** 717–722.

Mathews-Roth, M.M. (1987). β-Carotene therapy in erythropoietic protoporphyria and other photosensitivity diseases. In *Photomedicine,* vol. II, E. Ben-Hur and Rosenthal, eds. CRC Press, Boca Raton, Fla.

Mathews-Roth, M.M., and Krinsky, N.I. (1987). Carotenoids affect development of UV-B induced skin cancer. *Photochem. Photobiol.* **46,** 507–509.

Mathews-Roth, M.M., Wilson, T., Fujimori, E., and Krinsky, N.I. (1974). Carotenoid chromophore length and protection against photosensitization. *Photochem. Photobiol.* **19,** 217–222.

Mathis, P., and Rutherford, A.W. (1984). Effect of phenolic herbicides on the oxygen-evolving side of photosystem. II. Formation of the carotenoid cation. *Biochem. Biophys. Acta* **767,** 217–222.

Mathis, P., and Schenck, C.C. (1982). The functions of carotenoids in photosynthesis. In *Carotenoid Chemistry and Biochemistry,* G. Britton and T.W. Goodwin, eds. IUPAC, Pergamon Press, Oxford, pp. 339–351.

Matile, Ph. (1980). Catabolism of chlorophyll: Involvement of peroxidase. *Z. Pflanzenphysiol.* **99,** 475–478.

Matile, Ph., Dügellen, T., Schellenberg, M., Rentsch, D., Bortlik, K., Peisker, Ch., and Thomas, H. (1989). How and why is chlorophyll broken down in senescent leaves. *Plant Physiol. Biochem.* **27,** 595–604.

Matile, Ph., Ginsburg, S., Schellenberg, M., and Thomas, H. (1987). Catabolites of chlorophyll in senescent leaves. *J. Plant Physiol.* **129,** 219–228.

Matsuno, T., Ito, T., Taniguchi, T., Takeda, S., and Hirota, S. (1970). Constituents of *Trichosanthes cucumeroides*. Pigments of the fruits. *Yakugaku Zasshi* **90**, 248–250.

Matsuno, T., Tani, Y., Maoka, T., Matsuo, K., and Komori, T. (1986). Isolation and structural elucidation of cucurbitaxanthin A and B from pumpkin *Cucurbita maxima*. *Phytochemistry* **25**, 2837–2840.

Mattheis, J.R., and Rebeiz, C.A. (1977). Chloroplast biogenesis: Metabolism of protochlorophyllide and protochlorophyllide ester in developing chloroplasts. *Arch. Biochem. Biophys.* **184**, 189–196.

Matthews, R.F., Locascio, S.J., and Ozaki, H.Y. (1975). Ascorbic acid and carotene contents of peppers. *Proc. Fla. State Hortic. Soc.* **88**, 263–265.

Maudinas, B., Bucholtz, M.L., Papastephanou, C., Katiyar, S.S., Briedis, A.V., and Porter, J.W. (1977). The partial purification and properties of a phytoene synthesizing enzyme system. *Arch. Biochem. Biophys.* **180**, 354–362.

Maudinas, B., and Lematre, J. (1979). Violaxanthin, the major carotenoid pigment in *Zea mays* root cap during seed germination. *Phytochemistry* **18**, 1815–1817.

Maunders, M.J., and Brown, S.B. (1983). The effect of light on chlorophyll loss in senescing leaves of sycamore. *Planta* **158**, 309–311.

Maunders, M.J., Brown, S.B., and Woolhouse, H.W. (1983). The appearance of chlorophyll derivatives in senescent tissue. *Phytochemistry* **22**, 2443–2446.

Mayland, H.F., and Dean, L.L. (1971). Chlorophyll content of persistent green and normal snap bean pods (*Phaseolus vulgaris* L.). *J. Am. Soc. Hortic. Sci.* **96**, 362–365.

McMahon, R.W., and Gladon, R.J. (1984). Method for extraction and quantitation of chloroplast pigments in fruits and other organs by high performance liquid chromatography. *HortScience* **19**, 220–221.

Medina, P.V.L., Dostal, H.C., and Medina, R.M.T. (1981). Specific inhibition of pectinmethylesterase isozymes by ethylene during tomato fruit ripening. *Rev. Ceres* **28**, 435–443.

Mehwald, J. (1965). Möhrenbau und Karotinwerte. *Rhein. Gemüse-Obst-Gartenbau* **53**, 112–113.

Meier, D., and Lichtenthaler, H.K. (1981). Ultrastructural development of chloroplasts in radish seedling grown at high- and low-light conditions in the presence of the herbicide bentazon. *Protoplasma* **107**, 195–207.

Mejía, L.A., Hudson, E., Mejía, G.E., and Vasquez, F. (1988). Carotenoid content and vitamin A activity of some common cultivars of Mexican peppers (*Capsicum annuum*) as determined by HPLC. *J. Food Sci.* **53**, 1448–1451.

Meller, E., Belkin, S., and Harel, E. (1975). The biosynthesis of δ-aminolevulinic acid in greening maize leaves. *Phytochemistry* **14**, 2399–2402.

Meredith, F.I., and Purcell, A.E. (1966). Changes in the concentration of carotenes of ripening Homestead tomatoes. *Proc. Am. Soc. Hortic. Sci.* **89**, 544–548.

Michalik, H. (1985). Effect of fertilization with macro and micro elements on the dry matter, sugar and β-carotene contents of carrots. *Biul. Warzywniczy* **28**, 141–163.

Miki, N., and Akatsu, K. (1970). Effects of heating sterilization on color of tomato juice. *Nippon Shokuhin Kogyo Gakkaishi* **17**, 175–181.

Miki, N., and Akatsu, K. (1971). Stability of lycopene inside and outside of the flesh in tomato juice. *Nippon Shokuhin Kogyo Gakkaishi* **17**, 309–312.

Milborrow, B.V. (1982). Stereochemical aspects of carotenoid biosynthesis. In *Carotenoid Chemistry and Biochemistry,* G. Britton and T.W. Goodwin, eds. IUPAC, Pergamon Press, Oxford, pp. 279–295.

Miller, J.C., and Covington, H.M. (1942). Some of the factors affecting the carotene content of sweet potatoes. *J. Am. Soc. Hortic. Sci.* **40,** 519–522.

Miller, J.C., Melampy, R.M., Mikell, J.J., and Hernandez, T.P. (1949). Effect of storage on the carotene content of fourteen varieties of sweet potatoes. *Proc. Am. Soc. Hortic. Sci.* **54,** 399–402.

Mínguez Mosquera, M.I., and Fernández Díez, M.J. (1981). Ripe spanish red pepper *Capsicum* genus. Relationship between red and yellow carotenoid pigments. *Grasas y Aceites* **32,** 293–298.

Mínguez Mosquera, M.I., Fernández Díez, M.J., and Pereda Marin, J. (1984). Paprika (*Capsicum annuum*). Ratio between the red and yellow carotenoid pigments. *Grasas y Aceites* **35,** 4–10.

Mitchell, H.L., and Hauge, S.M. (1946). Enzymic nature of the carotene destroying system of alfalfa. *J. Biol. Chem.* **163,** 7–14.

Mizrahi, Y., Dostal, H.C., McGlasson, W.B., and Cherry, J.H. (1978). Effects of abscisic acid and benzyladenine on fruits of normal and *rin* mutant tomatoes. *Plant Physiol.* **56,** 544–546.

Mohr, W.P. (1979). Pigment bodies in fruits of crimson and high pigment lines of tomatoes. *Ann. Bot.* **44,** 427–434.

Monselise, J.J., and Berk, Z. (1954). Oxidative destruction of lycopene during the manufacture of tomato puree. *Bul. Res. Council Isr.* **4,** 188–191.

Moran, R. (1982). Formulae for determination of chlorophyllous pigments extracted with *N,N*-dimethylformamide. *Plant Physiol.* **69,** 1376–1381.

Moran, R., and Porath, D. (1980). Chlorophyll determination in intact tissues using *N,N*-dimethylformamide. *Plant Physiol.* **65,** 478–479.

Morgan, A.F., and Smith, L.L.W. (1928). Development of vitamin A in tomatoes. *Soc. Exp. Biol. Med. Proc.* **26,** 44.

Morris, M.M., Park, Y., and MacKinney, G. (1973). On the photodecomposition of chlorophyll *in vitro. J. Agric. Food Chem.* **21,** 277–279.

Moss, G.P., and Weedon, B.C.L. (1976). Chemistry of the carotenoids. In *Chemistry and Biochemistry of Plant Pigments,* 2nd ed. vol. 1, T.W. Goodwin, ed. Academic Press, London, pp. 149–224.

Moster, J.B., Quackenbush, F.W., and Porter, J.W. (1952). The carotenoids of corn seedlings. *Arch. Biochem. Biophys.* **38,** 287–290.

Muftugil, N. (1986). Effect of different types of blanching on the color and the ascorbic acid and chlorophyll contents of green beans. *J. Food Proc. Preserv.* **10,** 69–76.

Murphy, E.W., Criner, P.E., and Gray, B.C. (1975). Comparison of methods for calculating retentions of nutrients in cooked foods. *J. Agric. Food Chem.* **23,** 1153–1157.

Murray, H.C. (1948). Vitamin content of field-pea products; retention of the thiamine and riboflavin in cooking. *Food Research* **13,** 397–399.

Musser, R.L., Thomas, S.A., Wise, R.R., Peeler, T.C., and Naylor, A.W. (1984). Chloroplast ultrastructure, chlorophyll fluorescence and pigment composition in chilling-stressed soybeans. *Plant Physiol.* **74,** 749–754.

Mustardy, L.A., Machowitz, E., and Faludi-Daniel, A. (1976). Light induced structural changes of thylakoids in normal and carotenoid deficient chloroplasts of maize. *Protoplasma* **88,** 65–73.

Nagra, S.A., and Khan, S. (1988). Vitamin A (β-carotene) losses in Pakistani cooking. *J. Sci. Food Agric.* **46,** 249–251.

Naito, K., Ueda, K., and Tsuji, H. (1981). Differential effects of benzyladenine on the ultrastructure of chloroplasts in intact bean leaves according to their age. *Protoplasma* **105,** 293–306.

NAS-NRC (1974). *Recommended Dietary Allowances (1974),* 8th ed. Food and Nutrition Board, Washington, D.C., pp. 20–54.

Neamţu, G., Illyes, Gh., and Pescaru, E. (1982). Carotenoid content in some corn varieties. *Bul. Inst. Agric. Cluj* **36,** 55–59.

Neumann, K.-H., and Schwab, B. (1975). Investigation on the influence of gibberellic acid treatments on yield, root anatomy and carotene distribution of carrot plants. *Z. Pflanzenern. Bodenk.* **1,** 19–23.

Nickerson, J.T.R., Proctor, B.E., and Goldblith, A.S. (1956). Ionizing radiations in the processing of plant and animal products. *Food Technol.* **10,** 305–311.

Noack, K. (1943). Ueber den biologischen Abbau des Chlorophylls. *Biochem. Z.* **316,** 166–187.

Noack, K., and Kiessling, W. (1930). Zur Entstehung des Chlorophylls und seiner Beziehung zum Blutfarbstoff. *Hoppe-Seyler's Z. Physiol. Chem.* **193,** 97–137.

Noble, A.C. (1975). Investigation of the color changes in heat concentrated tomato pulp. *J. Agric. Food Chem.* **23,** 48–49.

O'Connor, R.T., Heinzelman, D.C., and Jefferson, M.E. (1946). Determination of β-carotene in sweet potatoes and sweet potato products: an improved method. *Ind. Eng. Chem. (Anal.)* **18,** 557–562.

Odland, D., and Eheart, M.S. (1975). Ascorbic acid, mineral and quality retention in frozen broccoli blanched in water, steam and ammonia steam. *J. Food Sci.* **40,** 1004–1007.

Ölze-Karow, H., and Mohr, H. (1978). Control of chlorophyll *b* biosynthesis by phytochrome. *Photochem. Photobiol.* **27,** 189–193.

Ogunlesi, A.T., and Lee, C.Y. (1979). Effect of thermal processing on the stereoisomerisation of major carotenoids and vitamin A value of carrots. *Food Chem.* **4,** 311–318.

Ogura, N., Ueno, M., Matsunaga, M., Sato, T., and Nakagawa, H. (1987). Chlorophyllase activities and color change of fresh green leaves on blanching. *Nippon Nogeikagaku Kaishi* **61,** 451–456.

Ohad, I., Friedberg, F., Neeman, Z., and Schramm, M. (1971). Biogenesis and degradation of starch. I. The fate of the amyloplast membrane during maturation and storage of potato tubers. *Plant Physiol.* **47,** 465–467.

Omata, T., and Murata, N. (1980). A rapid and efficient method to prepare chlorophyll *a* and *b* from leaves. *Photochem. Photobiol.* **31,** 183–185.

Omata, T., and Murata, N. (1983). Preparation of chlorophyll *a,* chlorophyll *b* and bacteriochlorophyll *a* by column chromatography with DEAE-sepharose CL and sepharose-CL-6B. *Plant Cell Physiol.* **24,** 1093–1100.

Onyewu, P.N., Ho, C.-T., and Daun, H. (1986). Characterization of β-carotene thermal degradation products in a model food system. *JAOCS* **63,** 1437–1441.

Orthoefer, F.T., and Dugan, L.R., Jr. (1973). The coupled oxidation of chlorophyll *a* with linoleic acid catalyzed by lipoxidase. *J. Sci. Food Agric.* **24,** 357–365.

Oser, B.L., Melnick, D., and Oser, M. (1943). Influence of cooking procedure upon retention of vitamins and minerals in vegetables. *Food Res.* **8,** 115–122.

Oszmianski, J., and Lee, C.Y. (1990). Inhibitor effect of phenolics on carotene bleaching in vegetables. *J. Agric. Food Chem.* **38,** 688–690.

Ouyang, J.M., Daun, H., Chang, S.S., and Ho, C.-T. (1980). Formation of carbonyl compounds from β-carotene during palm oil deodorization. *J. Food Sci.* **45,** 1214–1217.

Packer, J.E., Mahood, J.S., Mora-Asellano, V.O., Slater, T.F., and Wolfenden, B.S. (1981). Free radicals and singlet oxygen scavengers: reaction of a peroxy-radical with β-carotene, diphenyl furan and 1,4-diazobicyclo(2,2,2)-octane. *Biochem. Biophys. Res. Commun.* **98,** 901–906.

Panalaks, T., and Murray, T.K. (1970). The effect of processing on the content of carotene isomers in vegetables and peaches. *J. Inst. Can. Technol. Aliment.* **3,** 145–151.

Papageorgiou, G. (1975). Chlorophyll fluorescence. In *Bioenergetics of Photosynthesis,* Govindjee, ed. Academic Press, New York, pp. 319–371.

Papastephanou, C., Barnes, F.J., Briedis, A.V., and Porter, J.W. (1973). Enzymatic synthesis of carotenes by cell-free preparations of several genetic selections of tomatoes. *Arch. Biochem. Biophys.* **157,** 415–425.

Pardo, A.D., Chereskin, B.M., Castelfranco, P.A., Franceschi, V.R., and Wezelman, B.E. (1980). ATP requirement for Mg chelatase in developing chloroplasts. *Plant Physiol.* **65,** 956–960.

Park, Y.W. (1987). Effect of freezing, thawing, drying and cooking on carotene retention in carrots, broccoli and spinach. *J. Food Sci.* **52,** 1022–1025.

Park, Y., Morris, M.M., and MacKinney, G. (1973). On chlorophyll breakdown in senescent leaves. *J. Agric. Food Chem.* **21,** 279–281.

Parker, R.S. (1988). Carotenoid and tocopherol composition of human adipose tissue. *Am. J. Clin. Nutr.* **47,** 33–36.

Parkes, K.E.B., Pattenden, G., Baranyai, M., Molnar, P., Szabolcs, J., and Toth, G. (1986). Novel carotenoid 3,6-epoxides from red paprika (*Capsicum annuum*). *Tetrahedron Lett.* **27,** 2535–2538.

Patil, B.C., Salunkhe, D.K., and Singh, B. (1971). Metabolism of solanine and chlorophyll in potato tubers as affected by light and specific chemicals. *J. Food Sci.* **36,** 474–476.

Pavisa, A.Cs., Hajdu, F., Hoschke, A., Bodnar, J., and Biacs, P. (1987). High-performance liquid-chromatographic control of standard specification of ground paprika. *Acta Aliment.* **16,** 129–142.

Paynter, V.A., and Jen, J.J. (1976). Comparative effects of light and ethephon on the ripening of detached tomatoes. *J. Food Sci.* **41,** 1366–1369.

Pendlington, S., Dupont, M.S., and Trussel, F.J. (1960). The carotenoid pigments of *Solanum tuberosum. Biochem. J.* **94,** 25 p.

Pennington, F.C., Strain, H.H., Svec, W.A., and Katz, I.J. (1964). Preparation and properties of pyrochlorophyll *a,* methyl pyrochlorophyllide *a,* pyropheophytin *a* and methyl pyropheophorbide *a* derived from chlorophyll by decarbomethoxylation. *J. Am. Chem. Soc.* **86,** 1418–1426.

Penteado, M.V.C., and Almeida, L.B. (1988). Occurrence of carotenoids in roots of five cultivars of cassava (*Manihot esculenta* Crantz) from São Paulo state. *Rev. Farm. Bioquim. Univ. São Paulo* **24**, 39–49.

Penteado, M.V.C., Minazzi, R.S., and Almeida, L.B. (1986). Carotenoids and provitamin A activity of green leaves consumed in North Brazil. *Rev. Farm. Bioquim. Univ. São Paulo* **22**, 97–102.

Pepkowitz, L.P., Larson, R.E., Gardner, J., and Owens, G. (1944). The carotene and ascorbic acid concentration of vegetable varieties. *Plant Physiol.* **19**, 615–626.

Pepping, F., Vencken, C.M.J., and West, C.E. (1988). Retinol and carotene content of foods consumed in East Africa determined by high performance liquid chromatography. *J. Sci. Food Agric.* **45**, 359–371.

Perrin, P.W., and Gaye, M.M. (1986). Effects of simulated retail display and overnight storage treatments on quality maintenance in fresh broccoli. *J. Food Sci.* **51**, 146–149.

Pesek, C.A., and Warthesen, J.J. (1987). Photodegradation of carotenoids in a vegetable juice system. *J. Food Sci.* **52**, 744–746.

Peterson, C.E., Simon, P.W., Rubatzky, V.E., and Strandberg, J.O. (1988). Cultivar germplasm releases. Beta III carrot. *HortScience* **23**, 917.

Peto, R., Doll, R., Buckley, J.D., and Sporn, M.B. (1981). Can dietary β-carotene materially reduce human cancer rates? *Nature* **190**, 201–208.

Petzold, E.N., and Quackenbusch, F.W. (1960). Zeinoxanthin, a crystalline carotenol from corn gluten. *Arch. Biochem. Biophys.* **86**, 163–165.

Petzold, E.N., Quackenbusch, F.W., and McQuistan, M. (1959). Zeacarotenes, new provitamins A from corn. *Arch. Biochem. Biophys.* **82**, 117–124.

Pfündel, E., and Strasser, R.J. (1988). Violaxanthin de-epoxidase in etiolated leaves. *Photosynth. Res.* **15**, 67–73.

Philip, T., and Chen, T.-S. (1988). Separation and quantitative analysis of fatty acid esters of fruits by liquid chromatography. *J. Chromatogr.* **435**, 113–126.

Philip, T., and Francis, F.J. (1971). Oxidation of capsanthin. *J. Food Sci.* **36**, 96–97.

Philip, T., Nawar, W.W., and Francis, F.J. (1971). The nature of fatty acids and capsanthin esters in paprika. *J. Food Sci.* **36**, 98–100.

Picha, D.T. (1985). Crude protein, minerals and total carotenoids in sweet potatoes. *J. Food Sci.* **50**, 1768–1769.

Pinsky, A., Grossman, S., and Trop, M. (1971). Lipoxygenase content and antioxidant activity of some fruits and vegetables. *J. Food Sci.* **36**, 571–572.

Platenius, H. (1943). Effect of oxygen concentration on the respiration of some vegetables. *Plant Physiol.* **18**, 671–684.

Poenicke, E.F., Kays, S.T., Smittle, D.A., and Williamson, R.E. (1977). Ethylene in relation to postharvest quality deterioration in processing cucumbers. *J. Am. Soc. Hortic. Sci.* **102**, 303–306.

Polgár, A., and Zechmeister, L. (1942). Isomerization of β-carotene. Isolation of a stereoisomer with increased adsorption affinity. *J. Am. Chem. Soc.* **64**, 1856–1861.

Porra, R., Klein, O., Dörnemann, D., and Senger, H. (1980). A simple method for the rapid determination of 4,5-dioxovaleric acid in the presence of 2-oxoglutarate and other 2-oxocarboxylic acids. *Hoppe-Seyler's Z. Physiol. Chem.* **361**, 187–190.

Porter, J.W. (1969). Enzymatic synthesis of carotenes and related compounds. *Pure Appl. Chem.* **20,** 449–481.

Porter, J.W., and Lincoln, R.E. (1950). I. *Lycopersicon* selections containing a high content of carotenes and colorless polyenes. II. The mechanism of carotene biosynthesis. *Arch. Biochem.* **27,** 390–403.

Porter, J.W., and Spurgeon, S.L. (1979). Enzymatic synthesis of carotenes. *Pure Appl. Chem.* **51,** 609–622.

Porter, J.W., Strong, F.M., Brink, R.A., and Neal, N.P. (1946). Carotene content of the corn plant. *J. Agric. Res.,* 169–187.

Porter, J.W., and Zscheile, F.P. (1946). Carotenes of *Lycopersicon* species and strains. *Arch. Biochem.* **101,** 537–545.

Possingham, J.V., and Saurer, W. (1969). Changes in chloroplast number per cell during leaf development in spinach. *Planta* **86,** 186–194.

Prebble, J., and West, S. (1975). Carotenoids of cauliflower bud tissue. *Ann. Bot.* **39,** 1097–1102.

Premachandra, B.R. (1984). A simple TLC method for the determination of provitamin A content of fruits and vegetables. *Intl. J. Vit. Nutr. Res.* **55,** 139–147.

Pressman, E., and Palevitch, D. (1973). Cytokinin-like activity of benomyl as a senescence inhibitor of broccoli heads. *HortScience* **8,** 496–497.

Purcell, A.E. (1958). Partition separation of carotenoids by silica-methanol columns. *Anal. Chem.* **30,** 1049–1051.

Purcell, A.E. (1962). Carotenoids of Goldrush sweet potato flakes. *Food Technol.* **16,** 99–102.

Purcell, A.E., and Walter, M.W., Jr. (1968). Carotenoids of Centennial variety sweet potato, *Ipomea batatas* L. *J. Agric. Food Chem.* **16,** 769–770.

Purcell, A.E., Walter, M.W., and Thompkins, W.T. (1969). Relationship of vegetable color to physical state of the carotenes. *J. Agric. Food Chem.* **17,** 41–42.

Quackenbush, F.W. (1963). Corn carotenoids: effects of temperature and moisture on losses during storage. *Cereal Chem.* **40,** 266–269.

Quackenbush, F.W. (1987). Reversed phase HPLC separation of cis and trans-carotenoids and its application to β-carotenes in food materials. *J. Liq. Chromatogr.* **10,** 643–653.

Quackenbush, F.W., Firch J.G., Brunsen, A.M., and House, L.R. (1963). Carotenoid, oil and tocopherol content of corn inbreds. *Cereal Chem.* **40,** 250–259.

Quackenbush, F.W., Firch, J.G., Rabourn, J.W., Mcquistan, M., Petzold, E.N., and Kargl, T.E. (1961). Composition of corn. Analysis of carotenoids in corn grain. *J. Agric. Food Chem.* **9,** 132–135.

Quackenbush, F.W., and Smallidge, R.L. (1986). Vitamins and other nutrients. Non aqueous reverse phase liquid chromatographic system for separation and quantitation of provitamins A. *JAOAC* **69,** 767–772.

Qureshi, A.A., Andrewes, A.G., Qureshi, N., and Porter, J.W. (1974a). The enzymatic conversion of *cis*-[^{14}C] phytofluene, *trans*-[^{14}C] phytofluene and *trans*-ζ-[^{14}C] carotene to more unsaturated acyclic, monocyclic and dicyclic carotenes by a cell free preparation of red tomato fruits. *Arch. Biochem. Biophys.* **162,** 93–107.

Qureshi, A.A., Qureshi, N., Kim, M., and Porter, J.W. (1974b). The enzymatic conversion of *cis*-[^{14}C] phytofluene, *trans*-[^{14}C] phytofluene and *trans*-ζ-[^{14}C] carotene to poly-*cis*-acyclic carotenes by a cell-free preparation of tangerine tomato. *Arch. Biochem. Biophys.* **162,** 108–116.

Rabourn, W.J., and Quackenbush, F.W. (1953). The occurrence of phytoene in various plant materials. *Arch. Biochem. Biophys.* **44,** 159–164.

Rabourn, W.J., Quackenbush, F.W., and Porter, J.W. (1954). Isolation and properties of phytoene. *Arch. Biochem. Biophys.* **48,** 267–274.

Radmer, R.J. (1966). Study on the Biosynthesis of Chlorophyll in *Phaseolus vulgaris.* M.Sc. Thesis, University of Chicago.

Radmer, R.J., and Bogorad, L. (1967). (-)S-Adenosyl-L-methionine-magnesium protoporphyrin methyltransferase, an enzyme in the biosynthetic pathway of chlorophyll in *Zea mays. Plant Physiol.* **42,** 463–465.

Rahman, F.M.M., and Buckle, K.A. (1980). Pigment changes in capsicum cultivars during maturation and ripening. *J. Food Technol.* **15,** 241–249.

Ramakrishnan, T.V., and Francis, F.J. (1980). Autoxidation of carotenoids and their relative polarity. *J. Food Qual.* **3,** 25–34.

Ramaswamy, N.K., and Nair, P.M. (1973). δ-Aminolevulinic acid synthetase from cold-stored potatoes. *Biochim. Biophys. Acta* **293,** 269–277.

Ramaswamy, N.K., and Nair, P.M. (1974). Temperature and light dependency of chlorophyll synthesis in potatoes. *Plant Sci. Lett.* **2,** 249–256.

Ramaswamy, N.K., and Nair, P.M. (1976). Pathway for the biosynthesis of delta-aminolevulinic acid in greening potatoes. *Indian J. Biochem. Biophys.* **13,** 394–397.

Ramos, D.M.R., and Rodriguez-Amaya, D.B. (1987). Determination of the vitamin A value of common Brazilian leafy vegetables. *J. Micronutr. Anal.* **3,** 147–155.

Randerath, K. (1963). *Thin Layer Chromatography.* Academic Press, New York.

Raymundo, L.C., Chichester, C.O., and Simpson, K.L. (1976). Light dependent carotenoid synthesis in tomato fruit. *J. Agric. Food Chem.* **24,** 59–64.

Raymundo, L.C., Griffiths, A.E., and Simpson, K.L. (1967). Dimethyl sulfoxide and biosynthesis of carotenoids in detached tomatoes. *Phytochemistry* **6,** 1527–1532.

Raymundo, L.C., Griffiths, A.E., and Simpson, K.L. (1970). Biosynthesis of carotenoids in the tomato fruits. *Phytochemistry* **9,** 1239–1245.

Raymundo, L.C., and Simpson, K.L. (1972). The isolation of a poly-*cis*-ζ-carotene from the Tangerine tomato. *Phytochemistry* **11,** 397–400.

Rebeiz, C.A. (1982). Chlorophyll: anatomy of a discovery. *Chemtech.* **12,** 52–63.

Rebeiz, C.A. (1988). Biocides from first principles. *Chemtech.* **18,** 90–93.

Rebeiz, C.A., Abou-Haidar, M., Yaghi, M., and Castelfranco, P.A. (1970). Porphyrin biosynthesis in cell-free homogenates from higher plants. *Plant Physiol.* **46,** 543–549.

Rebeiz, C.A., Tripathy, B.C., Wu, S.M., Montazer-Zouhoor, A., and Carey, E.E. (1986). Chloroplast biogenesis 52: demonstration *in toto* of monovinyl and divinyl monocarboxylic chlorophyll biosynthetic routes in higher plants. In *Regulation of Chloroplast Differentiation,* vol. 2, G. Akoyunoglu and H. Senger, eds. Alan R. Liss, New York, pp. 13–24.

Reddy, N.N., and Sistrunk, W.A. (1980). Effect of cultivar, size, storage, and cooking method on carbohydrates and some nutrients of sweet potatoes. *J. Food Sci.* **45,** 682–684.

Renqvist, U.H., De Vreeze, A.C., and Evenhuis, B. (1978). The effect of traditional cooking methods on carotene content in tropical leafy vegetables. *Ind. J. Nutr. Diet.* **15,** 154–158.

Richter, M.L., and Rienits, K.G. (1980). The synthesis of magnesium and zinc protoporphyrin IX and their monomethylester in etioplast preparations studied by high pressure liquid chromatography. *FEBS Lett.* **116,** 211–216.

Richter, M.L., and Rienits, K.G. (1982). The synthesis of magnesium protoporphyrin IX by etioplast membrane preparations. *Biochim. Biophys. Acta* **717,** 255–264.

Rinno, G. (1965). Die Beurteilung des ernährungsphysiologischen Wertes von Gemüse. *Arch. Gartenbau* **13,** 415–429.

Rodriguez-Amaya, D. (1989). Critical revue of provitamin A determination in plant foods. *J. Micronutr.* **5,** 101–225.

Rodriguez, D.B., Lee, T.-C., and Chichester, C.O. (1975). Comparative study of the carotenoid composition of the seeds of ripening *Momordica charantia* and tomatoes. *Plant Physiol.* **56,** 626–629.

Rodriguez, D.B., Raymundo, L.C., Lee, T.-C., Simpson, K.L., and Chichester, C.O. (1976). Carotenoid pigment changes in ripening *Momordica charantia* fruits. *Ann. Bot.* **40,** 615–624.

Rogers, L.J., Shah, S.P.J., and Goodwin, T.W. (1966). Intracellular localization of mavalonate-activating enzymes in plant cells. *Biochem. J.* **99,** 381–388.

Rouchaud, J., Moons, C., and Meyer, J.A. (1982). Effects of soil treatment with the insecticide chlorphenvinphos and of covering the culture with plastic film on the provitamin A content of early carrots. *J. Agric. Food Chem.* **30,** 1036–1038.

Rouchaud, J., Moons, C., and Meyer, J.A. (1983). Effects of selected insecticides and herbicides on the carotene content of summer carrots. *Sci. Hortic.* **19,** 33–37.

Rouchaud, J., Moons, C., and Meyer, J.A. (1984a). Effects of pesticide treatment on the carotenoid pigments of lettuce. *J. Agric. Food Chem.* **32,** 1241–1245.

Rouchaud, J., Moons, C., Meyer, J.A., Benoit, F., Ceustermans, N., and van Linden, F. (1984b). Effects of soil fertilization and covering of the culture with plastic film on the provitamin A content of early lettuce. *Plant. Soil.* **80,** 139–141.

Rouchaud, J., Moons, C., and Meyer, J.A. (1985). The effects of selected herbicide and fungicide treatments on the carotenes and xanthophylls in lettuce. *J. Hortic. Sci.* **60,** 245–249.

Rousos, P.A., Harrison, H.C., and Palta, J.P. (1986). Reduction of chlorophyll in cabbage leaf disks following Cu^{2+} exposure. *HortScience* **21,** 499–501.

Rudolph, H. (1934). Über die Einwirkung des farbigen Lichtes auf die Entstehung der Chloroplastenfarbstoffe. *Planta* **21,** 104–155.

Rüdiger, W. (1986). Regulation of chlorophyll biosynthesis. In *Regulation of Chloroplast Differentiation,* vol. 2, C. Akoyunoglu and H. Senger, eds. Alan R. Liss, New York, pp. 3–12.

Rüdiger, W., Benz, J., and Guthoff, C. (1980). Detection and partial characterization of activity of chlorophyll synthetase in etioplast membranes. *Eur. J. Biochem.* **109,** 193–200.

Rüdiger, W., Benz, J., Lempert, U., Schoch, S., and Stefens, D. (1976). Inhibition of phytol accumulation with herbicides. Geranylgeraniol and dihydrogeranylgeraniol containing chlorophyll from wheat seedlings. *Z. Pflanzenphysiol.* **80,** 131–143.

Rüdiger, W., Hedden, P., Köst, H.-P., and Chapman, D.J. (1977). Esterification of chlorophyllide by geranylgeranyl pyrophosphate in a cell-free system from maize shoots. *Biochem. Biophys. Res. Commun.* **74,** 1268–1272.

Rüttimann, A. (1982). Synthesis and stereochemistry of red pepper carotenoids. In *Carotenoid Chemistry and Biochemistry, Proc. 6th Int. Symp. on Carotenoids,* G. Britton and T.W. Goodwin, eds. Pergamon Press, Oxford, pp. 71–86.

Russo, L., Dougherty, R.H., Dostal, H.C., Wilcox, G.E., and Nelson, P.E. (1975). Quality of processed tomato products from ethephon-ripened fruit. *HortScience* **10,** 138–139.

Rygg, G.L. (1949). Changes in carotenoid content of harvested carrots. *Proc. Am. Soc. Hortic. Sci.* **54,** 307–310.

Rymal, K.S., and Nakayama, T.O.M. (1974). Major carotenoids of the seeds of three cultivars of the tomato. *J. Agric. Food Chem.* **22,** 715–717.

Sabater, B., and Rodriguez, M.T. (1978). Control of chlorophyll degradation in detached leaves of barley and oat through effect of kinetin on chlorophyllase levels. *Physiol. Plant.* **43,** 274–276.

Sadowski, R., and Wojcik (1983). Chromatography of chloroplast carotenoids on magnesium oxide thin layer. *J. Chromatogr.* **262,** 455–459.

Sahlberg, I., and Hynninen, P.H. (1984). Thin-layer chromatography of chlorophylls and their derivatives on sucrose layers. *J. Chromatogr.* **291,** 331–338.

Samuels, G., and Landrau, P. (1952). The influence of fertilizers on the carotene content of sweet potatoes. *Agron. J.* **44,** 348–352.

Sanger, J.E. (1971). Quantitative investigations of leaf pigments from their inception in buds through autumn coloration to decomposition in falling leaves. *Ecology* **52,** 1075–1089.

Santamaria, L., Bianchi, A., Arnaboldi, A., Ravetto, C., Bianchi, L., Pizella, R., Andreoni, L., Santagati, G., and Bermond, P. (1988). Chemoprevention of indirect and direct chemical carcinogenesis by carotenoids as oxygen radical quenchers. *Ann. N.Y. Acad. Sci.* **534,** 584–596.

Sapozhnikov, D.J., Krasovskaya, T.A., and Mayerskaya, A.N. (1957). Changes observed in the relation between the main carotenoids in the plastids of green leaves exposed to light. *Dokl. Akad. Nauk. U.S.S.R.* **113,** 465–467.

Sato, N., and Murata, N. (1978). Preparation of chlorophyll *a,* chlorophyll *b* and bacteriochlorophyll *a* by means of column chromatography with diethylaminocellulose. *Biochim. Biophys. Acta* **501,** 103–111.

Sayre, C.B., Robinson, V.B., and Wishnetzky, T. (1953). Effect of temperature on the colour, lycopene and carotene content of detached and of vine-ripened tomatoes. *Proc. Am. Soc. Hortic. Sci.* **61,** 381–387.

Schaffer, A.A., Boyer, C.D., and Gianfagna, T. (1984). Genetic control of plastid carotenoids and transformation in the skin of *Cucurbita pepo* L. fruit. *Theor. Appl. Genet.* **68,** 493–501.

Schaffer, A.A., Paris, H.S., and Ascarelli, I. (1986). Carotenoid and starch content of near-isogenic B^+B^+ and B,B genotypes of cucurbita. *J. Am. Soc. Hortic. Sci.* **111,** 780–783.

Schanderl, S.H., and Lynn, D.Y.C. (1966). Changes in chlorophylls and spectrally related pigments during ripening of *Capsicum frutescens. J. Food Sci.* **31,** 141–145.

Schanderl, S.H., Marsch, G.L., and Chichester, C.O. (1965). Color reversion in processed vegetables. I. Studies on regreened pea puree. *J. Food Sci.* **30,** 312–316.

Schillinger, A., and Zimmermann, G. (1965). Über die Stabilität verschiedener Vitamine in Trockenkartoffeln und Trockengemüsen. *Deutsch. Lebensmittel.-Rundsch.* **61,** 144–147.

Schlottmann, H., Mühlendyck, E., and Schuphan, W. (1961). Wertstoffverluste bei Gemüse und Obst zwischen Ernte und Verzehr. *Deutsch. Lebensm. Rundsch.* **57,** 270–276.

Schmid, L., and Lang, R. (1939). Die Farbstoffe an Gelbfleckigkeit erkrankter Kartoffelknollen. *Monatsh. Chem.* **72,** 322–326.

Schneegurt, M.A., and Beale, S.I. (1988). Characterization of the RNA required for biosynthesis of δ-aminolevulinic acid from glutamate. Purification by anticodon-based affinity chromatography and determination that the UUC glutamate anticodon is a general requirement for function in ALA biosynthesis. *Plant Physiol.* **86,** 497–504.

Schneider, H.A.W. (1966). Eine einfache Methode zur dünnschichtchromatographischen Trennung von Plastidenpigmenten. *J. Chromatogr.* **21,** 448–453.

Schneider, H.A.W. (1970). Anreicherung und Eigenschaften von δ-Aminolävulinat-dehydratase aus Spinat (*Spinacia oleracea*). *Z. Pflanzenphysiol.* **62,** 323–342.

Schneider, H.A.W. (1973). Changes in activity of four consecutive enzymes in the porphyrin and chlorophyll biosynthesis chain during development and illumination. *Z. Naturforsch.* **28C,** 45–48.

Schoch, S. (1978). The esterification of chlorophyllide *a* in greening bean leaves. *Z. Naturforsch.* **33c,** 712–714.

Schoch, S., Blank, M., and Schmidt, B. (1984). The influence of temperature on chlorophyll degradation in tobacco cell culture. In *Advances in Photosynthesis Research, Proc. Int. Congress in Photosynthesis Sixth Meeting,* C. Sybesma, ed., vol. 4, pp. 259–262.

Schoch, S., and Brown, J. (1987). The action of chlorophyllase on chlorophyll-protein complexes. *J. Plant Physiol.* **126,** 483–494.

Schoch, S., Lempert, U., and Rüdiger, W. (1977). On the last steps of chlorophyll biosynthesis. Intermediates between chlorophyllide and phytol containing chlorophyll. *Z. Pflanzenphysiol.* **83,** 427–436.

Schoch, S., Lempert, U., Wieschhoff, and Scheer, H. (1978). HPLC of tetrapyrrole pigments. Pheophytin esterified with different diterpene alcohols, isomeric biliverdins and synthetic bilins. *J. Chromatogr.* **157,** 357–364.

Schoch, S., Rüdiger, W., Lüthy, B., and Matile, Ph. (1984). 13^2-Hydroxychlorophyll *a* the first product in the reaction of chlorophyll oxidase. *J. Plant Physiol.* **115,** 85–89.

Schoch, S., and Schäfer, W. (1978). Tetrahydrogeranylgeraniol a precursor of phytol in the biosynthesis of chlorophyll *a*. Localization of the double bonds. *Z. Naturforsch.* **33c,** 408–412.

Schoch, S., and Vielwerth, F.X. (1983). Chlorophyll degradation in senescent tobacco cell culture. *Z. Pflanzenphysiol.* **110,** 309–318.

Scholz, B., and Ballschmiter, K. (1981). Preparation and reversed phase high-performance liquid chromatography of chlorophylls. *J. Chromatogr.* **208,** 148–155.

Siefermann-Harms, D. (1981). The role of carotenoids in chloroplasts of higher plants. In *Biogenesis and Function of Plant Lipids,* P. Mazliak et al., eds., pp. 331–340.

Siefermann-Harms, D. (1984). Evidence for a heterogeneous organization of violaxanthin in thylakoid membranes. *Photochem. Photobiol.* **40,** 507–512.

Siefermann-Harms, D. (1985). Carotenoids in photosynthesis. I. Location in photosynthetic membranes and light harvesting function. *Biochim. Biophys. Acta* **811,** 325–355.

Siefermann-Harms, D. (1988). High performance liquid chromatography of chloroplast pigments. One-step separation of carotene and xanthophyll isomers, chlorophylls and pheophytins. *J. Chromatogr.* **448,** 411–416.

Siefermann-Harms, D., Hertzberg, S., Borch, G., and Liaaen-Jensen, S. (1981). Lactucaxanthin, an ε,ε-carotene-3,3′-diol from *Lactuca sativa. Phytochemistry* **20,** 85–88.

Siefermann-Harms, D., Joyard, J., and Douce, R. (1978). Light-induced changes of the carotenoid levels in chloroplast envelopes. *Plant Physiol.* **61,** 530–533.

Siefermann-Harms, D., and Ninnemann, H. (1982). Pigment organization in the light-harvesting chlorophyll *a/b* protein complex of lettuce chloroplasts. Evidence obtained from protection of the chlorophylls against proton attack and from excitation energy transfer. *Photochem. Photobiol.* **35,** 719–731.

Siefermann-Harms, D., and Yamamoto, H.Y. (1974). Light induced de-epoxidation of violaxanthin in lettuce chloroplasts. III. Reaction kinetics and effect of light intensity on the de-epoxidase activity and substrate availability. *Biochem. Biophys. Acta* **357,** 144–150.

Sievers, G., and Hynninen, P.H. (1977). Thin-layer chromatography of chlorophylls and their derivatives on cellulose layers. *J. Chromatogr.* **134,** 359–364.

Simeonova, W. (1965). Veraenderungen der chemischen Zusammensetzung einiger Lebensmittel bei Einwirkung von Waerme bzw. von Pepsin. *Nahrung* **9,** 307–312.

Simon, P.W. (1987). Genetic improvement of carrots for meeting human nutritional needs. In *Horticulture and Human Health, ASHS Symposium Series* No. 1, Quebedeaux and Bliss, eds. Prentice-Hall, Englewood Cliffs, N.J.; ASHS, Va., pp. 208–214.

Simon, P.W., and Lindsay, R.C. (1983). Effects of processing upon objective and sensory variables of carrots. *J. Am. Soc. Hortic. Sci.* **108,** 928–931.

Simon, P.W., Peterson, C.E., and Lindsay, R.C. (1982). Genotype, soil and climate effects on sensory and objective components of carrot flavour. *J. Am. Soc. Hortic. Sci.* **107,** 644–648.

Simon, P.W., and Wolff, X.Y. (1987). Carotenes in typical and dark orange carrots. *J. Agric. Food Chem.* **35,** 1017–1022.

Simon, P.W., Wolff, X.Y., and Peterson, C.E. (1985). Selection for high carotene content in carrots. *HortScience* **20,** 586.

Simon, P.W., Wolff, X.Y., Peterson, C.E., Kammerlohr, D.S., Rubatzky, V.E., Strandberg, J.O., Basset, M.J., and White, J.M. (1989). High carotene mass carrot population. *HortScience* **24,** 174.

Simpson, D.J., Baqar, M.R., and Lee, T.H. (1977). Fine structure and carotenoid composition of the fibrillar chromoplasts of *Asparagus officinalis* L. *Ann. Bot.* **41,** 1101–1108.

Simpson, D.J., and Lee, T.H. (1976). The fine structure and formation of fibrils of *Capsicum annuum* chromoplasts. *Z. Pflanzenphysiol.* **77**, 127–138.

Simpson, K.L., and Chichester, C.O. (1981). Metabolism and nutritional significance of carotenoids. *Ann. Rev. Nutr.* **1**, 351–374.

Simpson, K.L., Lee, T.-C., Rodriguez, D.A., and Chichester, C.O. (1976). Metabolism in senescent and stored tissues. In *Chemistry and Biochemistry of Plant Pigments,* vol. 2, 2nd ed., T.W. Goodwin, ed. Academic Press, New York, pp. 780–842.

Simpson, K.L., Tsou, S.C., and Chichester, C.O. (1985). Carotenes. In *Methods of Vitamin A Assay,* J. Augustin, B.P. Klein, D. Becker, and P.B. Venugopal, eds. Wiley, New York, pp. 185–220.

Singh, B.P., and Dankhar, B.S. (1980). Effect on growth regulators and prepackaging on the storage of okra. *Haryana Agric. Univ. J. Res.* **X**, 398–402.

Sistrom, W.A., Griffiths, M., and Stanier, R.Y. (1956). The biology of a photosynthetic bacterium which lacks colored carotenoids. *J. Cell Comp. Physiol.* **48**, 473–515.

Sistrunk, W.A., Mahon, M.K., and Freeman, D.W. (1977). Relationship of processing methodology to quality attributes and nutritional value of canned spinach. *HortScience* **12**, 59–60.

Sitte, P. (1977). Chromoplasten-bunte Objekte der modernen Zellbiologie. *Biol. in unserer Zeit.* **7**, 65–74.

Sitte, P., Falk, H., and Liedvogel, B. (1980). Chromoplasts. In *Pigments in Plants,* 2nd ed., F.-Ch. Czygan, ed. Gustav Fischer, Stuttgart, pp. 117–148.

Škrbić, K. (1987). The effect of nitrogen nutrition on quality in some carrot cultivars. *Agrohemija* **4**, 263–273.

Sluiters-Scholten, C.M.T., van den Berg, F.M., and Stegwee, D. (1973). Aminolaevulinate dehydratase in greening leaves of *Phaseolus vulgaris. Z. Pflanzenphysiol.* **69**, 217–227.

Smith, A.G. (1988). Subcellular localization of two porphyrin-synthesis enzymes in *Pisum sativum* (pea) and *Arum* (cuckoo-pint) species. *Biochem. J.* **249**, 423–428.

Smith, B.B., and Rebeiz, C.A. (1977). Chloroplast biogenesis: detection of Mg-protoporphyrin chelatase *in vitro. Arch. Biochem. Biophys.* **180**, 178–185.

Smith, J.H.C., and Benitez, A. (1955). Chlorophylls. In *Modern Methods of Plant Analysis,* vol. IV, K. Paech and M.V. Tracey, eds. Springer, Berlin, pp. 142–196.

Soll, I., and Schulz, G. (1981). Phytol synthesis from geranylgeraniol in spinach chloroplasts. *Biochem. Biophys. Res. Commun.* **99**, 907–912.

Soll, J., Schulz, G., Rüdiger, W., and Benz, J. (1983). Hydrogenation of geranylgeraniol. Two pathways exist in spinach chloroplasts. *Plant Physiol.* **71**, 849–854.

Sood, R., and Bhat, C.M. (1974). Changes in ascorbic acid and carotene content of green leafy vegetables on cooking. *J. Food Sci. Technol.* **11**, 131–133.

Speek, A.J., Speek-Saichua, S., and Schreurs, H.P. (1988). Total carotenoids and β-carotene contents of Thai vegetables and the effect of processing. *Food Chem.* **27**, 245–257.

Speek, A.J., Temelilwa, C.R., and Shrijver, J. (1986). Determination of β-carotene content and vitamin A activity of vegetables by high-performance liquid chromatography and spectrophotometry. *Food Chem.* **19**, 65–74.

Spurgeon, S.L., and Porter, J.W. (1983). Biosynthesis of carotenoids. In *Biosynthesis of Isoprenoid Compounds,* J.W. Porter and S.L. Spurgeon, eds. Wiley, Chichester, pp. 1-122.

Spurgeon, S.L., Sathyamoorty, N., and Porter, J.W. (1984). Isopentenyl pyrophosphate isomerase and prenyltransferase from tomato fruit plastids. *Arch. Biochem. Biophys.* **230,** 446-454.

Sri Kantha, S., and Erdman, J.W., Jr. (1987). Legume carotenoids. *CRC Crit. Rev. Food Sci. Nutr.* **26,** 137-155.

Stahl, E. (1967). *Thin Layer Chromatography. A Laboratory Handbook,* 2nd ed. Springer, Berlin.

Stewart, I., and Wheaton, T.A. (1971). Continuous separation of carotenoids by liquid chromatography. *J. Chromatogr.* **55,** 325-336.

Stewart, J.K., and Uota, M. (1971). Carbon dioxide injury and market quality of lettuce held in controlled atmosphere. *J. Am. Soc. Hortic. Sci.* **96,** 27-30.

Stimson, C.R., Tressler, D.K., and Maynard, L.A. (1939). Carotene (Vitamin A) content of fresh and frosted peas. *Food Res.* **4,** 475-482.

Stobart, A.K., and Hendry, G.A.F. (1984). The turnover of chlorophyll in greening wheat leaves. *Phytochemistry* **23,** 27-30.

Stoll, A., and Wiedemann, E. (1938). Chlorophyll. In *Fortschritte der Chemie Organischer Naturstoffe,* vol. 1, L. Zechmeister, ed. Springer, Berlin, pp. 159-254.

Stone, M.B., Toure, D., Greig, J.K., and Naewbanij, J.O. (1986). Effects of pretreatment and dehydration temperature on color, nutrient retention and sensory characteristics of okra. *J. Food Sci.* **51,** 1201-1203.

Storey, R., and Beevers, L. (1978). Enzymology of glutamine metabolism related to senescence and seed development in the pea (*Pisum sativum* L.). *Plant Physiol.* **61,** 494-500.

Strain, H.H. (1939). Carotene. XI. Isolation and detection of α-carotene and the carotenes of carrot roots and of butter. *J. Biol. Chem.* **127,** 191-201.

Strain, H.H. (1941). Unsaturated fat oxidase: specificity, occurrence and induced oxidations. *J. Am. Chem. Soc.* **63,** 3542.

Strain, H.H. (1954). Chlorophyll reactions. Oxidation and isomerization reactions of chlorophyll in killed leaves. *J. Agric. Food Chem.* **2,** 1222-1226.

Strain, H.H. (1966). Fat soluble chloroplast pigments: their identification and distribution in various Australian plants. In *Biochemistry of Chloroplasts,* vol. I, T.W. Goodwin, ed. Academic Press, London, pp. 387-406.

Strain, H.H., and Manning, W.M. (1942). Isomerisation of chlorophylls *a* and *b*. *J. Biol. Chem.* **146,** 275-276.

Strain, H.H., and Sherma, J. (1972). Investigations of the chloroplast pigments of higher plants, green algae and brown algae and their influence upon the invention, modifications and applications of Tswett's chromatographic method. *J. Chromatogr.* **73,** 371-397.

Stransky, H. (1978). A method for the quantitative estimation of chloroplast pigments in the picomole-range by use of an isocratic HPLC system. *Z. Naturforsch.* **33c,** 836-840.

Straub, O. (1987). *Key to Carotenoids,* 2nd ed., H. Pfander, M. Gerspacher, M. Rychener, and R. Schwabe, eds. Birkhäuser, Basel.

Strmiska, F., and Holčiková, K. (1970). Die Veränderungen des Carotinoidgehaltes bei der Lagerung und Verarbeitung von Obst und Gemüse. *Fluess. Obst.* **37**, 46–53.

Stute, R. (1981). Die Bedeutung der Carotinoide für die Stabilität und zur Stabilisierung von Kartoffelpureeprodukten. *Arbeitsgemeinschaft Kartoffelforsch.* **3**, 42–47.

Subbarayan, C., Kushwaha, S.C., Suzue, G., and Porter, J.W. (1970). Enzymatic conversion of isopentenyl pyrophosphate-4-^{14}C and phytoene-^{14}C to acyclic carotenes by an ammonia sulfate precipitated spinach enzyme. *Arch. Biochem. Biophys.* **137**, 547–557.

Sugano, N., and Hayashi, K. (1967). Dynamic interrelation of cellular ingredients relevant to the biosynthesis of anthocyanin during tissue culture of carrot aggregen. Studies on anthocyanins. LVII. *Bot. Mag. Tokyo* **80**, 440–449.

Suginome, H., and Ueno, K. (1931). On the carotenoids of cucurbits. I. The pigments in the fruit of *Cucurbita maxima* Duch. *Bull. Chem. Soc. Japan* **6**, 221–228.

Sumner, J.B., and Dounce, A.L. (1939). Carotene oxidase. *Enzymol.* **7**, 130–132.

Sumner, J.B., and Sumner, R.J. (1940). The coupled oxidation of carotene and fat by carotene oxidase. *J. Biol. Chem.* **134**, 531–533.

Sundqvist, C., and Ryberg, H. (1979). Structure of protochlorophyll-containing plastids in the inner seed coat of pumpkin seeds (*Cucurbita pepo*). *Physiol. Plant.* **47**, 124–128.

Susuki, N., Saitoh, K., and Adachi, K. (1987). Reversed-phase high-performance thin-layer chromatography and column liquid chromatography of chlorophylls and their derivatives. *J. Chromatogr.* **408**, 181–190.

Svec, W.A. (1978). The isolation, preparation, characterization and estimation of the chlorophylls and the bacteriochlorophylls. In *The Porphyrins,* vol. V, D. Dolphin, ed. Academic Press, London, pp. 341–399.

Sweeney, J.P., Gilpin, G.L., Staley, M.G., and Martin, M.E. (1959). 1. Ascorbic acid and carotene. Effect of cooking methods on broccoli. *J. Am. Diet. Assoc.* **35**, 354–358.

Sweeney, J.P., and Marsh, A.C. (1970). Vitamins and other nutrients. Separation of carotene stereoisomers in vegetables. *JAOAC* **53**, 937–940.

Sweeney, J.P., and Marsh, A.C. (1971a). Effect of processing on provitamin A in vegetables. *J. Am. Diet. Assoc.* **59**, 238–243.

Sweeney, J.P., and Marsh, A.C. (1971b). Effects of selected herbicides on provitamin A content of vegetables. *J. Agric. Food Chem.* **19**, 854–856.

Sweeney, J.P., and Martin, M.E. (1958). Determination of chlorophyll and pheophytin in broccoli treated by various procedures. *Food Res.* **23**, 635–647.

Sweeney, J.P., and Martin, M.E. (1961). Stability of chlorophyll in vegetables as affected by pH. *Food Technol.* **15**, 263–266.

Swirski, M., Allouf, R., Guimard, A., and Cheftel, H. (1969). A water-soluble, stable green pigment, originated during processing of canned Brussels sprouts picked before the first autumn frosts. *J. Agric. Food Chem.* **17**, 799–801.

Szabolcs, J., and Rónai, A. (1969). The structure of α-cryptoxanthine and the identity of zeinoxanthin with α-cryptoxanthin. *Acta Chim. Acad. Sci. Hung.* **61**, 309–313.

Szabolcs, J., Rónai, A., and Toth, Gy. (1970). The identification of petaloxanthin. *Acta Chim. Acad. Sci. Hung.* **66**, 229–233.

Takagi, S. (1979). Effect of β-carotene on spinach leaf lipoxygenase activity. *Okayama Daigaku Nogakusu Gakujutsu Hokoku* **53**, 29–35.

Takagi, S. (1985). Determination of green leaf carotenoids by HPLC. *Agric. Biol. Chem.* **49**, 1211–1213.

Takagi, S., and Matsugami, M. (1977). Effect of carotenoids on leaf lipoxygenase activity. II. Studies on the inhibition of lipoxygenase activity by lutein. *Nippon Nogei Kagaku Kaishi* **51**, 489–495.

Takagi, S., Matsugami, M., and Moriloki, T. (1977). Effect of carotenoids on lipoxygenase activity. I. Purification of spinach chloroplast lipoxygenase and the inhibition by lutein of its activity. *Okayama Daigaku Nogakubu Gakujutsu Hokoku* **49**, 35–44.

Tămaş, V., Săăjan, Gh., and Bodea, C. (1970). Effect of corn pollen in poultry feed on egg yolk pigmentation. *Stud. Cerc. Biochim.* **13**, 423–429.

Tan, C.T., and Francis, F.J. (1962). Effect of processing temperature on pigments and color of spinach. *J. Food Sci.* **27**, 232–241.

Tauber, H. (1940). Unsaturated fat oxidase. *J. Am. Chem. Soc.* **62**, 2251.

Taylor, R.F. (1983). Chromatography of carotenoids and retinoids. In *Advances in Chromatography,* vol. 22, J.C. Giddins, E. Grushka, J. Cazes, and P.R. Brown, eds. Dekker, New York, pp. 157–213.

Teixeira Neto, R.O., Karel, M., Saguy, I., and Mizrachi, S. (1981). Oxygen uptake and β-carotene decoloration in a dehydrated food model. *J. Food Sci.* **46**, 665–669, 676.

Temple, N.J., and Basu, T.K. (1988). Does beta-carotene prevent cancer? A critical appraisal. *Nutr. Res.* **8**, 685–701.

Terpstra, W. (1981). Identification of chlorophyllase as a glycoprotein. *FEBS Lett.* **126**, 231–235.

Tevini, M., and Bergthaller, W. (1985). Veränderungen der Carotinoide während der Herstellung und Lagerung von Trockenkartoffeln. *Veröff. Arbeitsgem. Kartoffelforsch.* **7**, 26–32.

Tevini, M., Schönecker, G., Iwanzik, W., Riedman, M., and Stute, R. (1984a). Analyse, Vorkommen und Verhalten von Carotinoidestern in Kartoffeln. *Arbeitsgem. Kartoffelforsch.* e.V., pp. 42–47.

Tevini, M., Iwanzik, W., and Schönecker, G. (1984b). Analyse, Vorkommen und Verhalten von Carotinoiden in Kartoffeln und Kartoffelprodukten. *Jahrbuch Forschungskreis Ernährungsindustrie* e.V., pp. 36–53.

Tevini, M., Iwanzik, W., and U. Thoma (1981). Some effects of enhanced UV-B irradiation on the growth and composition of plants. *Planta* **153**, 388–394.

Tevini, M., Schönecker, G., and Stute, R. (1986). Das Verhalten von Carotinoiden bei der Herstellung und Lagerung von Kartoffelgranulat. *Veröff. Arbeitsgem. Kartoffelforsch.* **8**, 86–90.

Tevini, M., and Steinmüller, D. (1985). Composition and function of plastoglobuli. II. Lipid composition of leaves and plastoglobuli during beech leaf senescence. *Planta* **163**, 91–96.

Theorell, H., Holman, R.T., and Akeson, A. (1947). A note on the preparation of crystalline soybean lipoxidase. *Arch. Biochem.* **14**, 250–252.

Thimann, K.V., and Satler, S. (1979). Relation between senescence and stomatal opening: senescence in darkness. *Proc. Natl. Acad. Sci. Wash.* **76**, 2770–2773.

Thomas, J.R., and Gausman, H.W. (1977). Leaf reflectance vs. leaf chlorophyll and carotenoid concentrations for eight crops. *Agron. J.* **69,** 799–801.

Thomas, R.L., and Jen, J.J. (1975a). Red light intensity and carotenoid biosynthesis in ripening tomatoes. *J. Food Sci.* **40,** 566–568.

Thomas, R.L., and Jen, J.J. (1975b). Phytochrome-mediated carotenoid biosynthesis in ripening tomatoes. *Plant Physiol.* **56,** 452–453.

Thornber, J.P., Markwell, J.P., and Reinmann, S. (1979). Plant chlorophyll-protein complexes: recent advances. *Photochem. Photobiol.* **29,** 1205–1216.

Tichenor, D.A., Martin, D.C., and Wells, C.E. (1965). Carotenoid content of frozen and irradiate sweet corn. *Food Technol.* **19,** 106–109.

Tomes, M.L. (1963). Temperature inhibition of carotene synthesis in tomato. *Bot. Gaz.* **124,** 180–185.

Tomes, M.L. (1967). The competitive effect of the beta- and delta-carotene genes on alpha- or beta-ionone ring formation in the tomato. *Genetics* **56,** 227–231.

Tomes, M.L. (1969). Delta-carotene in tomato. *Genetics* **62,** 769–780.

Tomes, M.L., and Quackenbush, F.W. (1958). Caro-Red, a new provitamin A rich tomato. *Econ. Bot.* **12,** 256–260.

Tomes, M.L., Quackenbush, F.W., and Kargl, T.E. (1956). Action of gene B in biosynthesis of carotenes in tomato. *Bot. Gaz.* **117,** 248–253.

Tomkins, R.G., Mapson, L.W., and Wager, H.G. (1946). Drying of vegetables. V. The effect of scalding in starch or sulfite solutions on the loss of carotene and deterioration in culinary quality of dried carrots during storage. *J. Soc. Chem. Ind.* **65,** 384–385.

Tóth, G., and Szabolcs, J. (1981). Occurrence of some mono-*cis*-isomers of asymmetric C_{40}-carotenoids. *Phytochemistry* **20,** 2411–2415.

Trombly, H.H., and Porter, J.W. (1953). Additional carotenes and a colorless polyene of *Lycopersicon* species and strains. *Arch. Biochem. Biophys.* **43,** 443–457.

Trudel, M.J., and Ozbun, J.L. (1971). Influence of potassium on carotenoid content of tomato fruit. *J. Am. Soc. Hortic. Sci.* **96,** 763–765.

Tsukida, K., Saiki, K., and Sugiura, M. (1981). Structural elucidation of the main *cis*-β-carotenes. *J. Nutr. Sci. Vit.* **27,** 551–561.

Tsukida, K., Saiki, K., Takii, T., and Koyama, Y. (1982). Separation and determination of *cis/trans*-β-carotenes by high-performance liquid chromatography. *J. Chromatogr.* **245,** 359–364.

Tswett, M. (1906). Adsorptionsanalyse und chromatographische Methode: Anwendung auf die Chemie des Chlorophylls. *Ber. Deutsch. Bot. Ges.* **24,** 384–393.

Tswett, M. (1911). Über den makro- und mikrochemischen Nachweis des Carotins. *Ber. Deutsch. Bot. Ges.* **29,** 630–636.

Turrell, F.M., Weber, J.R., and Austin, S.W. (1961). Chlorophyll content and reflection spectra of citrus leaves. *Bot. Gaz.* **123,** 10–15.

Ueno, T., Maekawa, A., and Susuki, T. (1982). Effect of lipoxygenase on determination of carotene in vegetables (Japanese). *Vitamins* **56,** 83–89.

Umiel, N., and Gabelman (1971). Analytical procedures for detecting carotenoids of carrot (*Daucus carota* L.) roots and tomato (*Lycopersicum esculentum*) fruits. *J. Am. Soc. Hortic. Sci.* **96,** 702–704.

Umiel, N., and Gabelman, W.H. (1972). Inheritance of root color and carotenoid synthesis in carrot, *Daucus carota* L.: orange vs. red. *J. Am Soc. Hortic. Sci.* **97,** 453–460.

Valadon, L.R.G., and Mummery, R.S. (1967). Carotenoids of some apples. *Ann. Bot.* **31,** 497–503.

Valadon, L.R.G., and Mummery, R.S. (1969). The effect of light on carotenoids of etiolated mung bean seedlings. *J. Exp. Bot.* **20,** 732–742.

Valadon, L.R.G., and Mummery, R.S. (1977). Carotenoids of lilies and of red pepper: biogenesis of capsanthin and capsorubin. *Z. Planzenphysiol.* **82,** 407–416.

Van Blaricom, L.O., and Martin, J.A. (1951). Retarding the loss of red color in cayenne pepper with oil antioxidants. *Food Technol.* **5,** 337–339.

van Buren, J.P., Lee, C.Y., and Massey, L.M., Jr. (1982). Variation of vitamin concentration and retention in canned snap beans from three processing plants during two years. *J. Food Sci.* **47,** 1545–1548.

Van Hasselt, Ph.R. (1972). Photooxidation of leaf pigments in Cucumis leaf discs during chilling. *Acta Bot. Neerl.* **21,** 539–548.

Van Hasselt, Ph.R., and Strikwerda, J.T. (1976). Pigment degradation in discs of thermophilic *Cucumis sativus* as affected by light, temperature, sugar application and inhibitors. *Physiol. Plant.* **37,** 253–257.

Vecchi, M., Englert, G., Maurer, R., and Meduna, V. (1981). Separation and characterization of the *cis*-isomers of β, β-carotene. *Helv. Chim. Acta* **64,** 2746–2758.

Veloso de Almeida, P.G., Brune, W., and Kling de Moraes, G.H. (1977). Chemical content fluctuation of β-carotene and lycopene in stored tomatoes. *Rev. Ceres.* **24,** 187–199.

Vernon, L.P. (1960). Spectrophotometric determination of chlorophylls and pheophytins in plant extracts. *Anal. Chem.* **32,** 1144–1150.

Virgin, H.I. (1955). The conversion of protochlorophyll to chlorophyll *a* in continuous and intermittent light. *Physiol. Plant.* **8,** 389–403.

Virgin, H.I. (1981). The physical state of protochlorophyll(ide) in plants. *Ann. Rev. Plant. Physiol.* **32,** 451–463.

Vogele, A.C. (1937). Effect of environmental factors upon the colour of the tomato and the watermelon. *Plant Physiol.* **12,** 922–955.

von Euler, H., Ahlstroem, L., Hoegberg, B., and Tingstam, Sv. (1943). Vitamin content of foodstuffs. I. *Ark. Kemi. Mineral. Geol. Ser. B* **16,** No. 11, 8 pp.

Wackenroder, H. (1831). Über das Oleum radicis Dauci aetherum, das Carotin, den Carotenzucker und den officiellen sucus Dauci; sowie auch über das Mannit welches in dem Möhrensafte durch eine besondere Art der Gährung gebildet wird. *Mag. Pharm.* (*Geiger*) **33,** 144–172.

Wagenknecht, A.C., and Lee, F.A. (1956). The action of lipoxidase in frozen raw peas. *Food Res.* **21,** 605–610.

Wagenknecht, A.C., and Lee, F.A. (1958). Enzyme action and off-flavor in frozen peas. *Food Res.* **23,** 25–31.

Wagenknecht, A.C., Lee, F.A., and Boyle, F.P. (1952). The loss of chlorophyll in green peas during frozen storage and analysis. *Food Res.* **17,** 343–350.

Wagner, G.R., and Elstner, E.F. (1989). Oxidative xanthophyll degradation is not a major biosynthetic pathway for xanthoxin formation. *J. Plant Physiol.* **134,** 47–53.

Wai, K.N.T., Bishop, J.C., Mack, B.P., and Cotton, R.H. (1947). The vitamin content of soybeans and soybean sprouts as a function of germination time. *Plant Physiol.* **22,** 117–126.

Walker, G.C. (1964a). Color deterioration in frozen French beans (*Phaseolus vulgaris*). *J. Food Sci.* **29**, 383–388.

Walker, G.C. (1964b). Color deterioration in frozen French beans (*Phaseolus vulgaris*). 2. The effect of blanching. *J. Food Sci.* **29**, 389–392.

Walles, B. (1972). An electron microscope study on photodestruction of plastid ribosomes in β-carotene-deficient mutants of *Helianthus annuus* L. *Protoplasma* **75**, 215–227.

Walter, W.M., and Purcell, A.E. (1974). Lipid autoxidation in precooked dehydrated sweet potato flakes stored in air. *J. Agric. Food Chem.* **22**, 298–302.

Walter, W.M., and Schadel, W.E. (1982). Effect of lye peeling conditions on sweet potato tissue. *J. Food Sci.* **47**, 813–817.

Wang, C.Y. (1977). Effect of CO_2 treatment on storage and shelf life of sweet peppers. *J. Am. Soc. Hortic. Sci.* **102**, 808–812.

Wang, C.Y. (1979). Effect of short-term high CO_2 treatment on the market quality of stored broccoli. *J. Food Sci.* **44**, 1478–1482.

Wang, C.Y. (1983). Postharvest responses of Chinese cabbage to high CO_2 or low O_2 storage. *J. Am. Soc. Hortic. Sci.* **108**, 125–129.

Wang, S.S., Haard, N.F., and Di Marco, G.R. (1971) Chlorophyll degradation during controlled-atmosphere storage of asparagus. *J. Food Sci.* **36**, 657–661.

Wang, W.-Y., Gough, S.P., Kannangara, C.G. (1981). Biosynthesis of δ-aminolevulinate in greening barley leaves. IV. Isolation of three soluble enzymes required for the conversion of glutamate to δ-aminolevulinate. *Carlsberg Res. Commun.* **46**, 243–257.

Watada, A.E., Aulenbach, B.B., and Worthington, J.T. (1976a). Vitamins A and C in ripe tomatoes as affected by stage of ripeness at harvest and by supplementary ethylene. *J. Food Sci.* **41**, 856–858.

Watada, A.E., Norris, H.K., Worthington, J.T., and Massie, D.R. (1976b). Estimation of chlorophyll and carotenoid contents of whole tomato by light absorbance technique. *J. Food Sci.* **41**, 329–332.

Watanabe, T., Kobayashi, M., Hongu, A., Nakazato, M., Hiyama, T., and Murata, N. (1985a). Evidence that a chlorophyll *a′* dimer constitutes the photochemical reaction center I(P700) in photosynthetic apparatus. *FEBS Lett.* **191**, 252–256.

Watanabe, T., Nakazato, M., Mazaki, H., Hongu, A., Konno, M., Saitoh, S., and Honda, K. (1985b). Chlorophyll *a* epimer and pheophytin *a* in green leaves. *Biochim. Biophys. Acta* **807**, 110–117.

Weber, E.J. (1987). Carotenoids and tocols of corn grain determined by HPLC. *JAOCS* **64**, 1129–1134.

Weckel, K.G., Santos, B., Hernan, Ed., Laferriere, L., and Gabelman, W.H. (1962). Carotene components of frozen and processed carrots. *Food Technol.* **16**, 91–94.

Weedon, B.C.L. (1967). Carotenoids: recent advances. *Chem. Br.* **3**, 424–432.

Weedon, B.C.L. (1971). Stereochemistry. In *Carotenoids,* O. Isoler, ed. Birkhäuser, Basel, pp. 267–323.

Weinstein, J.D., and Castelfranco, P.A. (1978). Mg-Protoporphyrin-IX and δ-aminolevulinic acid synthesis from glutamate in isolated greening chloroplasts. δ-Aminolevulinic acid synthesis. *Arch. Biochem. Biophys.* **186**, 376–382.

Weinstein, J.D., Mayer, S.M., and Beale, S.I. (1987). Formation of δ-aminolevulinic acid from glutamic acid in algal extracts. Separation into an RNA and three required enzyme components by serial affinity chromatography. *Plant Physiol.* **84,** 224–250.

Weits, J., Van der Meer, M.A., Lassche, J.B., Mettivier Meyer, J.C., Steinbuch, E., and Gersons, L. (1970). Nutritive value and organoleptic properties of three vegetables fresh and preserved in six different ways. *Int. J. Vit. Res.* **40,** 648–658.

Went, F.W., LeRosen, A.L., and Zechmeister, L. (1942). Effect of external factors on tomato pigments as studied by chromatographic methods. *Plant Physiol.* **17,** 91–100.

Werner, H.O. (1941). Dry matter, sugar and carotene content of morphological portions of carrots through the growing and storage season. *Proc. Am. Soc. Hortic. Sci.* **38,** 267–272.

Westcott, D., Livingston, G., Esselen, W., and Fellers, C. (1955). Nonenzymatic discoloration of green bean puree. *Food Res.* **20,** 149–159.

Whatley, J.M. (1974). Chloroplast development in primary leaves of *Phaseolus vulgaris. New Phytol.* **73,** 1097–1110.

White, J.W., Jr., Zscheile, F.P., and Brunson, A.M. (1942). The carotenoids of yellow corn grain. *J. Am. Chem. Soc.* **64,** 2603–2606.

White, R.C., Jones, I.D., and Gibbs, E. (1963). Determination of chlorophylls, chlorophyllides, pheophytins and pheophorbides in plant material. *J. Food Sci.* **28,** 431–436.

White, R.C., Jones, I.D., Gibbs, E., and Butler, L.S. (1972). Fluorometric estimation of chlorophylls, chlorophyllides, pheophytins and pheophorbides in mixtures. *J. Agric. Food Chem.* **20,** 773–778.

White, R.C., Jones, I.D., Gibbs, E., and Butler, L.S. (1977). Estimation of copper pheophytins, chlorophyll and pheophytins in mixtures in diethyl ether. *J. Agric. Food Chem.* **25,** 143–146.

Whitfield, D.M., and Rowan, K.S. (1974). Changes in the chlorophylls and carotenoids of leaves of *Nicotiana tabacum* during senescence. *Phytochemistry* **13,** 77–83.

Whu, M.T., and Salunkhe, D.K. (1976). Effect of vacuum packaging on quality and storage life of broccoli. *Lebensm. Wiss. Technol.* **9,** 251–254.

Wider de Xifra, E.A., Battle, A.M. del C., and Tigier, H.A. (1971). δ-Aminolaevulinate synthetase in extracts of cultured soybean cells. *Biochim. Biophys. Acta.* **235,** 511–517.

Williams, R.J.H., Britton, G., Charlton, J.M., and Goodwin, T.W. (1967a). The stereospecific biosynthesis of phytoene and polyunsaturated carotenes. *Biochem. J.* **104,** 767–777.

Williams, R.J.H., Britton, G., and Goodwin, T.W. (1967b). The biosynthesis of cyclic carotenes. *Biochem. J.* **105,** 99–105.

Wills, R.B.H. (1987). Composition of Australian fresh fruit and vegetables. *Food Technol. Austral.* **39,** 523–526.

Wills, R.B.H., Lim, J.S.K., and Greenfield, H. (1987). Composition of Australian foods. Vegetable fruits. *Food Technol. Austral.* **39,** 488–490.

Wills, R.B.H., Nurdin, H., and Wootton, M. (1988). Separation of carotenes and xanthophylls in fruit and vegetables by HPLC. *J. Micronutr. Anal.* **4,** 87–98.

Willstaetter, R., and Mieg, W. (1907). Untersuchungen über Chlorophyll. IV. Über die gelben Begleiter des Chlorophylls. *Ann. Chem.* **355,** 1–28.

Willstaetter, R., and Stoll, A. (1913). *Untersuchungen über Chlorophyll,* Springer, Berlin.

Winterstein, A., Studer, A., and Rüegg, R. (1960). Neue Ergebnisse der Carotinoidforschung. *Chem. Ber.* **93,** 2951–2965.

Wise, R.R., and Naylor, A.W. (1987). Chilling enhanced photooxidation. Evidence for the role of singlet oxygen and superoxide in the breakdown of pigments and endogenous antioxidants. *Plant Physiol.* **83,** 278–282.

Wolf, M.J., Buzan, C.L., MacMasters, M.M., and Rist, C.E. (1952). Structure of the mature corn kernel. I. Gross anatomy and structural relationships. *Cereal Chem.* **29,** 321–333.

Wong, Y.-S., and Castelfranco, P.A. (1984). Resolution and reconstitution of Mg-protoporphyrin IX monomethyl ester (oxidative) cyclase, the enzyme system responsible for the chlorophyll isocyclic ring. *Plant Physiol.* **75,** 658–661.

Wright, S.W., and Shearer, J.D. (1984). Rapid extraction and high-performance liquid chromatography of chlorophylls and carotenoids from marine phytoplankton. *J. Chromatogr.* **294,** 281–295.

Wrischer, M. (1974). Plastid transformation in carrot roots induced by different lights. *Acta Bot. Croat.* **33,** 53–61.

Wrischer, M. (1978). Ultrastructural changes in plastids of detached spinach leaves. *Z. Pflanzenphysiol.* **86,** 95–106.

Wu, M., Singh, M.T., Wu, M.T., Salunkhe, D.K., and Dull, G.G. (1970). Effects of certain soil fumigants on essential nutritive components and the respiratory rate of carrot (*Daucus carota* L.) roots. *HortScience* **5,** 221–222.

Wu, M.T., and Salunkhe, D.K. (1975). Effect of vacuum packaging on light-induced greening and glycoalkaloid formation of potato tubers. *J. Inst. Can. Sci. Technol. Aliment.* **8,** 185–187.

Wu, M.T., and Salunkhe, D.K. (1976). Effects of vacuum packaging on quality and storage life of broccoli. *Lebensm. Wiss. Technol.* **9,** 251–254.

Wu, S.-M., and Rebeiz, C.A. (1984). Chloroplast biogenesis 45: Molecular structure of protochlorophyllide (E433 F625) and chlorophyllide *a* (E458 F674). *Tetrahedron* **40,** 659–664.

Yadava, U.L. (1986). A rapid and nondestructive method to determine chlorophyll in intact leaves. *HortScience* **21,** 1449–1950.

Yakir, D., Sadovski, A., Rabinovitch, H.D., and Rudich, J. (1984a). Effect of high temperature on quality of processing tomatoes of various genotypes ripened off the vine. *Sci. Hortic.* **23,** 323–330.

Yakir, D., Sadovski, A., Rabinovitch, H.D., and Rudich, J. (1984b). Effect of high temperature on quality of processing tomatoes of various genotypes ripened on the vine. *Sci. Hortic.* **24,** 33–43.

Yamaguchi, M., and Sugiyama, T. (1960). The carotenoid contents of the Kintoki and Kokubu varieties of carrots grown in Japan. *J. Hortic. Assoc. Jpn.* **29,** 310–312.

Yamamoto, H.Y. (1979). Biochemistry of the violaxanthine cycle in higher plants. *Pure Appl. Chem.* **51,** 639–648.

Yamamoto, H.Y., Chichester, C.O., and Nakayama, T.O.M. (1962). Biosynthetic origin of oxygen in the leaf of xanthophylls. *Arch. Biochem. Biophys.* **96,** 645–649.

Yamauchi, N., Hamaguchi, S., and Ogata, K. (1980). Physiological and chemical studies on ascorbic acid of fruits and vegetables. VII. Mechanism of chlorophyll degradation and action of ascorbic acid in the inhibition of yellowing in harvested parsley leaves. *J. Jpn. Soc. Hortic. Sci.* **49,** 414–420.

Yamauchi, N., Iida, S., Minamide, T., and Iwata, T. (1985). Foliage yellowing by peroxides in stored spinach (mechanism of chlorophyll degradation in harvested leaf vegetables, Part II). *J. Jpn. Soc. Food Sci. Technol.* **32,** 814–817.

Yamauchi, N., Iida, S., Minamide, T., and Iwada, T. (1986). Polar lipids content and their fatty acid composition with reference to yellowing of stored spinach leaves. *J. Jpn. Soc. Food Sci. Technol.* **55,** 355-362.

Yamauchi, N., Iida, S., Minamide, T., and Iwada, T. (1987). Localization of lipolytic acyl hydrolase in chloroplast of stored spinach leaves. *J. Jpn. Soc. Hortic. Sci.* **55,** 510–515.

Yamauchi, N., and Minamide, T. (1985). Chlorophyll degradation by peroxidase in parsley leaves. *J. Jpn. Soc. Hortic. Sci.* **54,** 265–271.

Yang, Y.J., and Henze, J. (1987). Influence of CA-storage on external and internal quality features in broccoli (*Brassica oleacea* var. *italica*). I. Changes in external and sensory quality features. *Gartenbauwiss.* **52,** 223-226.

Yang, Y.J., and Henze, J. (1988a). Influence of CA-storage on external and internal quality features in broccoli (*Brassica oleacea* var. *italica*). II. Changes in chlorophyll and carotenoid contents. *Gartenbauwiss.* **53,** 41–43.

Yang, Y.J., and Henze, J. (1988b). Influence of CA-storage on external and internal quality features in broccoli (*Brassica oleacea* var. *italica*). III. Changes in carotenoid pattern. *Gartenbauwiss.* **53,** 55–57.

Yokoyama, H., Coggins, C.W., and Henning, G.L. (1971). The effect of CPTA on the formation of carotenoids in citrus. *Phytochemistry* **10,** 1831–1834.

Yokoyama, H., Debenedict, C., Coggins, C.W., and Henning, G.L. (1972). Induced color changes in grapefruit and orange. *Phytochemistry* **11,** 1721–1723.

Young, A.J., Britton, G., and Musker, D. (1989). A rapid method for the analysis of the mode of action of bleaching herbicides. *Pestic. Biochem. Physiol.* **35,** 244–250.

Zakaria, M., Simpson, K., Brown, P.R., and Krstulovic, A. (1979). Use of reversed-phase high-performance liquid chromatographic analysis for the determination of provitamin A carotenes in tomatoes. *J. Chromatogr.* **176,** 109–117.

Zechmeister, L. (1934). *Carotenoids.* Springer, Berlin.

Zechmeister, L. (1962). *Cis-trans isomeric carotenoids, vitamin A and arylpolyenes.* Springer-Verlag, Wien.

Zechmeister, L., and Cholnoky, L. (1927). Paprika coloring matters. I. *Ann. Chem.* **454,** 54–71.

Zechmeister, L., and Cholnoky, L. (1928). Paprika coloring matters. II. *Ann. Chem.* **455,** 70–81.

Zechmeister, L., and Cholnoky, L. (1934). Paprika coloring matters. VII. Adsorption analysis of the pigments. *Ann. Chem.* **509,** 269–287.

Zechmeister, L., and Cholnoky, L. (1935). Paprika coloring matters. VIII. Constitution of capsanthin and capsorubin. *Ann. Chem.* **516,** 30–45.

Zechmeister, L., and Cholnoky, L. (1936). Lycoxanthin und Lycophyll, zwei natürliche Derivate des Lycopins. *Ber. Deutsch. Chem. Ges.* **69,** 422–429.

Zechmeister, L., LeRosen, A.L., Schroeder, W.A., Polgár, A., and Pauling, L. (1943). Spectral characteristics and configuration of some stereoisomeric carotenoids including prolycopene and pro-γ-carotene. *J. Am. Chem. Soc.* **65**, 1940–1951.

Zechmeister, L., LeRosen, A.L., Went, F.W., and Pauling, L. (1941). Prolycopene, a naturally occurring stereoisomer of lycopene. *Proc. Natl. Acad. Sci. USA* **27**, 468–474.

Zechmeister, L., and Sandoval, A. (1945). The occurrence and estimation of phytofluene in plants. *Arch. Biochem.* **8**, 425–430.

Zechmeister, L., and Tuzson, T. (1934). Das Pigment des *Cucurbita maxima* Duch. (Riesenkürbis). *Ber. Deutsch. Chem. Ges.* **67**, 824–828.

Zhu, Y.S., Merkle-Lehman, D.L., and Kung, S.D. (1984). Light induced transformation of amyloplasts into chloroplasts in potato tubers. *Plant Physiol.* **75**, 142–145.

Ziegler, R., Dresken, M., Herl, B., Menth, M., Schneider, H., and Zange, P. (1982). Magnesium-dechelatase, ein Chlorophyll abbauendes. Enzym. Meeting, Deutsche Botanische Gesellschaft (Freiburg) Abstract 472.

Zimmermann, W.I., Tressler, D.K., and Maynard, L.A. (1940). Determination of carotene in fresh and frozen vegetables. I. Carotene content of green snap beans and sweet corn. *Food Res.* **5**, 93–101.

Zimmerman, W.I., Tressler, D.K., and Maynard, L.A. (1941). Determination of carotene in fresh and frozen vegetables by an improved method. II. Carotene content of asparagus and green Lima beans. *Food Res.* **6**, 57–68.

Zink, F.W. (1961). N^6-benzyladenine, a senescence inhibitor for green vegetables. *J. Agric. Food Chem.* **4**, 304–307.

Zinsou, C. (1971). Dégradation enzymatique et autooxydative du β-caroteène. *Physiol. Vég.* **9**, 149–167.

Zinsou, C., and Costes, C. (1973). Photodécoloration du β-carotène en solution. *Physiol. Vég.* **11**, 191–206.

Zonta, F., and Stancher, B. (1986). High-performance liquid chromatography (HPLC) of carotenes by alumina column. Applications to food analysis. *Riv. Soc. Ital. Sci. Aliment.* **15**, 17–22.

Zonta, F., Stancher, B., and Pertoldi Marletta, G. (1987). Simultaneous high-performance liquid chromatographic analysis of free carotenoids and carotenoid esters. *J. Chromatogr.* **403**, 207–215.

Zscheile, F.P., Beadle, B.W., and Kraybill, H.R. (1943). Carotene content of fresh and frozen vegetables. *Food Res.* **8**, 299–313.

Zsolt, J., Schneider, Gy., and Matkovics, B. (1963). Carotenoid changes in different maize varieties during ripening. *Can. J. Biochem. Physiol.* **41**, 481–486.

INDEX